Egyptian and Græco-Roman Wall Plasters and Mortars

Egyptian and Græco-Roman Wall Plasters and Mortars

A Comparative Scientific Study

Safaa A. Abd El Salam

BAR International Series 1319
2004

Published in 2016 by
BAR Publishing, Oxford

BAR International Series 1319

Egyptian and Græco-Roman Wall Plasters and Mortars

ISBN 978 1 84171 388 5

BAR Publishing is the trading name of British Archaeological Reports (Oxford) Ltd. British Archaeological Reports was first incorporated in 1974 to publish the BAR Series, International and British. In 1992 Hadrian Books Ltd became part of the BAR group. This volume was originally published by John and Erica Hedges Ltd. in conjunction with British Archaeological Reports (Oxford) Ltd / Hadrian Books Ltd, the Series principal publisher, in 2004. This present volume is published by BAR Publishing, 2016.

Printed in England

BAR
PUBLISHING

BAR titles are available from:

BAR Publishing
122 Banbury Rd, Oxford, OX2 7BP, UK
EMAIL info@barpublishing.com
PHONE +44 (0)1865 310431
FAX +44 (0)1865 316916
www.barpublishing.com

Contents

- **Except Where Stated All Photographs by the Author**

Illustrations: Part 1-Text

List of Figures

List of Figures: Part II

List of Maps and Plans: Part I

List of Plans: Part II

List of Tables: Part I

List of Graphs: Part I

List of Graphs: Part II

List of Photographs: Part I

Glossary

acqua sporca: (Italian) Light grey (greyish) water colour used to decrease the tonal contrast between the paint layer and the colour of the render in the lacunae.

aggregate*:* The coarse or inert ingredients of the mortar composition.

anhydrite*:* Crystalline calcium sulphate ($CaSO_4$).

binder: Material that binds together sand particles in a mortar, e.g. lime, clay, cement, etc.

buon fresco: True *fresco*, in which the pigments were mixed with water and applied to the damp lime surface (see Chapter 2, 3.8.2).

calcite*:* Crystalline calcium carbonate ($CaCO_3$).

cement*:* Is abbreviation of alumino silicate cement. Cement was used in the early 19th century, the first Portland cement was patented in 1824, and became a widely use in 1880s, which was made by burning in a kiln, using coal, a blend of limestone and clay containing silica and alumina, at 1450 °C. This produced a clinker, which was crushed to a fine powder.

cinnabar*:* Red mercury sulphide, natural vermilion called "*minium*" by Vitruvius (1960): Theophilus (1961) calls it "*cenobrium*". Many recipes for the preparation of cinnabar come from Latin manuscripts from the 8th century onwards. In the history of chemistry, Gaber/Jabir, the 8th–9th century Arab alchemist, speaks about a red compound formed by the union of sulphur and mercury. In 800 AD two descriptions for making artificial cinnabar were given (Mural Painting Conservation 1998).

dado: The lower part of the wall down to the floor level, often finished with a moulded rail called the dado rail, to prevent damage to the surface of lower part of wall.

distemper*:* A term employed to designate aqueous paints made with a simple glue-size or casein binder.

efflorescence*:* Some crystalline substances have the property of losing their water and crystallizing when exposed to dry air. In contrast**, hygroscopic** substances are capable of absorbing moisture from the air.

fresco: (Italian) "Fresh", used to describe the traditional *buon **fresco*** which is painted on a wet lime surface, the pigments being mixed only with water. In *fresco* ***secco,*** the pigments were mixed with lime or an organic medium and applied to dry plaster (see Chapter 2, 4.8.2 and 4.8.3).

gable: (Arabic) mountain.

gauging trowel: It is specialist trowel for applying plaster/mortar.

gesso: (Italian form of a Latin word) any kind of gypsum or gypsum plaster made with a purely aqueous binder of glue or gelatin solution (see Chapter 2, 4.4).

giornata (-e) di lavoro: Literally "day work" the medieval Italian term for the division of a painting into panels of wet plaster that can be completed before the plaster dried.

gypsum: (from the Greek *gypsos)* Calcium sulphate hydrate ($CaSO_4$ $2H_2O$) (see Chapter 2, 3.4 & 4.3.3).

hib: (Arabic) Natural mixture of clay and limestone that had been washed off mountain by the rainstorms.

intonaco: (Italian) Finishing layer of wet plaster. Dried unpainted *intonaco* was cut away and re-laid so that the work could be done on a new damp surface. Finer and more perfectly finished, sometimes polished, it is richer in lime, usually having a composition of two parts of finely sieved sand/marble dust to one part slaked lime and is applied up to 0.5 cm thick. It is possible to estimate approximately how many working days were spent in the execution of a *fresco* from an examination of the joints in the plaster surfaces.

knocking up: Kneading or pummeling the lime mixture to make it more mobile.

lacuna (-e): (Latin) refers to larger (deep) losses of one or more layers of the render.

lime putty: Traditional non-hydraulic lime putty pure lime, derived from limestone which do not contain clay or other reactive silicates and contain a very high proportion of calcium carbonate, about 95% it is made by the wet slaking of calcium oxide (quick lime). It is normally sieved and matured for at least three months before use to ensure all the materials are fully slaked. The lime putty particles will progressively reduce in size producing an increasingly higher quality of materials.

lime-wash: When water is added to lime putty in different proportions, it gives milk of lime or lime wash.

lime-water: This is formed when lime wash is thinned. Lime particles settle to the bottom of the container with limewater above it. Limewater is advisable for *fresco* painting because the very fine and thin film carbonate of lime on the surface of the water helps to fix the colour by chemical reaction.

mastaba: (Arabic) "bench" used as the term for free standing tombs of the early dynastic period and Old Kingdom. The basic form of the mastaba's superstructure is a rectangle with flat roof and vertical or slightly inclined walls. Lauer (1976) described the principles of the mastabas in the Old Kingdom and their decoration styles (see page 31-81 for more information).

medium*:* The liquid constituent of a paint, in which the pigments are suspended.

mortar*:* Used between stones as filler; mortar is usually rough and contains large pieces of stone/aggregate additives.

necropolis: (Greek) large cemetery, normally large and important burial areas that were in use for long periods, while "cemetery" usually described smaller and more homogenous sites, although cemeteries may be divisions of a necropolis.

oxgall*:* Traditional plaster additive from the gall bladder of the ox, cow, etc.

Paraloid B72*:* Acrylic resin.

plaster*:* Finishing layer or face layer often painted or used as a render.

Pontic wax (Unidentified wax from the Black Sea area).

popping or blowing: Surface disruption caused by expansion by the lime slaking in *situ*.

pozzolana: (Italian) natural volcanic earth, e.g., volcanic ash, volcanic sands, or manufactured products such as coal and wood ash, certain other vegetable ashes and soft fired clay products such as brick and tile. These are rich in silica, alumina and sometimes iron oxides. When mixed with lime and water it tends to strengthen and will even set hard to produce calcium silicate hydrate, which gives a chemical set to the mortar. Natural *pozzolana* was used with lime in Pompeii and widely in Roman Italy.

salt: A salt is a chemical product of the reaction between an acid and base. The term "salt" often relates to sodium chloride (NaCl) but in this work it could be any salt, e.g., sodium chloride, potassium chloride, magnesium chloride, etc., unless specified.

Saqqara: This name derives from Sokkar, Memphis god of the dead.

silt: Finely divided mineral particles, usually less than 75 µm. The smallest sizes, below 2 µm, stay in suspension in a liquid, in contrast to gravel, which is more than 2 mm, and sand, which is 0.1 mm to 2 mm.

stacco a massello: (Italian) detaching part of the wall as well as the plaster when removing painting.

stacco: (Italian) In which the fine (*intonaco*) as well as the coarse plaster (*arriccio*) layers are detached from the wall.

Stela (-e): (Greek) Slab of stone or sometimes wood with texts, reliefs or painting: such commemorative or votive stelae were placed in temples. The function of a stela is the decoration of a tomb.

strappo: (Italian) painted surface to be detached.

stucco: (Italian) Fine lime plaster used as a medium for decorative work in relief. For resistant coats, the renderings are put on the wet wall (see Chapter 2, 4.7).

tratteggio: (Italian) meaning hatching with vertical or horizontal painted lines following the design of the painting.

tempera: (Latin) derived from *temperare* (watercolour process), in which the pigments were mixed with organic materials, e.g., glue, gum and white/yolk of egg (see Chapter 2, 4.8.1).

Abbreviations

A: Anfushi.
EE: Emery Excavation.
EDS: Energy Dispersive Spectrometer (surface structure).
HCl: Hydrochloric acid.
ICCROM: International Centre for the Study of the Preservation and Restoration of Cultural Property.
ICOM: International Council of Museums.
ICP: Induction Coupled Plasma Spectroscopy (metal ions).
I.C.R: The Central Institute for Restoration.
M $CaCO_3$: Measurements of Calcium Carbonate.
MP: Mustafa Pasha.
NE: New Excavation.
PB72: Paraloid B72.
PCS: Polished Cross Section (identification of composition).
pH: Is a notation concentration of hydrogen ion ranging from 0-14 where 0-7 is acid, 7-14 is alkine, 7 being neutral.
PLM: Polarised Light Microscopy (internal structure).
ppm: Parts per million.
R: Room.
RH: Relative Humidity.
S: Sample.
SA: Standard Analysis (qualitative & quantitative analysis).
SEM Scanning Electron Microscope.
SS: Soluble Salts.
ST: Spot Test.
T: Tomb.
UV: Ultra Violet.
X-RD: X-Ray Diffraction (identification of single crystal.
X-RPD: X-Ray Powder Diffraction (identification of minerals).

Historical Dates; Egypt

Date	Period	Dynasty
3000 BC	Late pre-dynastic	-
2920-2575 BC	Early dynastic	(1-2)
2575-2134 BC	**Old Kingdom**	(3-8)
2134-2040 BC	First Intermediate	(9-11/1)
2040-1640 BC	**Middle Kingdom**	(14-17)
1640-1532 BC	Second Intermediate	(14-17)
1550-1525 BC	**New Kingdom**	(18)
1319-1307 BC		(19)
1196-1070 BC		(20)
1070-712 BC	Third Intermediate	(21-24)
712-332 BC	Late Period	(25-31)
332-30 BC	**Macedonian Ptolemaic**	
30 BC-AD 395	**Roman Period**	

(After Baines and Malek 1980).

Historical Dates; Britain	
43 AD	**Roman Invasion in Britain**
300-400 AD	Piddington
(4th century AD)	

Acknowledgements

My first debt of gratitude must go to my supervisor Dr. Graham Morgan. He has given his time, advice and enthusiasm freely, well beyond the level that I had a right to expect and has helped to shape my ideas without ever expecting his own view to prevail. He also gave generous encouragement and thoughtful comments throughout my publication and I have been very lucky to benefit from his support and experience. I am most grateful to the following staff School of Archaeological Studies and Ancient History at the University of Leicester. I would also like to express my gratitude to Prof. Graeme Barker for his great support since I started my research at the school, to Dr. Sarah Scott for her kind and supportive supervision and her useful comments through the first two years of my research and to Dr. Simon James for his reading my thesis and his useful comments throughout the writing up stage. My thanks also go to Dr. Marijke van der Veen and Mr. Tony Gouldwell (Environmental Laboratories Support) for their help in the identification of the seeds which I found throughout my research and also a great assistance from Mr. Gouldwell in the translation of some articles from Norway. Thanks also go to the Archaeological Technical staff, Mrs. Debbie Miles-Williams (Archaeological Illustration) and Ms. Lucy Farr (Technical Support, illustration) for their kind help. My thanks must also go to Alex Moseley (Computer officer, faculty of Arts) for his advice on computers.

Thanks also to the following people: Prof. John Baines, professor of Egyptology at the Oriental Institute, University of Oxford for his assistance with some Egyptian bibliography, Dr. Halen Whitehouse (Assistant Keeper) and Mr. Mark Norman at the Department of Antiquities Ashmolean Museum, in Oxford for providing me with some articles from their collections on Egyptian conservation, and Ms. Julie Dawson, Conservator at the Fitzwilliam Museum in Cambridge.

To Prof. Fekri Hassan, Petrie professor of Archaeology at University College London, I owe a debt of gratitude for his great assistance in contacting Dr. Shawki Nakhla, who was the retired director of the Egyptian museum in Cairo, and for his organization with others, which enabled me to collect samples from Egypt in 1997. I am much indebted to the Egyptian Department of Antiquities for granting permission to take samples under special supervision from the sites of Egypt. My thanks extend in particular to Mr. Mohamed Hagras, Chief Inspector of Saqqara and his team for their help and assistance in providing samples from the new excavations in Saqqara and Dahsur and for visits to the sites, to Dr. Mohamed es-Sughayir who was the Chief Inspector of Upper Egypt for his permission to visit and take photographs as well as to Dr. Mohamed Naser, Chief Inspector of Qurna. My sincerest thanks to Dr. Ahmed Abd El Fatiah, the Director of the Alexandria Museum and to Mrs. America Abu Baker, the Head of the Conservation Unit in the Alexandria Museum for her great help and providing all the information that was needed. I also wish to thank numerous people in Egypt for their contributions, direct and indirect, throughout the collection of the samples from Saqqara, Luxor and Alexandria. Thanks to Mr. Roy Friendship-Taylor for his assistance in my visit to the site at Piddington and in the collection of the fragments of wall paintings.

I have received great assistance from Prof. Mohamed El Sharkawi, who is a geologist and was the Cultural Counsellor of the Egyptian Educational Bureau in London. I would also like to thank the following members of staff in the University of Leicester's Geology Department, who have given practical and academic assistance: Dr. R. G. Clements, and Ms Emma Mansely for the ICP analysis, Mr. Kevin Sharkey for the X-RPD analysis and Mr. Rod Branson for his advice on the use of the SEM at the Geology Department. I also thank Mr. George McTurk of the University of Leicester electron microscope unit who conducted the scanning electron microscope and energy dispersive spectroscope analyses.

To the library staff at ICCROM in Rome for their great assistance in using the library during my attendance at the mural painting conservation course in 1998, and the following people: Dr. M.

Cataldi at Museo di Villa Guilia, who helped me to have a private permission to visit work sites in Italy in 1998, provided me with a catalogue on the Domenei tomb (in Italian) and arranged a visit to the work site of the Tifone tomb: Ms Loreneza Dillessandro, who is a conservator and restorer in private practice in Rome, and was in one of the groups who have been conserving the tomb of Nefertari in Luxor and was able to get me permission to visit the Vatican work site. In Florence, I had the good fortune to meet Dr. Mauro Mateini, who is a chemist and head of the scientific laboratories OPD in Florence. I discussed with him the scientific methods for analysing Egyptian materials and he explained and provided information on the cleaning methods using ammonium carbonate, not only on interior wall painting but also exterior and gave me examples to visit in Florence. My thanks go to the team working in Arezzo for allowing me to visit the work site and for explaining the methods used for conservation in the church painted by Piero della Francesca. In Venice, Dr. Loranzo Lazzarini, who is a geologist at I. U. A. V, Department of Architectural History in Venice allowed me to visit his laboratory and discussed with me the methodology of analysing pigments, which he had used on some Egyptian materials. He kindly provided me with two tiny samples of pigments as well.

In particular, I would like to thank the Egyptian Government for my scholarship to undertake my Ph.D. in Britain and to carry out my field research in Italy in 1998, as well as in attend the conference in the following year and also the University of Leicester for the grant and for their financial assistance for my field research in Italy and to attend my first conference in Birmingham 1999, and my third one in Ottawa in 2000. Finely, I am grateful to Miss. Vicky Roe for her great assistance for proof reading this research.

Part 1-Text

Egyptian and Græco-Roman Wall Plasters and Mortars: A Comparative Scientific Study

Chapter 1

Introduction

The main aim of this research was to determine the parameters that would help to distinguish wall plasters, mortars and related materials in Egyptian and the later Græco-Roman period in various areas in Egypt and Piddington Roman villa in Britain. This was done by analysing wall plasters and mortars from different sites and periods using different analytical methods. This identified the technologies used in the past, compositions, structures and mineral contents. A secondary aim was to define the main forms of deterioration and decay, which had affected the walls and paintings. These would both help in the investigative stages of the conservation and restoration processes.

In recent years, researchers have become increasingly interested in analysing the materials for the conservation and restoration of works of art. The science of conservation is a systematic and structured method of understanding the material world and scientists aim to describe materials in an objective manner. Scientific ideas and theories are continually evolving in the light of recent discoveries and observations.

There is a great tendency to analyse materials and concentrate on the methods of analysis, preferably non-destructive, used for the diagnosis of works of art. This approach has been especially important in the responsibility for caring for archaeological and historical works of art concerning the nature, sources and mechanism of the deterioration factors.

The technological analysis of the materials using different instruments provides us with different information about the objects, from simple investigation to sophisticated tools, in case the information produced would not be enough to draw an outline on all the materials and methods used. The development of analytical techniques also allows us to detect the causes of deterioration and to understand the ancient techniques used, which could assist future restoration.

Scientific analysis can show the methods of manufacture and preparation of the materials, constructional techniques and, to see what has happened to the materials since their manufacture investigating the structure, composition and decay factors, and what caused alteration to the materials as some of these conditions combine together to cause the archaeological materials to deteriorate. The geological location affects the structure of the mixture and the materials, which in turn affects the surface of the painted layer.

As a consequence, this has led to the development of methods of diagnosing the problems and provides scientific information about the materials, thus allowing understanding of the techniques used and leads us to determine the materials and methods to be employed for conservation. The systematic approach of examining the material's composition can be varied to give us all the information required. This depends on the methodology used and what we need to know about the materials of wall painting, their surfaces and compositions.

A previous study has been carried out on all the tools and instruments for conservation. A classification of these tools for the materials analysis of works of art is given by Mairinger & Schreiner (1982), Burnstock (1998) and Ferreti (1993). These tools need highly specialized workers and analysts each of whom can deliver a limited set of data, which will answer some of the questions. It is my view that we should combine techniques as necessary, but first use simple methods without expensive instrumentation, where these methods would be enough to identify all the problematic materials within works of art. The majority of the analytical methods can be applied, but a question addressed to the methodology could be used to examine and analyse wall plasters and mortars. From this point of view, the methodology involved in this research has been designed in order to obtain the different information about the plasters and mortars. Figure 5 (page 42) shows an outline of all the instruments, which the methodologies have required for the analysis of wall paintings (plasters, mortars and pigments). However, the author saw that by using different instruments one could obtain all the information required, understanding the characterisation of the materials used for wall paintings and their techniques as well. The methods of examination and analytical techniques allow us to identify and determine not only the original materials that were used, but also to define the causes of the alteration factors, which affected both the painted layer and the rendering.

The classical and literary sources provided important information on all the materials used. The most important were; Vitruvius, *The ten books on architecture* (1960), and Theophilus (12 century) *The various arts* (1961). Vitruvius lived in 1st century BC, and described the techniques and materials employed in classical times.

Lucas, in two books "*Antiques their restoration and preservation*" (1932) and "*Ancient Egyptian materials and industries*" (1989) described all the materials used in Egyptian times as well as the analyses of different materials. These I consider to be the most useful studies done on Egyptian objects. However, there is Nicholson and Shaw (2000), "*Ancient Egyptian Materials and Technology*", but the book came in at the end of my study and could not be taken into account. A study by Laurie, *The Materials of The painter's craft in Europe and Egypt from early times to the end of the XVVII TH century, with some account of their preparation and use* (1910) and "*The pigments and mediums of the old masters*" (1914) details all the techniques of wall paintings and materials used, as well as other painting techniques, e.g. encaustic, dry oil etc. In addition he focused on the methods used by Cennino Cennini and Theophilus. Finally Davey's "*A history of building materials*" (1961) was one of the most valuable sources to the present study.

However, the present research is concentrated on Græco-Roman wall painting in Alexandria. The studies carried out on the wall paintings in Greece itself were limited to the art historical techniques of wall paintings, most actual wall paintings having been lost. However, fortunately some studies have been done on the classical wall paintings in Alexandria. The most important, by Adriani (1936), concentrated on the wall painting of the Mustafa Pasha tombs, compared with some examples from other places in the Greek world. He has given us an explanation of the architectural style as well as the decoration executed. It is considered an important study in terms of art history, because it includes the decorative style in addition to techniques. Whilst Brown (1957) has reported on the Ptolemaic paintings and mosaics in Alexandria, he also reports on the styles of Alexandrian painting in terms of Stela, as well as the developments of the Alexandrian style. In addition, he pointed out the similarities between those and the Greek world.

Noshy (1937) in *The arts in Ptolemic Egypt: A study of Greek and Egyptian influences in Ptolemaic architecture and sculpture*, studied the decorative styles used and noticed the similarities between the architectural style of the Alexandrian tombs and those of Macedonia, as well as Greek Hellenistic houses. There was also an article by Rostovtzeff (1919), "Ancient decorative wall painting", important because Rostovtzeff has referred not only to the decorative style used in the Greek world, but also points out the general scheme of mural decorative painting of the Greek and Roman period. It is noteworthy to mention that the most interesting decorated vaults, as he states, belong to the earlier Hellenistic period in Alexandria and its neighbourhood. Studies of the history of vase painting during the Hellenistic period show a change from monochrome to polychrome, and the history of painted glass vessels shows a similar change. He explained how decorative style developed in the Greek world and discussed other techniques such as incrustation style and mosaics.

A wide range of studies has been carried out in recent decades on the analysis of materials in wall painting, in particular analysis of pigments, methods included; X-ray diffraction, polarized light microscope (PLM), electron probe micro-analysis (EPMA) and basic chemical testing. Numerous studies have been carried out on Egyptian and Græco–Roman samples (Profi *et al.* 1974, Filippakis *et al.* 1979, Lazzarini 1998, Rees-Jones 1990, Calamiotou *et al.* 1983, Green, 1995 & Corbeil *et al.* 1996). The pigments have been widely studied (Gettens & Stout 1966, Gettens & FitzHugh 1966, Gettens *et al.* 1972, Gettens, FitzHugh 1974, Gettens *et al.* 1974 and Baines 1985). However, the most interesting study done of Egyptian pigments was by Schiegl *et al* (1989 and 1992) who analysed more than 1000 samples from different parts of Egypt. Another study by Weatherhead and Buckley (1989) and Weatherhead (1995) was carried out on samples from Amarna. Egyptian blue has attracted most study, not only for its history but also for the study of ancient technology.

Several studies have been carried out on the analysis of mortars, using various methods (Jedrzejewska 1960, Malaguzzi-Valerj 1970, Penkala & Bralewska 1981, Reller Jedrezejewska 1982, Stewart & Moore 1982, Lewin 1982, Perander 1982, Preusser 1988, Schafer & Hilsdorf 1993, Martinez-Ramirez *et al.* 1996 & Wild 1992). In addition more recent research on the technology and analysis of Roman wall painting has been published by Béarat *et al.* 1997 and Burnstock (1998).

I saw the need for a systematic study into the scientific aspects as yet only touched on in the examination of wall painting, using analytical techniques in combination to analyse wall painting in terms of plasters and/or mortars. Technologically, analysis can show the methods of manufacture, the preparation of materials and also construction techniques. Analysis can show the quantities of aggregate and lime used, and also show the various sizes of materials and kinds of rocks, natural deposits or crushed materials which were used to produce such plaster or mortar.

The strategy of examination and analysis of these materials is divided into three stages:
1-Preliminary examination: A microscope was used at 10X magnification and optical microscopy of polished cross-sections to identify the structure of the mortar and the painted layers as on initial examination.

2-Chemical analysis: Microanalysis (spot test). Standard methods were used to identify the quantitative and qualitative nature of the composition of the plaster and mortars. Measurements of calcium carbonate ($CaCO_3$) content of the layers and in some instances, the deposits, which covered the painted surface, and analysis of water-soluble salts. Analysis of pigments using micro-chemical tests was used to determine their composition.

3-Physical methods: X-ray diffraction and X-ray powder diffraction (X-RD & X-RPD) confirmed the mineralogical compounds in the plasters and pigments. Induction coupled plasma spectroscopy (ICP) analysis was for other metal ions present in the materials. Polarized light microscopy (PLM) was to assess for internal structure. Scanning electron microscopy and dispersive X-ray microanalysis (SEM & EDS) were used for surface structure and to define deterioration and decay of factors.

The point of focusing on the analysis of plasters and mortars is to obtain the quantitative and qualitative nature of the mixture used. These are considered the most important aspects in the restoration, in terms of re-plastering.

There are several reason for this. Firstly, once the restorer know what materials were originally used that can then consider what materials would be appropriate for the restoration and conservation.

Also they should be prepared and applied using methods as similar as possible to the original to prevent an imbalance between suitable and unsuitable ancient and modern materials as observed by analysing a few samples of the re-plastering (see Chapter 6.2, 2.2.3). This may help to prevent efflorescence and discolouration of both plaster and mortar and the painted layer as well. The passage of water borne salts from the sand or aggregate used can be avoided, if one uses the same locally obtained materials with the same ratio of the mixture that was used originally.

From this point of view my research involves the analysis of plaster or mortar using standard methods enabling a calculation of the materials used to make the rendering. Although the methods considered are destructive, they have an important role in showing the percentages of both sediment types and the acid soluble component which are essential for restoration work and historical compositions. The present research provided a great opportunity to collect samples of re-plastering from different tombs and different areas as well as to analyse and distinguish between the materials used for re-construction. It was possible to point out the similarity and the differences between the original rendering and new ones used for the conservation of wall painting in Egypt.

We need to study not only the methods, which were used, but also materials utilized for pigments and the supporting structure for painted lime mortar and plaster. The techniques are extremely important in order to understand colours, pigments, adhesives and the paint layer of plaster or mortar used, as well as the motifs and the subjects in mural painting which relate to the particular century of manufacture. This is why analysis is important before any decision is taken to restore or treat the murals. The analysis allows us to get more information about the materials used and their structures. Restorers should use the similar materials, but it is sometimes difficult to find them, in which case the nearest materials should be employed. The analysis of materials, is an important aspect for the restoration of works of art, allowing the restorer an insight into the materials used and informing them of possible choices of restoration processes.

The materials used for the manufacture of plasters and mortars varied according to the main sources available. For that reason different sites were selected from which to collect samples as the aim of study was not only different periods of plasters and mortars, but also different areas. The samples came from four sites: Saqqara, Dahshur, Luxor and Alexandria in Egypt, dated to different periods between 3100-332 BC. The samples were collected from tombs from different centuries as well as different areas to compare and distinguish between the materials used for rendering. The fifth site was Piddington Roman villa in Britain dated to 4th century AD. This is described in detail in Chapter 6. Piddington is a closely dated Roman site, although the geology is different, the manufacture techniques are the same as those described by Vitruvius and studied in detail by Graham Morgan who analysed mortars, plasters and pigments from different Roman sites in Britain (1992), in addition to Béarat *et al.* 1997.

Samples of stone and mud brick (as supporting structure) were also collected as they might have contained saline solutions and it was necessary in some instances to look at their structures. Plasters and mortars and pigments were collected as well. Different surface salts were also collected to identify the kinds of salts that covered the wall paintings. Samples of modern re-plastering were collected to identify the composition and to see to what extent they differed from the original ones.

The research is divided into two parts. The first eight chapters, begin with the general introduction. The second includes further studies on art history from Egypt up to Græco-Roman times, reporting on the materials used and their techniques in antique and classical times. As this study is aimed at the examination of the wall paintings and the materials used, it was necessary to indicate the materials and techniques which applied in those times.

The different kinds of plaster and mortar were mentioned by the classical writers to distinguish between the techniques and their preparation. I have illustrated a scheme for studying wall paintings shown in Figure 2 (see page 12).

Over time, changes occur in mural painting materials and cause the alteration of pigments and their decay with different processes and their supporting structures on which the mural was painted in each century. As illustrated in figure 3 (page 28) the causes of deterioration and decay are varied. The agents of deterioration and preservation depend on the nature of the materials and the environment. The causes of decay affect artefacts and resultant changes in composition affect the colours and cause physical deterioration and the breakdown of the structure of materials. Before we make any decision on conservation measures, we have to study the causes of alteration and the techniques which were used, in addition to the effects of the general environment on mural paintings.

This is one of the most necessary aspects to be considered in order to understand the decay of materials and their alteration in both the painted layers and the rendering, as these factors are varied and could be combined together under certain conditions. As a result it was found that there was an essential need to point out the main reasons for the decay and deterioration of wall paintings, which are described in Chapter 3. It has been noticed that the most common decay of wall painting is caused by soluble salts.

Chapter 4 provides an introduction to the analytical techniques and methodology used in the examination of Egyptian and Græco-Roman wall paintings, and outlines the analytical techniques employed in this research, beginning with literary information and a description of the methodology used to analyse and examine plasters or mortars and pigments where they are found on the plaster. Simple chemical tests were carried out to identify the kind of plaster or mortar, e.g. calcareous or non-calcareous aggregate, and to distinguish materials such as gypsum (calcium sulphate) or chalk (calcium carbonate). Effervescence, due to the formation of carbon dioxide gas with acid, indicated the presence of a carbonate.

Chapter 5 contains an introduction to the five sites: historical background, samples areas, the strategy of taking samples and nature of samples as well. In addition, it described the deterioration and decay that affected the plasters, mortars and the fragmentary wall paintings from Piddington, as an example from different environments.

Chapter 6 contains the case studies of the examination of materials from wall paintings (supporting structures, plasters, mortars, pigments and salts) from the four sites of Saqqara, Dahshur, Luxor and Alexandria in Egypt and Piddington in Britain, in addition to the summary of the results of the chemical and physical methods. It was extremely important to diagnose the main characteristics of the materials used in the wall painting, and to understand the techniques that were employed, in order to ensure the right methodology for conservation. It is necessary to treat, not destroy, objects of art.

As a result of examination and analysis, conservation, in terms of cleaning and consolidation, was carried out on some fragments of wall paintings from Piddington Roman villa. It seemed to be necessary to review the techniques of cleaning and consolidation and describe the methods used for the cleaning of wall painting fragments, to test the effectiveness of the analytical processes as a guide to better conservation these are described in chapter 7.

A final conclusion, recommendation and a plan for future research is contained in chapter 8.

The second part of the study contains all the appendices related to the research.

I would point out that there are important factors to be considered. The first concerns the study of the objects of art, which does not mean only scientific methods and the application of sophisticated tools to diagnose problems, but also consideration of the associated environmental conditions, which act as an integral part of the cultural composition. Secondly, the working environment has to be appropriate to the different backgrounds of the art historians, restorers, biologists etc. They have to work together to ensure that the scientific work achieves the correct application of conservation and restoration.

As a result, my research involved the examination of wall paintings using different techniques to pursue a strategy for the examination of wall paintings which could be carried out without expensive instrumentation. This first stage is to distinguish the materials and the techniques used, to draw an accurate outline of the objects and to determine the proper processes for the conservation and restoration of wall paintings.

Chapter 2

The Art History and Materials Used for Wall Paintings in Egypt up to Græco-Roman Times

1-Introduction

The study of the techniques of works of art is extremely important because it helps us to understand the relationship between structure and appearance. It also provides us with the information which identifies the origin and history of the works, as well as the techniques which were used. In addition, it helps to determine the technological methods for laboratory examination not only in terms of defining the problems of the objects, but also of making decisions concerning the conservation methodology to be used.

This chapter will explain the materials and their preparation, as well as the techniques, which were used in Egyptian and Græco-Roman wall paintings, following this will give a brief account of the history and style of wall paintings will be studying, in terms of the relationship between the style and techniques used.

2-The History of Wall Painting

It is known that the first simple murals were made at the beginning of the upper Paleolithic period (about 30,000 years ago), when according to Mora, people either used natural materials to make impressions, or alternatively, coated the rock surface with grease and blew pigments through a tube onto the rock without any preparation of the surface (Mora 1984: 69-73). However, as Abd El Salam explained, the beginnings of wall painting in Egypt is dated to the Pre-dynastic period (circa 3300 BC) during which a surface, which had been covered with mud plaster and another whitish plaster, was then painted in varying shades of red, blue, green, black, white and yellow (Abd El Salam. 1992: 91).

Lucas (1932) states that “in the history of mural painting, temple painting is dated to the early 18th dynasty (1500 BC) and continues through to Ptolemaic times (332-30 BC). However, at Knossos in Crete it is dated to 1500 BC and is even later at Tiryns, on the mainland opposite Crete. In Egypt, at Amarna murals date back to in the reigns of Amenophis 111 & 1V (Akhentenl) (411-1373 BC), as well as at the houses of Herculaneum and Pompeii, however, as Lucas notes, only to be destroyed in the first century AD” (Lucas 1932: 139). A brief outline of the beginnings of wall painting, and early examples of Egyptian painting, can be found in Appendix 1. (See also Davies & Gardiner 1936, Smith 1949 and Robins 1997).

3-The Decorative Style Used in Ancient and Classical Times

There is an integral connection between the mural and the structure of the wall in the history of wall painting. Rostovtzeff (1919) states that “the structure of the building affected the decoration style, which was dependent on the division of the wall according to the building materials used”. The most primitive system of mural decoration of houses and vaults in Greece were employed using a simple idea prompted by the structural system of the walls, which were built up of unbaked brick, while large stone blocks were used for the foundation. Wood, stone or twigs were used for joining together the foundations, the brick wall, as well as the top of the wall to connect with the roof. While the central part of the wall was painted red, the cornice and intermediate portion were painted with geometrical and other designs. In contrast, the base was painted in imitation of stone coloured, different from the central wall-space. This was the earliest stylistic decoration of the ancient unbaked brick wall, a primitive example of which has been found in Kertch, southern Russia, along with Vassiurin’s advanced style progressive in the decoration, which was found near Aropa on the

Black Sea. The division of the walls was dependent on the base, the intermediate portion, the central part and the cornice (see Appendix 2) (figures 1 A and B). The intermediate part and the cornice were adorned with painted ornaments, simulated *stucco* and painted stone. The imitation of coloured stone and marble was applied onto the lower part; Although until the end of the Hellenistic period marble and coloured stones were commonly used, in his study, Rostovtzeff states that the reasons for the use of imitation was due to economic reasons (Rostovtzeff 1919: 147, 148, 127).

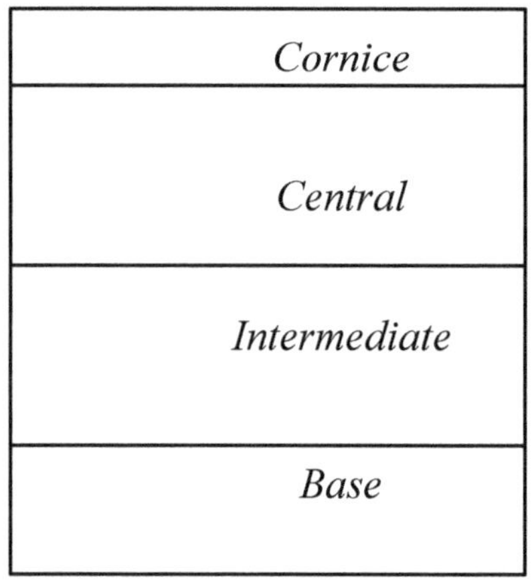

1 A

1 B

Figures 1 A & B. A, a simple diagram of the early division of the wall for painting.
B, a section of a painted wall in discovered a vault at Pantikapaion in 1908 (Rostovtzeff 1919: plate 1).

Brown in his book (1957) gives an explanation the decorative scheme of the Mediterranean world and he explains the remarkable significance of the development of Alexandrian styles. Where different, the colour techniques were limited to local colours, which were applied in a wash and less grandiose shades than those found in Pompeii and Rome during the first century BC (Brown 1957: 86). In Hellenic times another development was the notion that covering of walls with slabs of different coloured marble instead of *stucco*, but retaining thus, the structure of the Greek wall built of squared stone. According to Rostovtzeff this gave a richer harmony of colours throughout the Hellenistic world, which, in turn, was called the First style, or Incrustation style, or the Greek Structural style (Rostovtzeff 1919: 150) (see Appendix 2).

Vitruvius gave a description of the decorative elements used in wall paintings in Roman times (Vitruvius 7, 5, 2). Ling (1991) states these styles were divided by Augustus, of which the first style, which dated to between 200-80 BC, imitated *stucco*, which the exterior walls of public buildings, were decorated in or the marble veneering, imitation of veneer or alabaster. The second style, dated to between 90-10 BC accurately reproduces architectural elements and sculpture in paint, creating vistas and landscapes. The third style, dated to 30 BC-AD 50, introduces framed panels in the classical style, (which was popular in other forms of Augustan art) (see Barbet 1982). The fourth style, dated from about AD 50 onwards can be seen in the interior decoration of the golden house (see Ling 1991) of the emperor Nero.

Following of previous description on decoration style. I will consider now the techniques of wall paintings. In Figure 2, I have outlined the elements worth considering when studying wall paintings with regards to *fresco* and *tempera* techniques. This is divided into three parts; firstly, the supporting structures which were prepared with different kinds of rendering (plaster or mortar): Secondly, the wall paintings techniques which were applied in antiquity and thirdly, the pigments, adhesives and resins which were used in wall paintings. This should enable us to understand the history of monuments and their materials and provide technological data, before researching any laboratory work on the analysis of wall paintings.

Figure 2

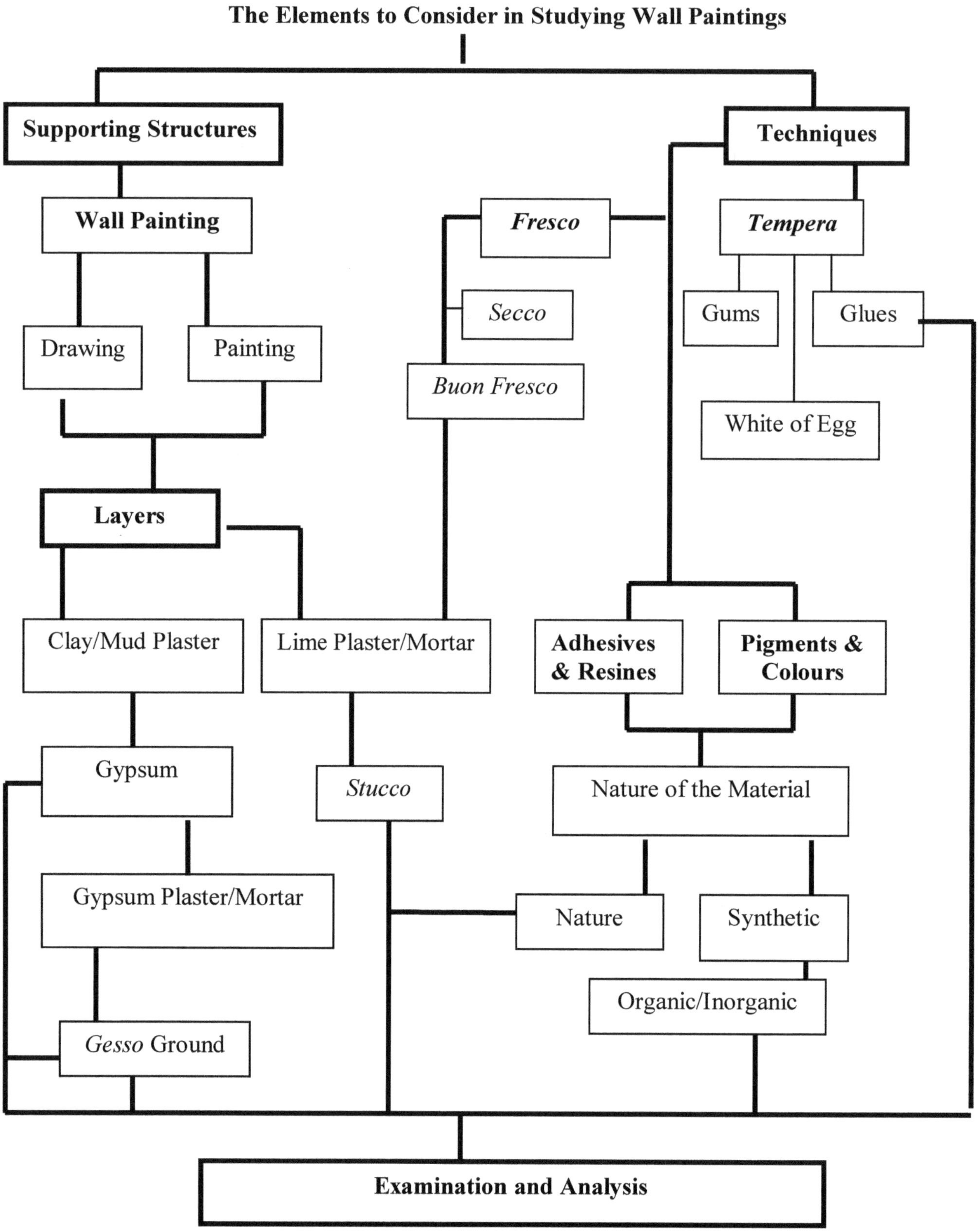

Wall painting, (as it is applied to and made integral with the surface of a wall or ceiling), is different to all other forms of pictorial art, in that, whatever the materials used, it is an integral part connected with architecture. The use of colour, design and theme treatment can radically affect the sensation of spatial proportion.

4-The Elements to Consider in Studying Wall Paintings

4.1-Supporting Structures

In Egypt, the building materials used were sun-dried brick, limestone and sandstone covered with plaster or mortar. Sun-dried brick was one of the oldest materials used in building and it has been discovered that the oldest bricks used in Egypt date back to the pre-dynastic. They consisted of a mixture of Nile alluvium and clay, as well as sand, which contained some impurities. Lucas notes that limestone consists of calcium carbonate and other minerals, e.g. silica, clay, and oxide of iron and magnesium carbonate, and differs not only in quality, good or poor, but also in texture and hardness, according to the geological areas where it was formed (Lucas 1989: 49, 52-55). Limestone was widely used from 2600 BC until 1550 BC, when they started to use sandstone in the middle of the 18th dynasty. While normal limestone was used inside the building, the best limestone was used to coat buildings, exteriors thus providing the best carving stone (Abd El Salam 1992: 360).

4.1.1-Rendering Materials and Their Manufacture

Up to classical times the renderings used were varied as shown in Figure 2. Davey (1961), Hodges (1974) and Lucas (1989) comment that different kinds of plasters or mortars were applied. While it is necessary to state the rendering compositions which were applied to form a suitable surface for painting, and point out the differences between them, it seems essential to analyse of the manufacturing methods by which they were produced. The basic materials for consolidating and surfacing walls were mud or clay, lime plasters or gypsum plasters (Hodges 1989: 170). Lucas states that the mortar used in ancient times was a type of clay used with sun-dried brick, while gypsum was to be used with stone. He also mentions that during the pre-dynastic period, the painting was executed onto clay plaster. At later periods, (e.g. during the 18th dynasty at El Amarna), clay plaster was used as a ground. In both palaces and houses, the walls were covered with clay plaster, which had a coat of white (gypsum) and then paint (Lucas 1989: 74, 353).

4.2-Clay Plaster

Clay, which is considered to be the best material for making brick, is a product of the natural decay and disintegration of igneous rock and shale, combined, at times with alluvium, wind-blown materials with sand and silt. As Davey notes, it is composed of both hydrate silicate of aluminum and hydrate oxides of both aluminum and iron (Davey 1961: 64). Lucas states that clay has high plasticity, and the cohesive properties of mud, enabling it to adhere, without the additives of adhesives. However, clay is rich in composition and, as a result, the brick not only dries slowly, but also shrinks and cracks, leading to loss of shape. A depiction of the brick-making of ancient Egypt was represented in the tomb of Ra-kh-Me-Re, (18th dynasty). Lucas mentions that chopped straw and animal dung was added to the mixture to increase both the strength and plasticity of the clay (1989: 49, 50). Clay also contains impurities; such as humus (an organic material formed from decayed leaves and plant material which is dark in colour), calcium carbonate, pyrites and gypsum. In addition, clay can also be re-deposited by water or wind and become contaminated during transportation (Davey 1961: 64).

4.2.1-The Use of Clay or Mud Plaster

Clay plaster was used to cover both the mud brick and limestone walls of tombs. However, it was not always coated with a thin layer of white plaster before painting (Lucas 1932: 184). As mentioned before clay was used from the pre-dynastic period to early dynastic period (as well as in the 18th dynasty). Lucas reports the use of two kinds of clay plaster; coarse clay plaster (Nile alluvium) a mixture of clay and sand varying in proportion mixed with calcium carbonate and a small amount of gypsum, whereas the clay plaster, better known as "*hib*," was a natural mixture of clay and limestone that had been washed out by the rainstorms (Lucas 1989: 79). Hodges argues that "the high ratio of clay as an insoluble residue suggested a calcareous mud or clay plaster, while a low ratio indicated a true plaster (1974: 171)". Clay plaster was not only applied in antiquity (Egypt and Mesopotamia) as a rendering preparation, which tended to be coarser in the late eras, but it has also been discovered to be the basic rendering for mural paintings of the classical Greek and Mycenaean age. Two thick layers consisting of a mixture of mud and rubble with impurities of calcium carbonate were covered with a layer of clay. Mora says this was the transition between ancient peoples and Greeks way of preparing the walls and is seen in the *fresco* technique. He also mentions another example of Coptic painting in Nubia, but this was executed in *fresco secco*. The surface was consisted of mud sand and straw, and was covered with a thin layer of kaolin, thereby proving that despite the Greek and Roman influence these traditional methods continued (Mora 1984: 84, 75).

4.3-Gypsum

Gypsum is a natural material, varying in colours and composition (white, pink, grey, and light brown). Chemical changes can occur in the iron compound, which is incorporated in the gypsum mixture, and are capable of change the colour of the plaster to pink. Such an event occurred in the tomb of Tutankhamun. However, the grey colour is due to the presence of small particles of burnt fuel. Gypsum occurs in two conditions: firstly, as a surface rock formation, which is found in the Mariout region to the west of Alexandria in the district between Ismailia and Seuz, in the Fayum and near the Red Sea coast. Secondly, it is found below the surface of limestone desert; and it is this later that was and still is, used for making plaster (Lucas 1989: 77, 78). It can be either pure or contain impurities; clay can contain lime or gypsum, therefore, poor quality gypsum could be impure, containing calcium carbonate (Hodges 1989: 171).

4.3.1-The Preparation of Gypsum

By calcining gypsum at a low temperature (130-170°C) for three hours, upwards of ¾ of the water content in gypsum is driven off, to produce Plaster of Paris ($CaSO_4.½ HO_2$). When the temperature is increased to 400 °C and upward, calcium sulphate ($CaSO_4$) anhydrite is formed (Davey 1961: 92). Hodges suggests that most early plaster may have contained charcoal due to the crude kilns in which gypsum (calcium sulphate) was formed (Hodges 1989: 171). In other words, calcium sulphate can be formed or by chemical reaction from other sources under certain conditions.

4.3.2-Gypsum Plaster

Gypsum Plaster was well known from the early dynastic period and was used for the walls and ceilings of houses, temples and tombs in order to provide a suitable surface for painting to coat clay plaster and stone surfaces (Lucas 1989: 77) Gypsum plaster has been used for many purposes in Egypt and was used in both ancient Greece and Rome (Lucas 1932: 181). It was rarely used by the Romans in Britain, and was re-introduced to England in the 13th century. Therefore there is evidence of its fairly common use in areas where it developed naturally (Simpson and Brown 1994: 4). Mixed with calcium carbonate, it has also been used as an artificial mixture to provide a white plaster base for painting or as an application of a coat of calcium carbonate (which contains some traces of gypsum) because it adheres well to the surface of stone or clay plaster. While a coarse quality was used to cover faults or irregularities in the stone, a finer quality was used to provide a suitable surface, using a white wash, of whiting and glue, to fill the pores and smooth the surface before painting (Lucas 1989: 77, 353).

4.3.3-Gypsum Mortar

Lucas notes that there are two types of mortar were applied between the edges of the stone to protect them. One type which was a mixture of powdered limestone and unidentified organic adhesive, and another type which was of calcium carbonate (Lucas 1989: 75). Furthermore, according to Hodges, there was also gypsum mortar composed of a mixture of Plaster of Paris and sand (Hodges 1989: 171). It has been identified mixed with ceramics, within the walls of a theatre in Alexandria which is dated to Roman times (Penkala and Bralewska 1981). The reason for the use of gypsum in Egypt was due to the hot weather, which made it suitable for painting. Lime was used in both Greek and Roman plaster because gypsum is ineffective in a wet climate when it is applied onto exteriors doors because it is liable to weathering (Lucas 1989: 75).

4.4-*Gesso* Ground

In Medieval Italy and Spain, gypsum was mixed with glue water or size to form *gesso*, which was employed by the artists to produce a panel as a base coat for painting. Davey explains that glue was produced by cooking animal (horse) hair and hooves in a caustic soda solution. Such solution a material must be added to delay the setting process of the mixture of water and Plaster of Paris; otherwise it would set too quickly (Davey 1961: 95). Two different forms of *gesso* were produced. The first, applied in several coats is (*gesso gresso*) a thick gesso of plaster of Paris mixed with size, which would crystallize into a hard mass when used for the first layers. The second is *gesso sottile;* a thin *gesso* for the finishing layer, which is mixed with Plaster of Paris and a considerable amount of water in order to produce a smooth, silky *gesso* surface. Both were well known in Cennini's day in the 15th century (Constable 1954: 69, 70).

However, to Egyptologists, the term *"gesso"* means a plaster made with chalk (whiting, whitening) and glues, the reason that it gives additional strength and cohesiveness as a binder. It was widely used in ancient Egypt to apply to wood as a base for painting and gilding as well as for covering the *cartonnage* mummy masks and coffins (Lucas 1932: 182). *Gesso* was also used for architectural decoration, evidence of which was found in certain temples of the New Kingdom, as a coat was applied onto very fine linen and glued to the sandstone, then the *gesso* was carved in low relief and painted (Lucas 1989: 4).

To summarize, it is necessary to point out the differences between the two materials which have been used for the preparation of *gesso*, namely chalk (calcium carbonate) or gypsum (Plaster of Paris). Lucas argues that the name "*gesso*" was used irrespective of the basic materials or purpose. He states that Theophilus refers to the use of both gypsum and glue, or chalk and glue for coating skins as a ground for painting. Constable, as previously mentioned, refers to the *gesso* ground as a mixture of size and gypsum or whitening. Lucas also mentions that Church has stated this method (size and gypsum) was the ordinary ground in Italy and Spain (Lucas 1932: 182, 183). There is great evidence to prove that the *gesso* used by the Egyptians differed from the ones used by the Italian and Spanish artists.

4.5-Lime Plaster

Lime has been applied as a lime wash since the Neolithic period, and in the historic period, lime plaster features in the Minoan civilization (Torraca 1988: 67). The process of slaking lime after calcining is described by Vitruvius (1960), Hodges (1974) and Torraca (1988). Limestone loses a third of its weight, converting the calcium carbonate to calcium oxide (quick lime) at a temperature of 700-900 °C. When mixed with water, it produces calcium hydroxide (slaked lime) $CaCO_3$ 700-900 °C $CaO + CO_2 \rightarrow CaO + H_2O$ $Ca(OH)_2$. However, if the limestone is overheated it does not react well with the water and, as a result, it forms lumps. The reaction of carbon dioxide in the air produces a slow conversion of calcium hydroxide to calcium carbonate, by the absorption of carbon dioxide and the hardening of slaked lime $Ca(OH)_2 + CO_2 \rightarrow CaCO_3 + H_2O$.

However, lime of a good quality depends on the nature of the limestone; pure limestone produces a very fat lime and produces a soft consistency; Grey limestone is slaked very slowly and, if the lime contains clay and lumps, these further complete the slaking on the surface and cause it to crack. The process of burning lime over a wood fire is considered to be the best. If it is burnt with coal, the lime will absorb the sulphuric acid and turn to gypsum, which prevents the setting of the *fresco* and causes efflorescence.

4.5.1-The Use of Lime Plaster or Mortar

Lime plaster was employed for Neolithic statues based on lime plaster mixed with limestone fragments, quartz particles and clay (Tubb 1987). Lime plaster or mortar, is produced by mixing slaked lime with plant fibre, animal hair or sand to strengthen the mix. Vitruvius mentions different types of sands, used in the mixture of the plaster or mortar; "*pozzolana*"; (see glossary) "pit sand" (volcanic earth), river or sea sand. He describes the recipe that was used for the plaster mixture, which consisted of six coats made of lime. The first three coats are coarse and he notes different kinds of sand, in the case of *pozzolana* (3 parts + one part of lime) and river or sea sand (2 parts + one part of lime + one of burnt brick). However, when sea sand was used in the plaster and this was coated with *stucco*, a salty efflorescence was given out and spoiled the surface. The disadvantage of using sea sand is that it contains salt. The last three coats contained marble sand and marble dust instead of sand. The mixture's filler was used in order to avoid any shrinkage or cracks, because slaked lime was never used on it own as a plaster or mortar. Sand makes the plaster porous, thus allowing the transformation of the slaked lime to a carbonate. The solidity of the plaster depends on the quality of the composition (free of impurities or containing clay or humus) as well as the shape of sands used (Vitruvius 2, 5, 1-3, Hodges 1989: 170, Torraca 1988: 68, 69 and Doerner 1969: 268, 270).

Although the discovery of lime mortar was near Baghdad dated to 2500 BC proves that lime mortar was used in Mesopotamia, *tempera* techniques were not used (Mora 1984: 75). In the Hellenic period (around 4th BC) men discovered that earth of volcanic origin mixed with slaked lime formed mortars which hardened even under wet conditions. Torraca indicates the composition of mortar used in Roman times, which was a mixture of lime, pozzolana, broken brick or tile and pumice aggregate, which was preferred for use instead of terracotta, as it is very light in weight (Torraca 1988: 69, 67). The process of slaking lime is considered an important factor in producing a lime mortar, otherwise little un-slaked pieces could be concealed and that would spoil and fragment the surface of the *stucco*, completing the slaking on the surface. Vitruvius mentions the application of a hoe to the slaked lime as a test of its slaking; if it sticks partially to the hoe it is an indication that the lime has not been fully slaked, while if it is clean and dry it means that the lime is weak, whereas rich and well slaked lime will stick like glue and indicates that it is fully tempered (Vitruvius 7, 3, 2). Hydraulic lime is formed by combining calcium silicate and aluminate, which reacts with water to form hydraulic lime. The formation of silicates and aluminate of calcium is a result of the reaction of the lime, which is formed by the decomposition of limestone with silica and alumina contained in the clay (Torraca 1988: 72).

The issue of the use of lime in Egypt has been disputed by modern writers, although it has been found in Mesopotamia, it was apparently used in Egypt. Davey said that lime was not used before the Ptolemaic period; although limestone was suitable for burning to lime, there is little evidence to prove that lime plaster was used before the Ptolemaic period (1961: 94). However, Lucas states that all the plaster used in Egypt was gypsum plaster because the temperature required for converting calcium carbonate to lime is about 900 °C. On the other hand, a temperature of 130 °C is not enough to convert the calcium carbonate in the mixture of gypsum into lime. Therefore, the gypsum plaster used was a mixture of calcium carbonate and sand, which is naturally occurring, without artificial additives, and that has corrected the belief of gypsum being a mixture of lime and gypsum (Lucas 1989: 77-79).

4.6-The Preparation of Walls in Ancient and Classical Times

In the Neolithic period, the irregularities in walls were levelled and the rock surface strengthened by covering them with clay as a ground for painting (Mora 1984: 72-73).

4.6.1-In Egypt the product was similar to the first example, and was used to treat the irregularities in stone by covering them with clay or mud plaster as previously mentioned. Whatever the building material used, it was prepared with a suitable rendering for paintings. The preparation of the

Egyptian walls, as well as the techniques used, seem to differ greatly from the classical methods due to the materials which were applied. The preparation of surfaces firstly depended on the nature of the supporting walls, and, secondly, on the decorative technique to be used as in the case of good quality stone. If the walls were smooth, a layer of gypsum covered the surface. If the walls were irregular a layer of plaster, which was a mixture of silt and chopped straw, was applied to fill the irregular surface before a layer of gypsum was added. The two main techniques used for the decoration were carving and painting the limestone (coloured relief) and painting directly on to the plaster.

4.6.2-Classical Times. The methods given by Vitruvius were the application of three layers of sand mortar as a (rough coat) and then a mixture of marble dust applied in three or more coats. The second layer of the marble dust was coated in a medium substance and rubbed-down before the fine layer was put upon it. These methods helped to prevent any cracks and added to the solidity of the wall. Rubbing not only smoothed and hardened the surface, but also revealed the sheen of the marble. Whilst still wet, the wall was painted, and then polished when dry to enhance the colours (Vitruvius 7, 3, 4-9).

4.7-The *Stucco* was used by the middle of the early Minoan period in Crete, *stucco* was applied to rubble masonry as a protective coating. In Greek and Roman times marble was added to the *stucco* mixture for use in covering stone work. The coat of *stucco* was used on internal and external wall surfaces for architectural decoration. *Stucco* was applied as a rendering of lime, gypsum or cement mixture, and painted over whether wet or drying and decorated when required. The application of *stucco* involved numerous coats of which the undercoat was a coarse mixture of lime, small pebbles and crushed pottery, onto which three coats of lime *stucco* were applied, to a total thickness of about ¾ inch (Davey 1961: 112, 113).

Laurie described the most interesting paintings, namely the Mycenaean *frescoes* (Knossos Crete 1500 BC). The rendering contained three layers; a coarse rubble mixture of lime and clay (½ inch in thickness), then a fine plaster applied in coats (the first ½ inch and the second ¼ inch in thickness). The fine plaster was of lime without any sand or marble dust, which indicated that it had been slaked for a long time before use. The pigments were applied without any medium, which indicates the use of the *buon fresco* technique. The colours used were natural pigments; yellow ochre, red oxide, a red made by burning the yellow ochre, another shade of red, which was made by burning the yellow, a black made of shale and Egyptian blue (Mayer 1973: 334).

The methods used in classical times formed a firm compact surface with the use of sand and mortar, and also marble dust for rubbing the surface. Not only were Greek plasters applied using these methods, but also lime was added to the mixture of the mortar for the production of slabs. Roman plaster was very thick, as numerous coats were applied, sometimes before the first coat had dried, helping to hold the water contained in the plaster keeping the plaster damp while the painting proceeded on the surface (Laurie 1910: 3, 89, 93). The final layer of plaster consisted of pure lime unadulterated by other materials. Marble dust and sand were used in undercoats of rough plaster and for *stucco* in the coarser under coated plasterwork (Mayer 1973: 335).

4.8-Techniques of Wall Paintings Used in Ancient and Classical Times

The study of the techniques used is extremely important for the conservation of works of art, in order to aid the diagnosis and, through analysis, distinguish the techniques, not only because of the preparation methods used on the wall, but also because of the materials which were used for painting (figure 2). In antiquity, two techniques were used and developed; *fresco* and *tempera* (Hodges 1989: 160). Davey argues that the ground of a mural or ceiling was often true *fresco* and the decorative design was in *tempera* (1961: 177).

4.8.1-*Tempera* Technique

The pigments used were mixed with an organic medium of glues and gums. The use of egg yolk or size as a medium was used mainly in the western world as the introduction of drying oil appeared to be a late invention (Hodges 1989: 160). *Tempera* was used on the *gesso* ground. Whence the wall was dry it was painted with one or two coats of size (diluted glue) (see also 4-10.2, glues). Afterwards, the colours were mixed with the same substance used for coating, and a binding medium was used to secure their adhesion to the ground when drying. Vasari notes that the blue colours were mixed with size as the yellow of egg would turn blue to green, while neither size nor gum affected them (Vasari 1907: 224).

4.8.2-*Buon Fresco* (true *fresco*)

The painting was applied when the plaster was still wet or fresh (day work), because if left dry the plaster formed a certain slight crust, whether from heat or cold. The pigments used were mixed with pure water. However, if any organic materials were used, it would result in the darkening of the colours and an alteration of the pigments would occur (Vasari 1907: 221-222). The application of the pigments onto wet plaster raises an important point to consider; Laurie says the use of the *fresco* technique in classical times indicates that large areas were covered by *buon fresco,* however, that particular technique involved the treatment of limited areas at a time. Therefore, joining areas should be visible in the case of *buon fresco*. Laurie referred to Pliny, who stated that materials used for the pigments should resist wet lime (Laurie 1910: 86).

4.8.3-*Fresco Secco*

This technique relied on the finishing coat (*intonaco*), which was rubbed with pumice stone for a smooth surface and had to dry completely before decoration. Following this, the wall was washed with water which contained small proportions of hydrated lime. This process was repeated and the decoration commenced while the surface was still damp (Davey 1961: 176). Slaked lime and water, which converted into chalk then was mixed with the pigments and applied while the plaster was still damp (Constable 1954: 60). This technique was considered a middle way between *buon fresco* and dry *fresco* (called *secco* painting), in which the painting is applied onto dry *intonaco*, but organic materials were used as a binder to adhere the colours to the surface such as, glue, egg etc. However in the case of the *fresco secco,* the painting does not form a layer on top of the *intonaco* unlike in *buon fresco* and it is less durable (Malaguzzi-Valerj 1970: 165).

4.8.4-The Differences Between *Buon Fresco* and *Fresco Secco*

One can assume that the differences between the two techniques are that the visible joints between the areas of work in *buon fresco* and the use of organic material, which distinguishes it from *fresco secco*. It was suggested that it is possible to distinguish between the different techniques by imagining that in one case, the pigments penetrate into the *intonaco* and in the other are spread on top of the plaster to form a separate crust ((Malaguzzi-Valerj 1970: 165). However, there are difficulties in determining whether it is *buon fresco* or *fresco secco* since the original organic binder used in *secco* technique as a secondary carbonation may have taken place to fix the pigments as a result of deterioration. On the other hand, the transformation of proteins into calcium oxalates is another process of *secco* painting which makes it so difficult to distinguish between the two techniques (Schmid 1998a: 5). Whereas in *fresco secco*, the pigments are held in a thin film of chalk or marble dust, which are applied by the medium of limewater, with the same composition as the ground, and do not penetrate far into the plaster. In contrast, in *buon fresco*, the water borne pigment goes deeper into the wet layer of the plaster and the painting is incorporated in to the outer layer of the ground itself, which can be bound more firmly to the underlying layer (Constable 1954: 62, 63, 68).

When lime is burnt, carbon dioxide (CO_2) is driven out and, as a result of slaking lime by using water the materials became saturated with an aqueous solution of calcium hydroxide, which is capable of rising to the surface of the plaster. When the pigments applied adhere to the surface and absorb carbon dioxide from the air to produce calcium carbonate, this acts as an additional binder material. The pigments were then fixed in a crystalline film (calcium carbonate) which formed on the surface of the plaster (Vasari 1907: 288). It is worth noting that the pigments were fixed via the migration of calcium hydroxide and the precipitation of calcium carbonate and not by penetration, as some believe. Malaguzzi-Valerj noted that he never noticed any penetration into the *intonaco* due to the nature of the process. The pigment becomes bound to the *intonaco*; it is almost the reverse to the calcium hydrate of the mortar which comes to the surface and spreads into the paint layer so that pigment can not penetrate into the *intonaco* as was believed (Schmid 1998a: 5 and Malaguzzi-Valerj 1970: 165).

Although *buon fresco* can resist exposure to air which contains sulphur, it is harmed by the conversion of the carbonate of lime in the *fresco* into sulphate, resulting in the disintegration of the plaster. The major enemy of the *fresco* is dampness, less so on the surface than behind as it carries salt to the surface where crystallization occurs (see Chapter 3, salt crystallization). Calcium carbonate can not protect the pigments from light or atmosphere and some pigments are restricted from use in fresco technique (lead white, verdigris or vegetable sources) due to alkali in the lime (Constable 1954: 65-66).

4.8.5-Techniques Employed on Egyptian and Classical Wall Paintings
Classical writers have stated that wall painting in Egypt was executed using *tempera* for the interior and external decoration of their building. Greeks and Romans also used it for interior decoration in association with other processes (Davey 1961: 176-177). Lucas states that the paintings were applied directly to a thin layer of gypsum plaster, a thin coat of distemper or whitewash which consisted essentially of calcium carbonate, containing traces of gypsum as impurities. However, as Lucas notes that a pavement was discovered by Petrie, who states that the technique used was *fresco* indicating that the painting was laid onto wet plaster. But Lucas later determined that the painting was applied to a gypsum layer containing a large preparation of calcium carbonate and particles of burnt fuel (Lucas 1932: 140-141).

On the other hand, Laurie states that, while analysis of Greek wall painting has shown the presence of wax, no wax has been found in Pompeian or Roman *frescoes*. The technique used was an example of *buon fresco* and methods of polishing have been applied to the painted plaster. The distinction between the methods used in Greece and Rome may have been due to climatic considerations, where *buon fresco* was found to be more suitable in Italy's cooler climate (Laurie 1910: 84, 102).

4.9-The Pigments Used in Egyptian and Græco-Roman Wall Paintings
As mentioned previously, the study of wall painting is associated with the supporting structures, the rendering used and their manufactures, in conjunction with the techniques applied, and the applied painting materials. In Egyptian and Græco-Roman times, both the pigments and the adhesive were varied, and in terms of the studying of wall painting, it is an essential point to consider, when identifying the main sources of the materials which were used in paintings, as well as their preparation, which has an important role in the identification and analysis of the pigments.

The early **pigments** used in the Palaeolithic paintings were natural colours and were mainly white, black, red and yellow. In addition to these colours blue has been found and green (malachite) had been introduced to pre-dynastic Egyptian painting, (where these colours were available for use). The pigments sources are varied; obtained naturally or artificially from earth deposits, organic or inorganic sources (containing oxides, sulphides, carbonates, silicates of heavy metals) or from insects or vegetable sources (roots, wood plant etc.) as well as from mineral sources (e.g. cinnabar, orpiment, realgar, azurite malachite, and lapis lazuli) (Gettens *et al.* 1972: 138-140). By the

Græco-Roman periods (332 BC-ca AD 250) the colours used in the Old Kingdom and the later New Kingdom had changed and evolved compared with earlier times, they were different and heavily symbolic. The most commonly used colours were; blue, black, white, red, green and yellow; blue-grey and red as well as ochre for yellow shades (Baines 1985: 286). Egyptian and Græco-Roman pigments were varied. Laurie gives a statement of the pigments (table 1), which were known and used since Pliny's day (1914: 32).

Table 1, The Pigments Used in Egyptian and Græco-Roman Wall Paintings

White	Black	Yellow	Red	Blue	Green	Purple
Chalk Gypsum lead White	Charcoal Lamp black Bone black	Yellow ochre Lead oxides Orpiment Vegetable pigment	Red ochre Red lead Dragon's blood Kermes	Indigo Egyptian blue Azurite	Malachite Verdigris Terra verte Green Vegetable pigment	Tyrian Murex Purpura Vegetable purples

4.9.1-White used as a pigment derived from natural deposits of chalk, gypsum and manufactured lime (Hodges 1989: 156). Hydrated or slaked lime ($Ca(OH)_2$) or lime white, which was produced from slaked quick lime to putty, whiting (calcium carbonate) made by calcining and grinding mollusc shells and eggshells (Davey 1961: 178). In addition, the Greeks, as mentioned earlier, used white marble to prepare of the ground. Laurie said chalk was a base for pigments in classical times, while with Egyptian madder, gypsum was a base as was mordant lime (Laurie 1910: 255, 254). Chalk is a soft, white, greyish white or yellowish (iron oxide) white rock composed of the remains of minute sea organisms, such as foraminifera, which is prepared by grinding it in water and later sieving it to separate the coarse materials (Mural Painting Conservation 1998: 32). When heated they remain white, but if over heated they turn to quick lime (calcium oxide (CaO). Artificial chalk, carbonate lime, was whiter, more homogenous, more controlled in particle size and contained no fossils, unlike natural chalk (Gettens *et al.* 1974: 161, 162). The white pigments of pre-dynastic times were calcium carbonate or calcium sulphate (gypsum), which was not only used by Egyptians, but also by the Græco-Romans as discovered in Hawara (as a pigment) (Lucas 1989: 349). However, it unfortunately sometimes difficult to distinguish chalk from manufactured lime.

Vitruvius has described the process of preparing lead white. "A piece of lead sheet was placed on some wood shavings which had been soaked with concentrated vinegar (in a jar that is covered for a certain time, to prevent evaporation), converting the lead to lead white, which if heated, would change to sandarac (Vitruvius 7, 12, 1, 2)". Lead white was used, without grinding or heating, to prepare the flesh tone for bodies. However, if heated it turns yellow and is mixed with ordinary flake white and vermilion, depending on the tone required. Green was used for pallid hues (Theophilus 1, 1, 5). Lead white is a basic lead carbonate made from the best lead, whose use is dated to prehistoric times; (as well as in the Greek period) in fact, it was the earliest artificially produced pigment (Mayer 1973: 51). But, if it turns grey when heated it indicates the presence of sugar of lead, which is a metal pigment and very poisonous. Basic lead acetate ($3[Pb(CH_3COO)_2 . 2 Pb(OH)_2]$) decomposed with time to $PbCO_3$. A mixture of lead white and vermilion produces a black colour (Doerner 1969: 52-54). An important study has been carried out by Gettens *et al* on lead white, its composition, sources, as well as its relationship to medium, in addition to the various methods that were used to identify lead white (Gettens *et al.* 1967: 125-139).

4.9.2-Black is an inorganic pigment made artificially from carbon; lampblack and soot are produced from organic matter by collecting them from the flame on a cool surface (Hodges 1989:

157). The inferior grades of lampblack contain small amounts of greasy materials (Mayer 1973: 97). While charcoal black, or carbon black, is made by heating wood, then grinding and washing it to remove potash, bluish-black is made from carbonized vine thinnings or vine wood, burnt nut or shells, e.g., peach stones or coconut shell (Davey 1974: 178). Other shades of blacks such as bone black and bone brown were made from bone which had not been entirely charred, and graphite, which is a black form of crystalline carbon. Iron oxide black (Mars black) is ferroso-ferric oxide 1 Fe + 3 Fe_2O_3, producing a dense, opaque and heavy colour which, if burnt, turns a very dark-red brown (Doerner 1969: 89 and Mayer 1973: 42).

Vitruvius notes the method of producing soot charcoal; by using lees of wine which is mixed with gum for writing but, if mixed with size, would be was used for *fresco* painting (Vitruvius, 7, 10, 2). However, fine quality black pigment is produced from ivory or bones (Hodges 1989: 157). Bone black made from glue stock which was boiled to remove fat and glue, is blue-black in colour, smooth in texture and denser than carbon or lamp black, while charcoal is lighter and porous due to the fine structure of wood (Mural Painting Conservation 1998: 33). The black pigment used in Egyptian paintings was mainly carbon consisting of soot (fine or coarse soot) depending on the method of collection, scraped from the surface of cooking vessels, masonry etc. Lucas reveals the result of the analyses of black pigment used and indicates the use of charcoal and pyrolusite, which have been identified on Egyptian paintings (Lucas 1989: 339-340).

A Grey was made by mixing black with white, as in a mixture of gypsum and charcoal, or gypsum and lampblack. A mixture of pale yellowish earth and lampblack has been identified by Lucas (1989: 339, 346). Overall the use of greys as blue-grey as backgrounds to scenes appears to suggest as a substitution for both green and blacks (Baines 1989: 286). Theophilus, states that when used in classical times the dark grey (black with a little white) was mixed with green earth as well as burnt ochre to form a colour for a darker shadow, (Theophilus 1, 6, 7).

4.9.3-Yellow Ochre and golden ochre (Geothite, Fe_2O_3 H_2O, deeper than yellow ochre) are natural formations of earth consisting of clay and silica. The quantity of hydrated iron oxide determines the variety of shades in the colour (Davey 1961: 179). Yellow ochre is caused by the presence of various hydrated forms of iron oxide and may contain impurities of gypsum, magnesium and carbonate. Ochre is prepared by grinding, washing, levitation and then drying which consequently forms various yellows, to reddish brown. When burnt, yellow ochre turns to red as a result of the loss of crystallization (Mural Painting Conservation 1998: 34-35). Lucas mentioned that two yellow ochres were used in Egypt: the first was hydrated oxide of iron and the second was orpiment, a natural sulphide of arsenic which was imported from Persia/Anatolia (as it did not occur in Egypt).

Other yellows have been recorded: sandstone used at Kerma (Middle Kingdom); massicot, a yellow oxide of lead, which has been dated to 400 BC; and vegetable yellow (Lucas 1989: 349-351). Orpiment is a very brilliant yellow and has been associated with realgar, the orange sulphide of arsenic, which when exposed to light, tends to turn into orpiment (Laurie 1914: 31). Orpiment was not used on fresh plaster and was known in classical times being present in Syria. Massicot was prepared by roasting white lead at 300 °C. Long time exposure to damp air transformed it into red lead (Gettens *et al.* 1966: 129).

4.9.4-Red colours are natural oxides of iron. Red iron oxide (anhydrous oxides of iron) and red ochre, known as hæmatite, are amorphous earthy varieties of the materials. Both reds have been proven to occur naturally in Egypt near Aswan and in the oases of the western desert (Lucas 1989: 347-348). When calcined raw sienna is changed from hydrate ferric oxide to ferric oxide it is reddish-brown in colour, whilst calcined yellow ochre forms a light red (Davey 1961: 179), however, the detail of the preparation varies, from a few minutes up to several hours at temperature below 800 °C. The transformation of goethite (α-Fe OOH) to hæmatite (α-Fe_2O_3) in which the mineralogical differences in the x-ray diffraction pattern of the product hæmatite occur, depends on the temperature to which it has been heated (Helwig 1997: 183). However, dark red was prepared by mixing black and burnt yellow (Theophilus 1, 5, 8).

Pink is a mixture of red and white and has also been found to be a mixture of gypsum and ochre (Lucas 1989: 348, 346). However, Laurie says that pink colour was produced from madder, which was prepared on a gypsum base (Laurie 1910: 25). Pink was a distinctive colour unlike light red, and was a new innovation in both Middle and New Kingdoms (Baines 1989: 287). However, orange was formed by painting red over yellow or by mixing them together (Lucas 1989: 346).

4.9.5-Brown is an inorganic pigment made from raw umber, (an earth pigment similar to ochre) ($Fe_2O_3 + MnO_2 + H_2O$, clay), which contains manganese dioxide and hydrous ferric oxide if it is a reddish-brown colour. When burnt, (Fe_2O_3+ MnO_2) raw umber is turned from hydrous ferric oxide to ferric oxide to form a redder brown (Davey 1961: 179). Umber and sienna contain small quantities of black manganese dioxide, which produces a brown colour. Bitumen, charcoal and graphite were all used to artificially create brown pigments (Hodges 1989: 157). Lucas states that brown was used since dynastic periods (fourth dynasty, Amerna period and late period) (1989: 344). However, Baines mentioned that in the Middle Kingdom (2040-1640) several reds occur, including a dark one used to illustrate copper, possible browns are found in the other paintings of the period (Baines 1985: 287)

4.9.6-Green and Green Earth, (Fe Mg, Al, K hydro-silicate) (terra verte) containing iron was used in ancient times (Laurie 1910: 41); green earth varied in composition like so many of the complex silicates. Most green earth originated as marine clays. There are two greens; glauconite and celadonite. They are hydrous iron, magnesium and aluminum potassium silicates (Gettens & Stout 1966: 117). Malachite (Cu CO_3.Ca $(OH)_2$) is a naturally formed basic copper carbonate and similar to azurite (Davey 1961: 178), which contains less chemically bound water. Selected lumps of malachite were crushed, ground to a powder, washed and levigated to prepare it for pigment use. It was used in Egypt for eye-paint as early as predynastic times and has been found on tomb paintings from the 4th dynasty on (for more information see Gettens & FitzHugh 1974).

Other greens, Verdigris, described by Vitruvius, was formed by the action of vinegar on a copper base (Vitruvius, 7, 12, 1). Viridian was mixed with black, and was used as a green earth. It was obtained naturally by being dissolved in water then strained through a cloth, not ground by a stone (Theophilus 1, 2, 5).

4.9.7-Blue pigments included some natural compounds, e.g. azurite or mountain blue, a basic copper carbonate (2Cu CO_3 Ca $(OH)_2$) (Davey 1961: 178), found in Egypt, Sinai and the eastern desert, and is similar to green malachite (Lucas 1989: 340). Azurite is a natural pigment prepared by grinding, washing, levigation and sieving. When ground coarsely it produces dark blue colour, while fine grinding gives a lighter tone. Another artificial copper blue was verditer, or blue bice, a basic copper carbonate that has the same chemical composition as azurite (Gettens *et al.* 1966: 54-61).

Egyptian Blue was found as early as in the fourth dynasty. It spread to the Aegean and Mesopotamian areas as early as the third millennium BC and was used throughout the Greek and Roman periods until the ninth century AD (Lazzarini 1982). However, the secret of its manufacture was lost between 200-700 AD (Gettens 1966: 120). Egyptian blue is produced by being heated at 830-850 °C. It has a coarse crystalline form and is a pure blue in colour, which is similar to azurite. The crystals can vary in colour from deep blue to faint, it depending whether they contain some silicate and quartz as impurities (Gettens *et al.* 1966: 112, 113). The silica used in the form of quartz pebbles has a major role in the formation of the blue, which depends on the amount of particles that are free of iron compound. However, if iron present in more than traces, the colour tends to be of a greener hue rather than blue (Lucas 1989: 341). Egyptian blue (blue frit or Pompeiian blue CaO, CuO, $4SiO_2$) (Davey 1961: 178). Pompeiian blue is similar in chemical composition and in optical proportion to the ancient calcium copper silicate blue, however, it is purer and finer (Mural Painting Conservation 1998: 33).

4.9.8-Cinnabar, (HgS) the red made from mercury sulphide is natural and classical writers (see glossary) have cited the ground form of cinnabar using different names. Hodges states that vermilion was formed by heating up a mixture of mercury and sulphur (1974: 157). Vitruvius explains the process of producing cinnabar and the treatment protecting it from light as, as well as testing whether or not it is adulterated with lime (Vitruvius 7, 9-10). The main source was Sisapo in Spain (modern Almaden), but the pigment existed in China in prehistoric times, about the 2nd Millennium BC. It was used in Greek times as, early as the sixth century and it was used in Asia Minor long before. It was also used in Roman wall paintings, but there is no mention that it was used in dynastic Egypt or Mesopotamia.

Vermilion (HgS), synthetic form of cinnabar, made by a dry process which forms a darker and cooler tone, while a wet process produces the lighter German vermilion (Mural Painting Conservation 1998: 37 & Gettens *et al.* 1972: 47-53). Theophilus explains how to make vermilion, a mixture of sulphur and mercury (Theophilus 1, 35, 31). Which is more finely ground than cinnabar, but is sometimes confused with red lead. However, the difference can be tested by moistening it with sodium sulphide; while red lead will turn black immediately, vermilion will remain unaltered (Laurie 1914: 29).

Red Lead (Pb_3O_4) is an inorganic mineral produced by roasting white lead at 480 °C. It was not used in Egypt till the Græco-Roman period (Mural Painting Conservation 1998: 35). Laurie says the substance is not only dark brown/black, but curiously has a metallic lustre due to the action of hydrogen sulphide gas on the lead (Laurie 1914: 30). It is very poisonous, as are many of other pigments e.g., cinnabar, orpiment etc.. Under intense heat, red lead becomes a violet shade and when cooled again turns to a yellow-red (Doerner 1969: 74).

4.9.9-Purple various purples are a natural pigment organic and inorganic discovered in the Mediterranean area in the 2nd Millennium BC. These stable pigments derive from molluscs and are, similar to indigo (Lazzarini 1998). Davey noted that artificial purple is produced from calcined iron oxide (Fe_2O_3), which is formed naturally as hæmatite, and gives a dark purple or maroon pigment (1961: 178). Vitruvius notes that purples are obtained from a marine shellfish (Murex) and about a thousand were required to produce one gram of pigment. Sunlight affecting the apparent of the colour. A bluish shade is found from north to west, while a violet shade is formed from east and west, in contrast, in southern countries it is naturally red (Rhodes) and it is black in the northern most those countries (Pontus and Gaul). Purple is produced by dyeing chalk with madder root, while fine purple is made by heating it with water and, then squeezing it onto linen and mixing it with milk (chalk). Another method was heating yellow ochre to a glow and quenching it in vinegar to produce purple (Vitruvius 7, 12-13). Not only was Murex purple, used as a pigment, but also madder, smalt weld, certain boiled woods and leeks, to stain chalk or gypsum with these dyes. While, Kermes was produced from a kind of cochineal insect, some dyes were produced from vegetable matter, e.g. dried vine-leaves (Laurie 1910: 44-46 & 1914: 44-48).

4.10-Organic Materials Used With the Pigments

Adhesives were varied in ancient times, but as mentioned before the materials used as a medium for the pigments were glues, gums and the egg white. These adhesives were mixed with the pigments to adhere to the surface or to prepare the ground. This medium serves to protect the pigments more or less perfectly, from chemical and mechanical injury. Lucas mentioned that sometimes beeswax and resin were used to cover tomb paintings during the 18th dynasty (1989: 352-3). Glue and gum were used as an organic medium for Egyptian paintings, but there is doubt about the use of albumin. Both gelatin and glue have been identified as media in Egyptian paintings. However, Lucas says the presence of glue in a paint does not necessarily mean that it was employed as a binder, as it was used to fill the pores in the plaster, stone or other ground before painting. On the other hand, he states that there is not any specific test for glue, gum or albumin. In addition, it is important to consider whether the materials have been exposed for thousands of years, during which chemical changes may have taken place in the pigments and their adhesives (Lucas 1989: 1-6). Therefore, the recent analysis using gas chromatography and mass spectrometry (GC/MS) has shown the use of animal glue, carbohydrates (plant gums) and other glue, which included wax and natural resin. It has long been assumed that the binding of Egyptian paintings was mainly animal glue or plant gum (Newman & Halpin 1992).

4.10.1-Egg White (albumin) and egg yolk. Although it is uncertain if these were used as media with the pigments, Lucas notes some people have identified albumin. Lucas stated that albumin is a nitrogenous organic matter with a small proportion of sulphur. Therefore, animals and plants have the same occurrence of nitrogenous matter, as does glue and so the presence of phosphates is not enough to indicate albumin, unless it might have been in the form of calcium phosphate (Lucas 1989: 1-2). The identification of egg white was as a binding media in ancient times can be identified using immunofluorescence as described by Kockaert *et al.* (1988).

4.10.2-Glues were used for different purposes in ancient Egypt and were made by boiling hides, skins, cartilage and bones for the extraction of gelatinous materials (gelatin, size and animal glue etc. (Lucas 1989: 3). The quality of the glue depended on the purity of the materials; animal glue was extracted from sheep and cattle, while fine glue was produced from skin, bones and the offal of fish (Hodges 1989: 162). Adhesives were not only used as a medium, but as previous mentioned it was also employed to prepare the ground, in terms of *gesso* ground (see 3.2.3).

4.10.3-Gums are hydrophilic, forming gels which are insoluble in organic solvent due to the chemical compounds and carbohydrates, similar to sugars or starches (Hodges 1989: 164). Lucas studied extensively the use of gum as a medium for painting (1989: 6). Laurie notes that gum Arabic (acacia) consists of the calcium, magnesium or potassium salts of an organic acid, which was dissolved in water to form a convenient medium for painting (1910: 7). The process of dissolving the gum Arabic in water is slow but, eventually, it will dissolve completely and it has been identified in the tomb of Nefertari (Kleiner 1995: 50).

4.10.4-Resins and Varnishes, resins are insoluble in water and set by the drying of oils which are chemically altered by the presence of air. In contrast, gums are set by the evaporation of water (Hodges 1989: 164). Lucas states that the use of resin, had various different purposes uses, including a binder (1989: 7, 8). There are different kinds of resin, e.g., pine tree, mastic, *Pistacia lentiscus* (Syria) and sandarac are found on the Mediterranean coast of Africa. Depending on the location and availability Laurie notes that all gum Arabic and glue were largely used as media on a surface which was prepared in a solid condition, yet at the same time very smooth to execute *tempera* painting (Laurie 1910: 29, 21). These media (glue, gum and egg white) were previously employed as the medium, before the application of egg yolk or dry oil, which is mentioned by the art historians.

A colourless **varnish** was used to cover the painting in certain cases, as a treatment for certain colours such as yellow and red. Varnish was also used for other objects. Lucas mentions two different kinds of varnish used in Egyptian times; colourless and black varnish (Lucas 1989: 356-358). Pontic wax was used which mixed with a little hot oil and spread onto the surface with a brush. When dry, it was heated to sweating point by placing charred gall–apples near by (charcoal in Vitruvius's description 7, 9, 3-4). It was then rubbed with candle/tallow, followed by clean cloths and then made to shine like marble. This was the method used by the Greeks to prevent light penetrating the pigment. When applied on a wet surface, it resulted in a black colour (Laurie 1910: 97-99).

4.11-Summary

It can be concluded that wall paintings were applied, whether in ancient or classical times, followed a special preparation of the materials used to produce a firm and suitable surface for painting. The differences were not only between the techniques used and the surface preparation, but also the locally available materials which were suitable for use according to the climatic conditions. As a result, the techniques used varied from in-*tempera* painting to *fresco*. Depending on the preparation of the ground, the pigments and their adhesives were susceptible to attack by air or moisture, being easily removed by water if the *tempera* techniques were used. The alteration of pigments could be caused by the deterioration of the rendering, plaster/mortar or the supporting structure, causing the paint to flake and the materials to powder or alternatively it could be the result of other different sources combined together. In the analysis of materials, not only is it necessary to look at the surface, but also the wall painting to identify the composition, the structure and to discover the main source of decay, which affected the wall painting. In the next chapter will be discussed the possible causes of alteration in wall paintings, which could be from different sources, depending on the condition of the materials, focusing on the main problem of deterioration and decay.

Chapter 3

The Causes of Alteration and Deterioration in Wall Paintings

1-Introduction

In the analysis of materials it is necessary not only to look at the surface of the wall painting and identify its composition and structure, but also to find the main source of decay which has affected the wall painting. When the materials which were used are weak or contain soluble salts, as well as having a low porosity because low porosity materials do not allow good evaporation, there are unable to resist long exposure to the atmosphere.

The deterioration of wall paintings can be brought about by different factors in environmental conditions, as well as the materials used in conservation; both of these factors are responsible for rapid degradation and decay in murals. Materials such as stone, masonry or brick are responsible for blistering of the plaster and the paint layer, the flaking of pigments and the loosening of the ground. Various inorganic salts efflorescence is caused by the movement of water across the wall, carried by migration of the solution and its subsequent evaporation from the paint surface, which results in much damage to the painted plaster on account of the efflorescence (Lal 1996: 12).

Atmospheric factors and chemical compounds, causing physical and chemical damage, which affect the wall painting, are natural and artificial agents. The natural pollutants are carbon dioxide (CO_2), such as the respiration of humans, and the natural aerosols, which are stable and contain fine silica and calcium carbonate from the ground, as well as chlorides and sulphate of alkali metals from the sea, which are deposited on the wall surface (Mora 1974: 22). As a result, the reaction of nitrogen and oxygen produces oxides and nitric or nitrous acids; these are compounded with SO_2 in polluted environments to attack monuments. An increase of oxidative reactions (in air or water) will also give rise to a higher growth of photosynthetic organisms, resulting in different kinds of biological growth being produced (Caneva *et al.* 1991: 12, 13). Artificial pollution comes from materials containing sulphur (coal and mineral oils) which are converted into sulphuric acid by the moisture in the atmosphere. When the materials of wall painting are calcareous based and the rendering is lime based, transformation into calcium sulphate occurs and causes disintegration in the rendering (Mora 1974: 22).

Atmospheric influences are produced by different factors, such as variation in temperature, salt crystallisation, the mechanical action of wind-borne sand or the solvent action of water. The acids in the ground cause the disintegration of limestone in the presence of moisture. The water containing carbonic acid helps to dissolve the rocks slowly, and forms a dilute solution of calcium bicarbonate. This is deposited on the surface again as carbonate when the object is exposed to higher temperatures (Sneyers & De Henau 1968: 216). In addition, some other causes are produced in the manufacturing districts of large towns and others where rain is almost entirely absent. The agent in the former case is sulphuric acid, which is given off by coal and coal gas (Lucas 1932: 194, 195), and more recently diesel and heating oils, and in the later case is salt and salt grains, which access the stone from salty subsoil water or from salty ground and can accumulate in hot or dry weather.

The agents responsible for deterioration or preservation depend on the nature of the materials and the environment surrounding the materials. Much of the decay is related to temperature and humidity. As a result, two different types of reaction may occur, effecting either the deterioration or the preservation of artifacts. The first is **physical deterioration**, which breaks down the structures of the materials and the second is **chemical deterioration**, which causes the alteration of the chemical composition of materials. Examples of these are water and air, which corrode iron;

acid, which dissolves lime plaster; bacteria and other biological factors; the action of salts, and the effect of humidity (Cronyn 1990: 14).

From these points, it seems necessary to stress the fact that the alterations could be from a variety sources and in this chapter I will highlight these factors and focus on the mechanism of the deterioration and decay in wall paintings. Figure 3 illustrates the most important causes of decay; **Physical and chemical deterioration, bio-deterioration and decay and the intervention factors**, in terms of the deterioration of murals, which results in alteration of the materials used in wall painting, both rendering and pigments, as well as the supporting structure itself.

Figure 3

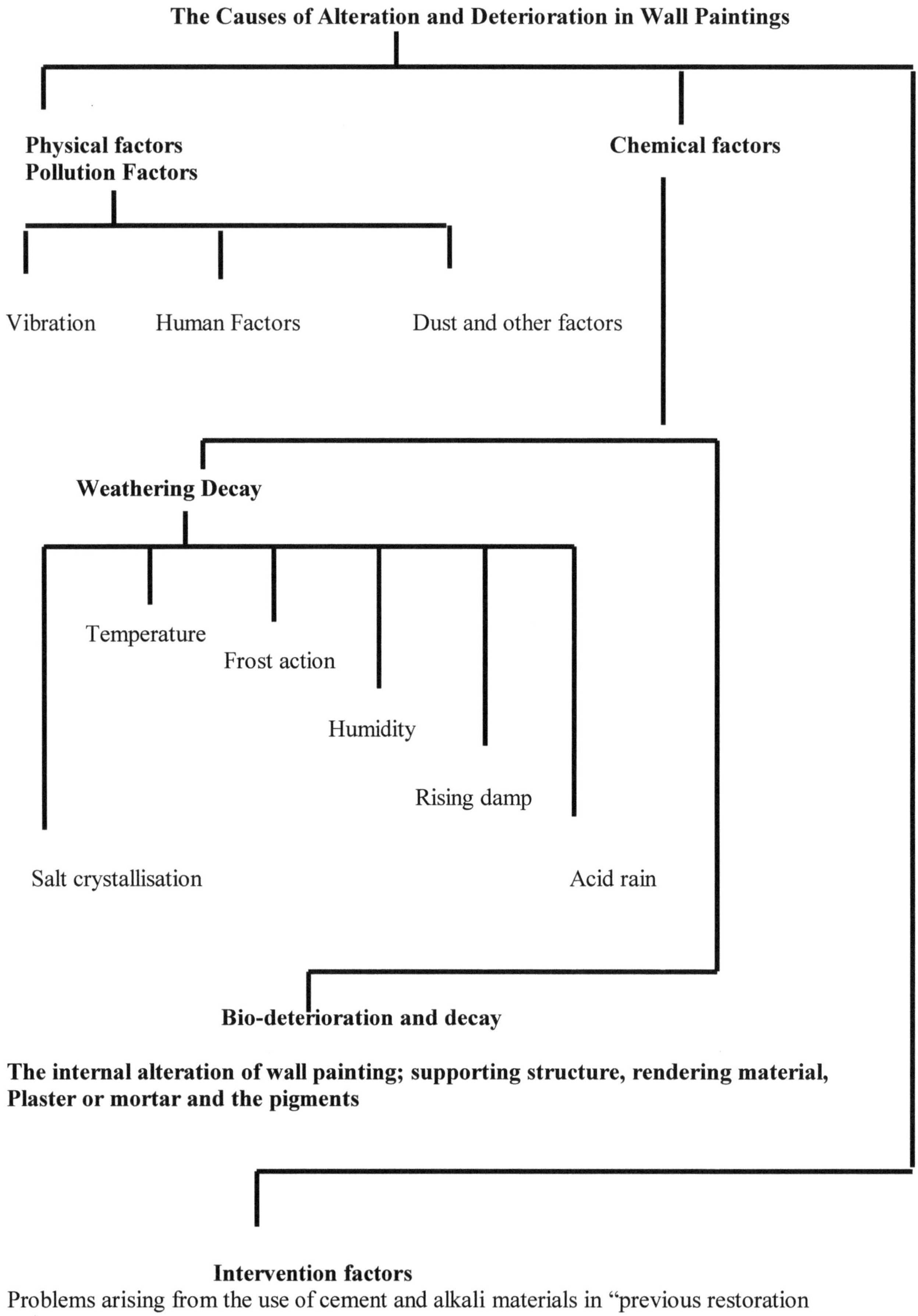
The Causes of Alteration and Deterioration in Wall Paintings
Physical factors
Pollution Factors
Chemical factors
Vibration
Human Factors
Dust and other factors
Weathering Decay
Temperature
Frost action
Humidity
Rising damp
Salt crystallisation
Acid rain
Bio-deterioration and decay
The internal alteration of wall painting; supporting structure, rendering material, Plaster or mortar and the pigments
Intervention factors
Problems arising from the use of cement and alkali materials in "previous restoration

2-Physical deterioration

2.1-Pollution Factors

2.1.1-Vibration; traffic causes vibration in the ground and in buildings, creating stresses which act on the structures and materials, promoting deterioration in the wall paintings and detachment of the plasters that are weak or not well connected to the surface. Vibration stress can interfere with the Physico-chemical processes of the deterioration and cause an increase in the rate of deterioration. As a result, the micro-cracks expanded by vibration stress can widen and give access to water, which causes destructive processes such as the crystallisation of salts. In addition, the type of building and its condition has great importance in determining the effect of the sonic waves on the structure (Torraca 1988: 55, 64).

2.1.2-Human Factors

One of the most damaging factors to monuments is the human factor. Although visitors are important for the economy, they are very dangerous to the preservation of the monuments. For example, Hawass states that each tourist who enters a tomb brings about 20 gm of water to the monument (Hawass 1993: 20). In addition tourists touching the monuments over a long period of time, has caused a deepening of the colours compared with the original, as if they were decayed by dampness. It is not dampness, but contamination; case studies (Arnold 1998b) have shown some examples that have been affected by tourism in Egypt. In addition, continuous inflow and outflow of air by visitors, allows the formation of incrustation due to evaporation or condensation of humidity over the surface of the murals (Torraca 1970: 172).

2.1.3-Dust, Animal Deposits etc.

There are some other factors responsible for the deterioration of wall painting. Such decay could be produced from dust, animal deposits, soot, insect excrement, as well as damage by bats, resulting in darkening of the surface, which is difficult to remove. In addition, the deposits are alkaline or acidic in nature, which promote the corrosion of the surface (Mora 1974: 24). However, in buried buildings or caves, deposits of dust can be cemented by condensation. When these are brought into contact with the external air, an extremely hard incrustation is formed over the mural painting, as happened in the tombs of Beni Hassen in Egypt. The incrustation was formed by dust cemented in place by calcium carbonate, which dissolved and re-precipitated, by a process of periodical condensation (Torraca 1970: 172).

3-Chemical Deterioration

3.1-Weathering Decay

The deterioration and decay caused by weathering is varied and has a bad effect on the other deterioration factors, not only in the building materials, but also in the rendering and painted layers, as well as the pigments. Some studies have been carried out on weathering decay and its effect on the monuments and numerous studies have been carried out on salt crystallisation and decay. The most important were by Arnold (1984, 1987 and 1998) and others by Hammer (1995) and Zehnder (1996).

3.1.1-Temperature is one of the most indirect factors, but has an important role in deterioration and decay. The extreme changes in temperature between day and night, which are common in tropical and desert areas, and the alternation of sunshine and rain have a destructive effect on the building materials. The result of such decay is the flaking of the surface as well as the interior layers, caused by the bad conduction of heat to the stone, exaggerated by the shaded areas of the building (Sneyers & De Henau 1968: 215, 216). The variation in temperature produces expansion and contraction of the building materials, which causes cracks and peeling of the outer layer of the surface. In addition, the variation of temperature allows the evaporation of the water, which helps the formation of salt crystals (Lucas 1915: 2). This has happened at Alexandria in the site of the Anfushi tombs as shown on Photographs 3 & 4.

3.1.2-Frost Action is another factor in weathering and can happen when structures are exposed to frost or to sudden changes of temperature. Frost action on damp walls is rapid and destructive because the rendering is weakened and disintegrates and becomes detached from the surface. The alternation between the night time and daily-time sunshine causes a slow disintegration of the surface of the wall painting (Mora 1974: 23). That is not the major problem in Alexandria, but could be in some other parts in Egypt, e.g., Luxor.

3.1.3-Humidity, is the main condition for all the chemical and biological decay, as well as physical damage. It is considered to be the primary cause of deterioration and decay in the porous materials. The notion of humidity refers to the kind of materials and their hygroscopicity in brick and mortar, but in both, the rate of evaporation is different. However, if the water is neither from the ground nor from the foundation, then moisture tends to descend. The surface water evaporates faster if the humidity arises from capillary action; as a consequence, a reduction in the capillaries is effective and even when moisture rises to the highest level, no evaporation results (Mora 1974: 15). There is a relationship between variations in humidity and the crystallisation and dissolution of salts on walls. Salt crystallisation increases when the relative humidity drops (less than 60%). When the relative humidity rises (more than 60%) previously crystallised salts re-dissolve (Arnold 1998a: 14).

3.1.4-Acid Rain as mentioned above has a bad effect on building materials as well as the mortar, reacting with its calcareous components and producing soluble salts. These will go into solution in rainwater or condensed moisture, and migrate through the porous matrix of the masonry. The effect of acid rain on mortar depends on the particular mixture. Most evaporation will take place if mortar has a very porous structure, and salt will concentrate in it, resulting in complete destruction. Lime-sand mortars are the most sensitive ones due to their composition; lime is attacked because of the small crystal size formed by the calcite, as well as calcium sulphate which can crystallise as a gypsum inducing mechanical stress into the matrix of the mortar (Charola & Lassarini 1986). If the rain falls on the top surface of the wall, it causes dampness in the wall. This affects the wall inside leading to flaking of the plaster. Sometimes rain may dissolve the binding media, as well as leaving a white stain of dissolve salt, when drying off (Arnold 1998b). A study has been carried out on the effect and the degradation as well as the behaviour of the acid rain on lime mortar and lime plus sepiolite (magnesium silicate hydrate). It concluded that the initial effect of the acid rain solution on the lime mortar resulted in a reaction with the binder ($CaCO_3$) to produce soluble substances, dependent on the reaction time (Martinez-Ramirez *et al.* 1996).

Therefore, acid rain may have a good role in distinguishing between two kinds of black crust; one produced with a pollutant factor, and the other by bio-deterioration, which are quite similar to each other in appearance. As I have observed on different examples, it was easy to determine between the two different crusts, because rain can clean off the pollution. This can be noticed on buildings or statues displayed in the open air. Bio-deterioration decay is not affected by rain, because the black crust may change to something else; such as from algae to fungi or fungi themselves. Analysis is essential, so that problems can be identified and the conservation process defined.

3.1.5-Salt Crystallisation

One of the most frequent and effective forms of weathering decay is the water-soluble salt, which has been known since antiquity and is still an unsolved conservation problem. It affects not only the building materials, but also causes serious deterioration both in the rendering and painted layers. Numerous studies have been carried out to establish the mechanisms of salt crystallisation, an many factors arise, e.g., ground water and rising damp, the original materials used, the materials used for conservation and repair, biological metabolism etc.

3.1.5-1-Water-Soluble Salts

The most characteristic process is migration to the wall surface and re-crystallisation. The process includes many aspects; surface evaporation and re-crystallisation, the process of disintegration by the crystallisation of salt and superficial incrustation, in addition to the nature as well as source of the salt (Mora 1974: 16). Salt accumulation in walls originated from the ions that have been leached out of different sources. In the case of building materials, the main sources have to be considered as; building materials, the stone and mortar, the soil, the materials used for conservation and repair, polluted atmosphere and organisms.

Salts are transported in an aqueous solution when the ground or wall surface contains water or could be precipitated in the case of rain, snow or dew, in addition to infiltration in the walls as a result of percolation. Rising damp in the walls has a great role in the transportation of the ions of salt in an aqueous solution, which causes deterioration and decay to the wall. Water running along a surface could also be responsible for transporting the ions (Arnold 1998b: 2 and Arnold & Zehnder 1987: 105). Salts dissolving from the wall and penetrating the surface, may be contaminated by water coming up from the soil with salt in a solution phase. However, this process is related to the immediate surroundings and depends on the temperature and relative humidity of the atmosphere, as well as the amount of ventilation. The structure of the materials, their porosity and density, also help the ingression of humidity by capillary action and infiltration (Mora 1974: 17). The growth of the salt crystals depends on the openings between the crystal surface and the internal walls of the capillaries, therefore, if the rendering is resistant, the crystals are forced out as an efflorescence, whilst, if the walls of the pores are weak, they disintegrate and decompose the rendering. Efflorescence appears in a variety of habits and aggregate forms, e.g. the structure of the crystals of sodium chloride ($NaCl$) is cubic or halite, which is a cubic form specific to crystalline sodium chloride. Other forms are prisms, long hair-like needles, due to the decreasing thickness of the solution film, while the substrata dried out, depending on the external environmental conditions, such as supersaturation, salt composition and the impurities of the solution. Certain crystals grow in anhydrous conditions, and the crystals increase, causing the wall to disintegrate. So the varied composition of the salts, in association with the temperature and relative humidity, result in different phenomenon of deterioration (Arnold 1998a: 11, 12 and Mora 1974: 18, 19).

The conditions of precipitation are created when the ambient relative humidity becomes lower than the equilibrium relative humidity of the saturated solution of the particular salt phase in the system of the evaporation process. Salt crystals grow in pores with a size of about 1-10 μm. The crystals are concentrated at the fissures because of the accelerated evaporation at the opening, as well as the widening of the structure. They continue to grow isometrically and because the solution may come from one or two sides of the fissures, they grow in a columnar habit. This enables them to be strong enough to widen the fissures, thereby increasing the growth of crystals. As a consequence the fissures expand and the solution is evaporated. Then the columnar crystals became whiskery, thinning out as a result of the reduction of the solution and the substrata drying out. They are able to lift up spalls of the paint layers, causing the crust, detached previously as the crystals grew, to turn whiskery. These crystals with different forms as well as habit may be formed on different places on a wall at the same time and/or on the same place at different times (Arnold 1998a: 12, 13 and Arnold & Zehnder 1987: 113, 119).

Arnold and Zehnder (1987) stated that, in the zones with ground water or moisture, three stages can be noticed in different zones, A, B and C (figure 4). There is a less deterioration in zone A above than zone B, as most of the paintings, mortars and stones are in zone B, more disintegration, crumbling and scaling could be found. The salts in the zone are mainly sulphates (sodium sulphates, magnesium sulphates, calcium sulphate and potassium nitrate and sodium carbonate). However, less deterioration can be found in zone C, although it looks dark and damp from several centimetres to several metres, where chlorides and nitrates are concentrated, as well as accumulated under a normal climate, due to the ion mixture, which is strongly hygroscopic. That could account for the difference between the deteriorations. If it is an advanced case, that occurred in zone A, it will be the same as in zone B, due to the ions activities (Arnold and Zehnder 1987: 111, 112).

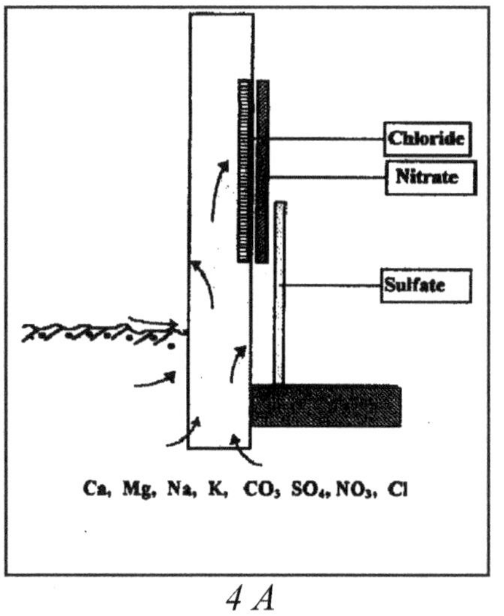

4 A

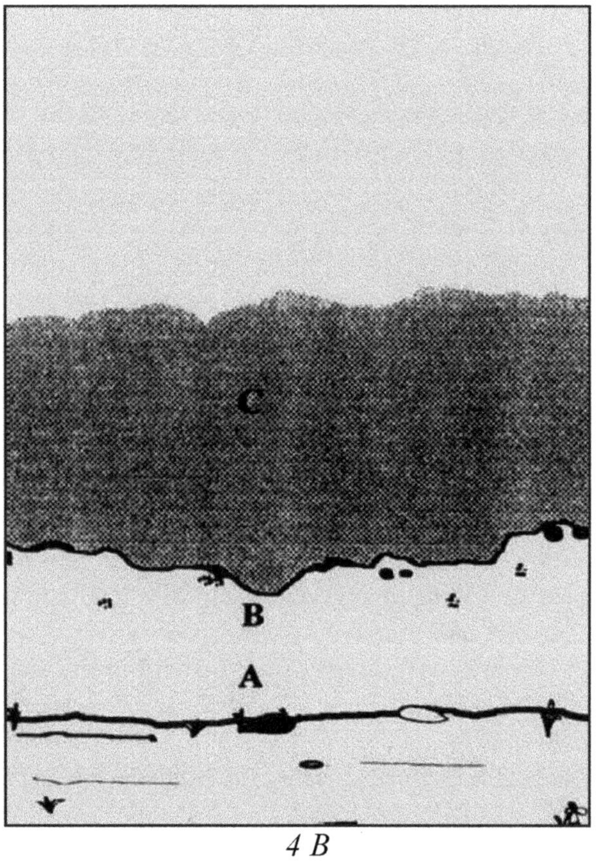

4 B

Figures 4, A & B, A is a model of the evolution in a wall of a salt system in relation to capillary rise of moisture. The ground is a diluted solution composed of Na, K, Mg, Ca, as the principal cations and No, C, So_4 and Co_3 as the principal anions.
B shows the different zones observed in the area of rising damp (from Arnold 1998).

The cause of gypsum efflorescence in the ground zone is a result of moisture; gypsum remains low down, whilst the chlorides and nitrates rise, even at moderate air humidity, so it could occur from a decrease or increase of the ambient humidity. It depends both on the way in which the gypsum solubility is reduced and the concentration of the solution. The crystallisation of gypsum occurs when the air humidity of the gypsum is low and if the solubility is reduced. It also happens when the air humidity is raised and the solution gets diluted. The decays produced take the form of powdery coatings, white dots and thin crust on the surface (Zehnder 1996).

Another phase of the crystallisation of salt can occur within the wall rather than on the surface, when the wind, focused towards the areas attacked, intensifies evaporation, which causes erosion to the exterior walls (Mora 1974: 23). Salts can be formed from the actual materials or as a result of the decomposition of calcium carbonate, sodium sulphate and sulphates of potassium, calcium and magnesium. The lime mortars of some buildings are made of dolomitic lime, and the compounds may react with sulphate ions in the water. Another source of salts is the soil, which contain sulphates, nitrates, chlorides, magnesium calcium, sodium, potassium and ammonium ions in a diluted solution. Soil near to human or other animal activity is enriched with nitrites, nitrates, chlorides and phosphates.

3.1.5.2-The Main Sources of the Salts and Their Effect on the Deterioration and Decay of the Wall Painting

The salts found in walls are known to be carbonates, sulphates, chlorides and nitrates and oxalates of sodium, potassium, calcium, magnesium and ammonia, which are precipitated from multi-components according to the materials present. The origin of salt is indicated by the type of salt present in the wall (Arnold & Zehnder 1987: 103). The walls, stones and mortar contain soluble salts presented as efflorescence forming different aggregates of crystals with different forms and habits on the surface, such as sub-florescence, which can form crystals behind the surface and as solutes in aqueous solution on and within the walls (Arnold 1998b: 2). Mechanical stress and chemical alteration are a result of cycles of crystallisation and dissolution of such salts, depending on the hygroscopic and deliquescent properties of the salts present. The type of support on which the salt is deposited has an important role in the kinetics of the water absorption process, and is of particular interest in wall paintings, as calcium nitrate is a deliquescent salt, which is commonly found in wall paintings (Pique *et al.* 1992).

The action of certain microorganisms on organic nitrogen compounds helps to transform them into ammonia, which is oxidized to form nitrous acid and then nitric acid, which attacks constituents of the soil converting them to nitrates. Sea air also contains sodium chloride, which is able to promote the disintegration of the surface by a process of hydration and dehydration. It also reacts with other salts and with humidity and temperature variation, as well as migrating towards the surface. Salt is deposited from the excreta of birds and bats (urine) which is transported by water (Paine 1993). The residue of the materials used in previous conservation could also include salts, as with the use of alkaline materials and chemical acid, in which the products are treated with water glass (sodium silicate, $Na_2\ SiO_3$). In addition, cement was used, injected with alkali, silicates and silicones etc. as a consolidant. These caused an increase in the rate of deterioration and decay, due to the reaction between the materials. Salts such as sodium sulphates, potassium, magnesium and calcium, as well as sodium chloride, are more damaging to the rendering and the paint layers, because, when they crystallise, they cause deterioration and decay on the surface and disintegration of the cohesion of the rendering used. Calcium sulphate and crystallisation within the rendering may cause a white veil over the surface by the sulphation of calcium carbonate. Thick efflorescence can occur as a result of the evaporation of solutions of sodium nitrate, potassium and calcium salts, which are soluble and easy to remove (Mora 1974: 19, 20 and Arnold & Zehnder 1987: 103-106).

A study has been carried out by Rossi-Manaresi and Tucci on different types on stone to show the correlation between the observed alteration process and the crystallisation pressure. They mentioned that the presence of soluble salts does not always lead to decay in the stone, depending on the pore structure and the crystallisation pressure. There is a relationship between the rate of the migration of the soluble salt and the rate of evaporation as well. If the evaporation rate is faster than the rate of drying, the liquid passes out onto the surface, then the crystals in solute are deposited on the top of the external surface without alteration of the stone, but visible efflorescence is a result. However, if evaporation is slower than the drying rate, the water evaporates by diffusion from within the pore space without any visible alteration of the stone. The alteration occurs when the deposition of all the solute within the pores of the stone is close to the surface. The growth of salt crystals is a result, which may lead to crumbling and powdering of the stone (Rossi-Manaresi & Tucci 1991).

3.2-Bio-deterioration and Decay of Wall Paintings

Micro-organisms and organisms play an important role in the mechanisms of weathering of organic and inorganic materials. The important acids and salts produced by micro-organisms are chlorides and nitrates and excreted sodium chloride. Micro-organisms produce nitrates from excrement and waste (Arnold 1998a: 5). Atmospheric factors like chemical components and climate (temperature, water and light) have an influence on their capacity for biological growth, which is more pronounced in warm and humid zones. Micro-organisms can resist very cold conditions, as they are able to dehydrate their biological structure. While low temperatures are favourable for biological growth, when the temperature rises, biological growth increases. However, high temperature is dangerous to biological growth because it gives rise to various negative phenomena and modification of the active sites of enzymes. Temperature is a limiting factor on biological growth, but heating can assist it if it is connected with other environmental factors such as humidity and light.

The absorption of water depends on many factors such as porosity, hygroscopicity of the materials, as well as ageing. Light is the most important factor in bio-deterioration decay and has a great role in biological growth, depending on the colour quality, quantity (intensity) and duration, which can be a favorable factor for some species of micro-organism. The light cannot be separated from the temperature, depending on the nature of solar radiation, which varies in reaching the soil according to latitudes and seasons, so that part of the energy is reflected and part absorbed. UV light is dangerous for all organisms, because it can induce rupture in the molecular links between cells and cause cell death. Other organisms protect themselves from this radiation by becoming darkly pigmented, such as the blackish patinas of cyanobacteria.

The dark colour induces light absorption and avoids its deep penetration so that the biological organisms grow under the surface. Red and infrared radiation may also promote biological colonization by creating physico-chemical stress. The climate, from the interaction of various factors such as temperature, rainfall and light, is considered the most important factor for biological growth (Caneva *et al.* 1991: 13-19).

3.2.1-Micro-organisms such as Fungi, Algae and Lichens grow in damp conditions with a relative humidity of over 65%. They are varied in colour and character (Mora 1974: 23). They are involved in the deterioration of building materials, stones, and mortars and degrade the mineral materials. The reasons for microbiological influenced corrosion are the production of organic and inorganic acids, which influence the physical and chemical weathering. While bacteria prefer to grow on stone at a pH value between 6-8, fungi grow better at a pH value between 5-6, as the growth conditions of these micro-organisms are different (Wilimzig & Bock 1995: 88).

Bacteria and Fungi are capable of producing chemical reactions, which form acids and oxidize the materials that are sensitive to acids. Some bacteria can produce a sulphur cycle in which sulphuric acid is formed from sulphur and sulphide, while other types of bacteria and fungi cause oxidation of organic materials, organic acids being a final product of their reaction (Torraca 1988: 49). Fungal cells grow as long strands or hyphae; the strands have a diameter about 5-9 μm and can be up to several metres long (Cameron *et al.* 1997: 4). Crust, black brown patinas, exfoliation, powdering and a whitish grey powder as an actinomyces decay, as well as white efflorescence are the results of bio-deterioration by bacteria, while fungi produce coloured stain patches, exfoliation and pitting (Caneva *et al.* 1991: 34). Bacteria are considered to be the main agent of deterioration and decay of murals in a number of churches in Italy and caused crumbling of the paint, blistering of the pictorial layer, and discolouration of the *frescoes* (Petushkova & Lyalikova 1986: 65).

Algae attack in very damp, light atmospheres, the superficial damage being caused by algal growth, in the case of painted or carved surfaces. The various pigments within the cells contain specialized chloroplasts, which make them appear as red, green, orange, and yellow. They can change colour as the environment changes, as well as when the different life cycles of the cells are reached. Algae and fungi form lichens, which cannot live in a polluted atmosphere. Algae excrete nutrients which fungi can absorb, and therefore provide an environment for the algae to protect themselves against extreme environmental conditions, as well as supplying them with the basic minerals derived from the atmosphere. They cause a white crust, bleaching and blistering the surface, the damage being done slowly.

Lichens grow on the materials and are able to produce organic acids; they vary in size from a few millimetres to several centimetres and differ in colours as well as the physical appearance. Depending on the species, they are very slow growing compared with other organisms, due to the availability of nutrients and the presence of water. The growths are in the form of discrete, rounded clumps, being often dark green or reddish in colour. They require a much higher water content and a certain amount of soil, so they cause more damage because they can hold moisture.

Other organisms such as **Liverworts**, are small primitive plants consisting of a green, flat, lobed thallus, similar to mosses, but they grow as individual plants on horizontal surfaces and require a lot of moisture for growth. **Higher plants**, e.g. roots of weeds and trees, cause disruption to the masonry materials. Insects, marine borers, snails and the deposition of the excrement of birds have a corrosive effect and cause halo scratches (Caneva *et al.* 1991: 39, Torraca 1988: 50, 51 and Cameron *et al.* 1997: 4-6).

As an example of decay of other counties, I noticed that in some tombs at Cerveteri in Italy, different kinds of bio-deterioration can be found on the site. In some little tombs underground, water damage had taken place and, more over, a layer of yellow colour had covered the water and turned it a fluorescent yellow, as shown in Photographs 1 A-C. In the site of Ostia Antica in Italy, the main problem was the roots of the trees, which forced underneath the floor to break up the mosaics. This has had a bad effect by moving the tessereae and producing different levels in the floor. In addition, micro-organisms have grown between the bonding as well.

There is some other bio-deterioration caused by different biological growths such as mosses, which developed on alkaline surfaces such as cement, concrete or lime mortar. The deterioration of the wall painting changed the appearance of the painted layer and the rendering. After cleaning, marks were left in the form of little halos and a certain amount of the surface was lost (Mora 1974: 23).

I have found similar marks on the cleaned fragments of Roman wall painting from Piddington (see Chapter 7 photograph 52 C), and tried to remove them, but without success.

The growth of microflora is often slow, but when biological populations attack organic materials their development is easy to detect and they enable changes to the structure of the materials (Caneva *et al.* 1991: 35). Microflora and bacteria have had a bad effect on the wall painting of the tomb of Tutankhamun. The deterioration and decay has spoiled the wall paintings all over with black patches (Fahd 1994). In the case of organic substances on the surface exposed to the open air, a small residue of previous biological colonization or old treatment favours the development of heterotrophic microflora.

Some studies have been carried out on preventing the growth of different microorganisms, such as fungi, bacteria etc. Mora mentioned the use of sodium silicofluoride, zinc and magnesium chloride for removal of vegetable growth such as moss and lichens (Mora 1968: 172). However, another study involved supplementing the mortar with a low concentration of copper and zinc for decreasing the cell growth of both bacteria and fungi (Wilimzig & Bock 1995). The durability of a new lime mortar with biocidal (apentachlorophenol) properties for reconstruction has been studied by Martinez-Ramirez *et al.* (1996).

It has been said by Petersen *et al* that methods for the control of microbial activity must be analysed and developed. It is important to consider whether the methods would be applied for short or long term techniques; in the first case these might be to kill the micro-organism, but not prevent re-colonization, but in the second case to inhibit biodecay for prolonged periods of time. The control of the environment is an important factor; it should be stabilized before using any materials for treatment, as in some instances it may encourage the growth of micro-organisms. The increase in microbial decay would result in an increase in salt decay, which is very harmful to both the rendering and the painted layer of the wall paintings. The control of the microbial activity depends on the materials used and the methods of application. Petersen *et al* have tested and described different methods on the effective inhibition of microbial decay, e.g. the use of lime water, alcohol and different kinds of biocides. The methods have been tried in the laboratory and on sites as well. They pointed out that in the case of using biocides, preliminary analyses should be carried out on their effectiveness because it is not possible to use biocides for the whole painting without these tests, although organotin compounds showed a more or less sufficient inhibition in agar diffusion. Some biocides have shown a great effect in inhibiting microbial activity, e.g. in bacteria, but are less effective against fungi (Petersen *et al.* 1993).

3.3-The Internal Alteration of Wall Paintings; Supporting Structures, Rendering Materials; Plaster & Mortar and Pigments.

3.3.1-Supporting Structures

The alteration of stone as the supporting structure of wall paintings is related to the structure, the shape and the nature of the stone. Carbonate rocks are those in which calcium carbonate is the common element and vary in their structures, colouration and hardness. Sedimentary rocks are derived from the disintegration or chemical weathering of pre-existing rocks. Saline rocks are deposited by water and glaciers; these rocks are sandstone, limestone, gypsum, travertine, and volcanic tuffs. Metamorphic rocks such as marble, quartzite and slate are derived from the transformation of pre-existing rocks of sedimentary or eruptive origin, due to the pressure and the temperature, which affect the composition and the appearance. The percentage and composition of the grains influence the resistance of the stone to alteration and determine its properties. According to the nature and size of the grains they might be classified as light coloured minerals or dark coloured, which both contain compounds of iron and magnesium (Sneyers & De Henau 1968: 212, 214 and Caneva *et al.* 1991: 87).

3.3.2-Rendering Materials; Plaster and Mortar

The internal deterioration of a wall painting can be caused by different factors and affect the plasters, pigments and binding media, on account of defects in their composition or the preparation of murals and the techniques used. When the rendering has a clay base, the composition of the clay

tends to be very sensitive to water and moisture. These materials used as a basis for plaster are sensitive because of their hygroscopic nature. The reaction of water and soluble salts causes alteration of the rendering depending upon the characteristic of the different types of rendering used. Renderings containing organic fibres are sensitive to moisture and they decompose physically, becoming softened and hollow as well as being washed away by the action of water. In the case of plaster, which consists of a lime rendering, sulphation is the result of the transformation of calcium carbonate into calcium sulphate under the influence of sulphur dioxide gas, a frequent impurity of contaminated atmospheres. However, in a very dry ambience, a hard crust of calcium carbonate is formed on the surface (see Chapter 5, deterioration of wall painting from Piddington). That can happen when the water has evaporated from the surface, which has been exposed to air. The water was absorbed, before all the calcium hydrate had time to be carbonated on the surface.
The quality of the rendering materials is important for wall painting. Since the wall painting and all coatings of architecture consist of different, heterogeneous layers, salt tends to crystallise in different zones according to the porosity and of the layers' cohesion, in addition to the environmental conditions, as mentioned above. The pressure of crystallisation can degrade the rendering and the binding media, physically breaking the mechanical cohesion of the carbonate crystals. In terms of *fresco* techniques, the rendering must be neither too friable nor too strong. In the first case, it will not adhere well to the surface and in the second it will cause shrinkage on drying. Organic materials such as straw, hair and other vegetable fibres were used to give cohesion to a rendering, as mentioned before. However, these materials enable moisture to penetrate the surface, producing good conditions for biological attack, because of the tendency to absorb water. As a result, the paint layer, in the case of *tempera* painting could flake, blister or wrinkle on account of the oxidation of the organic binding media in hot or humid environments (Mora 1974: 27-29, Hammer 1995: 578 and Lal 1996: 12).

The effect known as efflorescence is due to the materials used in the mixture of the plaster. The main sources of efflorescence are gypsum (calcium sulphate), potassium sulphate and carbonate and calcareous earth developed from caustic lime. The building materials have a great role in transference of the salt to the plaster or rendering layers. When salts have been dissolved by moisture and seeped through the wall, a film appears on the surface after drying (Mayer 1973: 337 and Doerner 1969: 305).

3.3.3-Pigments

The causes of specific changes in pigments are humidity and light, which affect the colour quality and although most pigments tend to discolour, organic pigments are very sensitive to humidity and light. Humidity promotes chemical action and results in changes in the composition of the pigment, which causes alteration in the colours, such as the blue basic carbonate, azurite, becoming greenish and partially converted into malachite. Salts e.g., ($NaCl$) are extremely dangerous to pigments and works of art, and may affect the colour by the action of the chloride ion. Copper colours (azurite and malachite) may change to copper sulphate if exposed to moisture for a long period and sulphur dioxide can produce black stains and if exposed to air, decomposed carbonate is the result. Light can cause some alteration to the pigment as a result of differences in the thermal expansion of the material composition of pigments. Ultra violet light is responsible for bleaching of organic pigments as well as the oxidation of the binding media (Mora 1974: 23, 24 and Mora 1984: 66). Another phase of the alteration of pigments is dependent on their reflective index and the fineness of the particles. The stronger reflective index gives better covering power, and the binder media, which reacts with the pigments, converts them into different phases according to whether or not they are fully compatible to the technique used, and how much the binder is attached to the pigments. This affects the action of the hiding power on the pigments (Mural Painting Conservation 1998: 30). Acids, alkalis and heat may affect pigments, and in the case of organic dyestuffs, light causes a definite degradation and fading to colourless products. The effect of light is accelerated by heat and moisture (Gettens & Stout 1966: 139). Lighting candles can also cause the alteration of pigments, as the heat was able to change the pigments because of the air contaminated through the oxidization of the atmosphere (Rickerby 1991). The fire not only altered the pigments but also converted calcium carbonate to quick lime in differing amounts. As a result of this, yellow and green oxides became red or brown and dark halos could be produced as well as the rendering breaking away.

Some examples of **the alteration of pigments** have been noted by Mora, especially some pigments which are not stable in *fresco* technique, such as azurite, cinnabar, and lead white (Mora 1974: 24, 29). Azurite ($2Cu(CO_3$ $Cu(OH)_2$, basic carbonate of copper-blue) is changed by the action of humidity transferred into green carbonate (malachite, $CuCO_3Cu(OH)_2$). In the presence of moisture and sulphuric acid, certain copper blue is changed to black copper sulphide. That happens when water gets into walls and is catalysed by alkaline materials. In the case of cleaning a surface painted in azurite great care is required in order to avoid using materials such as ammonia, ammonium salt or organic amines which react with copper.

With cinnabar (mercury sulphide, α-HS), the alteration of colour is caused by the reaction to the light wavelength between 400-570 µm, converting it into black (metacinnabar, α_1-HgS), both in water media (*fresco* and *tempera*), and oil painting. Cinnabar can be distinguished from metacinnabar, as it is in a hexagonal system in an alkaline solution, whilst, metacinnabar is a cubic system in an acid solution. Treatment with alkaline polysulphates converts it back to cinnabar (Gettens *et al.* 1972: 53, 54).

With white lead (see Chapter 2, page 26), alteration is caused by oxidization or chemical action. Sulphides (H_2S) form irregular red-black spots with sharp borders, but when the alteration occurs and is completed on the pigment, all the painted surface is converted into a red-black colour. However, Matteini has tested some samples with different media and found the oxidation by alkaline substances occurs in the presence of lime, animal glue, egg and casein with exposure to UV light at RH of 80% and treated with hydrogen peroxide sodium hypochloride chlorine gas (H_2O_2 NaCl O, CL_2), the sample that was treated with chlorine converted into the black lead oxide PbO_2 (Tabasso 1998). The process of this transformation into lead dioxide (PbO_2), its cause, the nature of the products, and possible methods of reconverting the altered products to their original colour have been described by Giovannoni *et al.* (1990).

Another study reported how oxidation can also be induced by biological growth in terms of bacteria that do not require organic substances for their development, and were revealed to participate in the conversion of white lead to lead sulphide, which produce hydrogen peroxide and was capable of causing various conversions of minerals which are present in pigments (Petushkova & Lyalikova 1986).

4-Intervention Factor

4.1-The Problems of Cement and Alkaline Materials Used in Previous Restorations

Cement (calcium alumina silicates) was one of the most common materials used for the conservation and restoration of works of art in the 1950s and 1960s. It contains several soluble salts, such as sulphate, nitrites and nitrates. When cement has been used in a building, where there is some moisture in the walls, the soluble salts present in the cement can migrate towards the original plaster or the mortar, which causes destructive efflorescence or crystallisation upon evaporation (Teutonico 1988: 62). Portland cement contains up to 1% of soluble alkalis, so when the ions are leached out, it forms efflorescence of alkali carbonate salts. Arnold has given as example of 100 kg of Portland cement with a content of 0.1% of soluble sodium oxide (Na_2O), which may produce about 460 g of carbonate (Na_2CO_3 $10H_2O$: Arnold 1998b: 3).

The problem of using cement is that it has a very low porosity and does not allow easy evaporation of water. In addition, the rainwater penetrates behind the plaster through cracks and causes tension in the cement at low or high temperatures. As a result, the plaster is detached from the wall or rising damp is increased because cement has a high desinty and is a thermally conductive material (Torraca 1988: 80). Cement may also form cracks in the painted layer or form zones of unseen and unequal condensation on the surface, varying in density beneath it. It is a matter of the fact that staining and disintegration may occur in the rendering, as Mora said, because the ambient humidity tends to condense on the cement and to be absorbed by the painting rather than the cement which is carrying the salt with it (Mora 1974: 27).

In addition to Portland cement, plaster of Paris has been used in previous restoration on objects of art (Tubb 1987). This a was source of soluble salts as has been investigated on Egyptian wall paintings (Miller *et al.* 1987). There are also alkaline cleaning substances and waterglass-like materials. Where acids such as hydrochloric or acetic acid or alkaline solutions such as caustic soda or ammonium carbonate are used for cleaning, they form more or less soluble salts. These materials can be used, but with some consideration that the possibility of a secondary effect has always to be taken in account. Waterglass, which is used for consolidation and still applied to wall paintings, can form a hard surface layer leading to rapid damage (Arnold & Zehnder 1987: 106, 107). The materials used in conservation such as varnishes, waxes or fatty substances were capable of altering the condition of the paint layer and changing the appearance of the colours, darkening both *fresco* and distemper pigments as well as changing the texture and the tonality, due to the fatty substances. In addition, bad application can cause contracting and cleavage of the paint on drying and allow the moisture to penetrate between the paint layer and the rendering and help to prevent evaporation of the water in the wall. The use of organic substances (gum Arabic or white of egg) attracts the growth of micro-organisms in the case of humid environments (Mora 1974: 26, 27), but are stable in dry environment.

I have noticed the problem of the use of cement in many areas in Italy, a few examples in the tombs of southern Etruria, Roman and Pompeiian wall paintings. The main problem was caused by the use of cement mortar for restoration in the 1950s. Damage has occurred to wall paintings and case studies have shown its effects. The same problem of previous restoration in the 1950s and 1960s was found in Egypt, Rome and other places.

5-Weathering in Egypt and the Main Types of Deterioration of Wall Paintings

There are several factors responsible for the deterioration and decay of the monuments in Egypt. Lucas mentioned that the main source of disintegration of the building stones in Egypt are variation extremes of temperature and the mechanical action of wind-borne particles of sand which produce abrasion. In addition, the solvent actions of water containing carbon dioxide and other substance in solution have a effect marked in the case of limestone. It is noteworthy to mention that the amount of rain in Egypt is very slight compared with European countries, where it will wash off all salts, whilst in Egypt the dryness allows salt to accumulate in the buildings.

Salts in Egypt come from the ground, the reason that the zone of disintegration in Egypt is so frequently found close to the ground level (Lucas 1915: 2, 5). Other elements affecting the Egyptian monuments are natural disasters such as floods and earthquakes, biological factors and the action of gravity and habitation. In addition, there is the pollution (industry), water-pollution, dust, fires and smoke produced by the people who are living in and around the monuments, as well as tourism.

Salt crystallisation is one of the most important aspects of decay in Egypt. It has been observed all over Upper and Lower Egypt. The tomb of Nefertari is one of the best known examples of the deterioration and decay of wall painting caused by salt crystallisation which has grown between the plaster and rock in large sized crystals and needles as well as fibres, penetrating through both the mortar and painted layers and causing disintegration by pulverization of the pigment itself (Rakowski 1973 & Getty Museum 1990).

The materials used in the past for conservation often enabled the deterioration of the wall paintings, due to the use of Plaster of Paris as well as cement. These are the main sources of soluble salts and are responsible for the decay of wall paintings, as mentioned before. These materials were used in conservation in Egypt both in stone buildings and wall paintings. The chemical materials or the adhesives used also have a great role in the deterioration of the painted layers. I have noticed that salts in different areas outside and within Egypt are a very serious problem, especially in Alexandria at Anfushi in both tombs 1 and 2 as well as bio-deterioration decay (see Chapter 5, photographs 3 & 4).

It can be concluded that the deterioration and decay of wall painting and the building materials could be from different factors, including not only the environment surrounding them as well as weathering decay, but also the composition, quality and the porosity of such materials. The processes of deterioration and decay however seem to be very complicated, as they can be produced by either chemical reaction or by the action of other factors, under certain conditions. The ability of the materials to resist or deteriorate, is due to the decay mechanisms which are not working in a linear, but a dynamic way.

The observation of the deterioration and decay processes is extremely important before analysing the materials of wall painting. This helps to draw an accurate outline of how the resulting deterioration and decay had affected them, to understand the mechanism and changes to the structures and their composition, and to determine the methods which could be used for preservation and the prevention of such decay. The analysis of materials is an important aspect of the conservation and restoration of objects of art. The instrumental tools used to analyse the materials are varied, but that analysis has an important role in the diagnosis of the main deterioration and decay principles.

In the next chapter, I have illustrated the methods involved in the analysis of the Egyptian and Græco-Roman wall paintings, in terms of the analysis of the plaster and mortar to identify main composition and structure, as well as determining the main source of the deterioration processes and decay.

Chapter 4

Methodology: The Analytical Techniques and Methods Used in the Present Examination and Analysis of Egyptian and Græco-Roman Wall Paintings: Plaster/Mortar & Pigments

1-Introduction

Analytical techniques provide important information in the identification of materials from which the artefacts were made, and can be used to establish whether the materials employed in the manufacture of the artefacts were local or imported. They also help us to understand more about the socio-economic context. Examination provides us with information about the manufacturing techniques which were used. Chemical and physical analysis attempts to define the composition of the ground sealant and surface plaster, as well as the composition of the range of pigments and organic plaster media.

The systematic approach to analysis of the material in the composition of works of art can be varied. Through these analyses we may get all the information required, depending on the methodology used and what we need to know about the materials, their structure and composition. The methods of examination and analysis have developed a great deal in recent years, particularly in the field of chemical and physical analysis. The use of analysis helps conservators to choose the right kind of specific techniques. This is extremely important as the types of materials used in antiquity vary considerably. Each mural painting is unique. Analysis of materials can provide us with the structure and composition of materials such as plaster and mortar, as well as pigments and colours (organic or inorganic) in the painted layer.

The various instruments for the analysis of the material in works of art are given by Mairinger and Schreiner (1982). Recent studies on the physical principles of scientific methods in works of art are reported by Ferretti (1993) and Burnstock (1998), with illustrations of the instrumental processes, and some notes on the features and limitations of each. In addition, the literature contains many examples of the use of established non-invasive methods for studying objects as a whole. These methods are widely used to analyse the materials. Numerous other studies describe experimental processes, with a consideration of each individual laboratory (e.g., Tite 1972; Jones 1987; Kempe & Templeman 1983).

I have chosen to use simple methods, without expensive instrumentation, as often these would be enough to identify most of the problems of materials in works of art and could be carried out by conservators using classical methods such as micro-chemical tests, which are required to identify the structure of the materials. The use of non-destructive techniques should be remembered as the need is to treat, not destroy, objects of art.

An outline of the strategy and plan for the analysis of materials involved in my research, is shown in Figure 5, which also lists the instruments used for the analysis of the materials in wall paintings. Three stages were applied to analyse materials and samples of Egyptian and Græco-Roman wall paintings from Saqqara, Dahshur, Luxor, and Alexandria in Egypt and Piddington Roman villa in Britain.

Figure 5

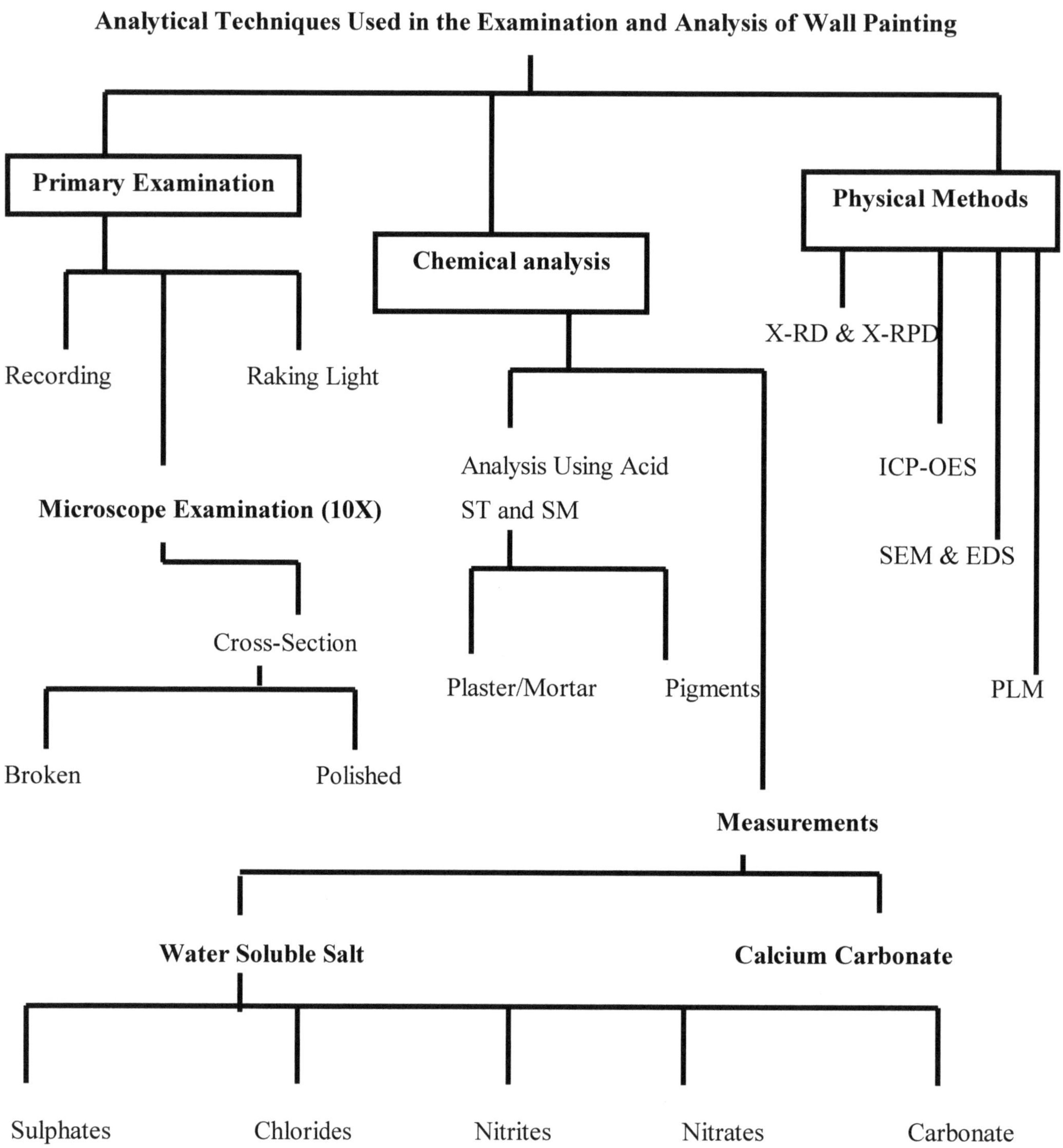

-ST: Spot Test **-SA:** Standard Analysis (British standard)
-X-RD: X-ray Diffraction and
-X-RPD: X-Ray Powder Diffraction (looking at pigments and Minerals)
-ICP-OES: Induction Coupled Plasma Optical Emission Spectrometry (Metals ions)
-PLM: Polarized Light Microscope (internal structure)
-SEM: Scanning Electron Microscope-**EDS:** Energy Dispersive Spectroscopy (Surface structure and Elements)

2-The Methods Used in the Present Examination of Wall Paintings

2.1-Primary Examination: The First Stage

The examination of excavated materials, as Cronyn said, is fundamental to archaeological conservation. It aims to identify the nature of artefacts, both for recording and treatment purposes. There is a need for more information during examination than can be obtained by naked eye alone, thus a binocular microscope, which has facilities for photographic recording, etc. is used in the first stage (Cronyn 1990).

Examination in the first stage, includes; recording and photographing the object or samples, microscope (10X) examination of cross sections, including; broken section structures, and polished sections for measurement of the wall plasters/mortars and the use of raking light.

The samples were examined under a binocular microscope (10X). Cross-sections were used to identify the shapes of the aggregates inside the mixture. Polished-sections were prepared using embedding resin (mixed with a liquid hardener at room temperature in proportion 98% resin and 2% catalyst), then the samples were ground down using silicon carbide paper from 220-1000 μm, and alumina powder for fine polishing. Cross-sections can provide information about the structure and identify the materials, as well as the condition of the paint present on the surface. This requires a binocular microscope. Cross-sections can be used for the identification of the painted layer using other techniques include; staining techniques, reagents for resin as well as protein (such as Amido black), fluorescent antibody technique and thin-layer chromatography for amino-acids, in addition to immunofluorescences (Hey 1957-8, Jones P. 1962, Johnson & Packard 1971, Elzinga-Tar 1971, Martin 1977 and Kockaert *at al.* 1988).

2.2-Chemical Analysis

The most widely known methods of painting on plaster were *fresco* and *tempera*. Plaster was made of sand and cementing materials, such as lime or gypsum mixed with water. Organic materials or natural fibres may be used, being added to strengthen the consistency, as previously mentioned in Chapter 2 (see 4.2). In my research the mortars or plasters were tested for their quantitative and qualitative mixtures, and measurements of water-soluble salts and of calcium carbonate content were taken.

2.2.1-Analysis Using Acid (spot tests)

Where necessary, micro-chemical tests were carried out using concentrated hydrochloric acid (HCl) for the identification of carbonates and diluted hydrochloric acid for pigments. In addition a test for iron was carried out in some cases. Hydrochloric acid (HCl) and acetic acid (CH_3COOH) were used for the identification of mortar and plaster with calcareous or non-calcareous aggregates. One should be careful, because some acid is stronger than others, e.g. hydrochloric acid (HCl) is stronger than acetic acid. If the mortar is lime based, it will dissolve with HCl, including any limestone aggregate. In such cases acetic acid should be used for chemical analysis. Limestone may not dissolve very quickly in acetic acid, and sometimes this may be used to disaggregate calcareous mortar or plaster. Other tests were carried out using warm hydrochloric acid to distinguish whether the plaster was lime or gypsum; in the case of pure lime, it will show effervescence. If some gypsum remains in the mixture, barium chloride is added to identify the gypsum, which in hydrochloric acid becomes milky and shows a white precipitate. Doerner described a method to determine the gypsum in the lime using 1 part of lime to 20 parts of hydrochloric acid. If the solution becomes milky and white precipitate is shown, then it is heated in a spoon until it becomes very hot and dry. A little water is poured into the palm of the hand and then added to the precipitate. If it turns hard quickly and warms, this indicates the presence of more than 5% of gypsum. If any clay or marl is present in the mixture, which would not dissolve, this indicates the impurity of the lime (Doerner 1969: 268). In addition, other tests were carried out where required throughout the research. Microchemical tests seemed to be very necessary, to distinguish the materials before carrying out full analyses. Microchemical tests indicated the reaction between the materials, which could be detected by the changes in colour, shape or by precipitation of particles, which can be observed by the naked eye or with the help of the microscope. Although it is a destructive technique, a quick result can be obtained and it can be useful for identification of the materials without expensive instrumentation.

2.2.2-Standard Analysis (SA)

There are a small number of chemical analyses, such as specific gravity, X-Ray diffraction or fluorescence which do not require interventionist samples, but the use of methods,which require a sample to be taken is more usual (Morgan 1992: 25-28). The size of the sample allows different levels of analysis to be undertaken. A small sample may not be representative of the whole and sometimes only a full section of plaster can give an overall view of the materials by analysis. Chemical analysis is sometimes not enough to determine the composition of materials, thus microscopic analysis is essential in determining their composition. The size of sample is a very important consideration. The scientific practice to use at least three samples for each analysis this is more likely to produce an accurate results to show what the plaster is composed of now, not what it was made from. Changes and alteraition may have taken place, as well as ageing (Morgan 1998). Some considerations on comparative analysis and basic methodology on ancient mortars also stated by Jedrzejewska (1982: 312-322).

Chemical and micro-chemical analyses give a fuller picture of the materials being examined, which would be essential if conservation and restoration were to take place. The standard methods (B. S. 890, 4551 and 812 part 119 & 103), were carried out on the analysis of mortars and plasters to identify the mixture of the plasters or mortars or “supporting structures” used in ancient and classical times. The standard methods required at least 100 g, as well as three different samples from the same area, but in some instances this was very difficult to obtain as it could be judged to destructive to heritage. The weight of samples, therefore, varied according to what was available.

Analyses were carried out on different samples from different sites, and different periods. They were treated with dilute acids, some with hydrochloric acid and some with acetic acid (10-25%), to separate the carbonated lime from the aggregate mixed with sand and other ingredients. Sieving analysis was carried out on each mortar after chemical separation of the binder and the aggregate. The stages of chemical analysis were; separation of the lime followed by washing (four times or more till pH 7), drying and sieving, (drying not in an oven because this can make the silt and clay hard, which would then cause some difficulty during the sieving), and then grading. Many sizes were used to sieve the samples, starting from 150 μm to 16 mm or larger for concrete. The final stage in grading was to identify the materials used in the aggregate. Graphs were constructed to compare the percentage in the aggregate grades of the mortar or plaster used from the different sites as well as the different periods.

Although this method is destructive, it allows us to study not only the grading and other characteristics of the aggregate, but also to determine the approximate proportion of the lime binder that was used in the original mixture. This information is useful and necessary for the reconstruction of the wall painting, in order to ensure the accurate proportions of the mixture to be used and to avoid any problems that could occur in the future. This method can provide us not only with the percentages of the mixture of the mortars and its lime, but also with the quantity of each grain size that was used.

2.2.3-Measurements

2.2.3.1-Water-Soluble Salts

Salt decay appeared to be one of the most important deterioration factors present in wall paintings, as was mentioned previously (see Chapter 2, 3.1.5). The methods for the identification of salts were carried out using different instruments, starting with micro-chemical qualitative analysis to identify sulphates, chlorides, nitrites, nitrates and carbonates as these salts were the most important to determine. Teutonico (1988), Hammer (1995), Bläuer (1996) have established the methods of analysis of water-soluble salt using different chemicals and reagents.

Microchemical slide tests can be used to identify the salts in situ; Arnold mentioned that a pH test paper allows the detection of alkaline salts on a wall. Highly hydrated salts, will dehydrate and dissolve in their own crystallisation water when held for a while between thumb and finger, which permit the detection of natron by its alkaline reaction and its capacity to dehydrate. However, the detection of alkali carbonates, where they were found, resulted in the use of alkali materials, such as Portland cement or water glass products, for restoration. Most of the mineral salts may be determined with a combination of optical and micro-chemical tests. However, physical methods, such as x-ray diffraction can be used as a complementary method, in particular where the results are un-certain. Analytical work should commence with the consideration of the building or stonework, since the occurrence of salt is related to specific weathering phenomena, and to particular micro-climates, as well as hydrological conditions (Arnold 1984: 129, 130 & 136). It has been found that thermogravimetry and differential thermal analysis (TG-DTA) are useful to identify salts such as calcium, nitrate, tetrahydrate, both as pure compounds and in mixture with other minerals. These techniques were also used in the preliminary investigation of mortar (see Pique *et al.* 1992).

I used standard weight of sulphates, chloride and nitrates prepared by dissolving 1g in one litre of deionised water. 1-2 g samples (plasters, mortar, salt and stone), were put in 20 ml of distilled water, shaken, and left to precipitate. Baloüer mentioned different views about the extracting time, e.g., 30, 60, 90 minutes assuming that longer contact times lead to the partly dissolution of sparingly soluble constituents (Bläuer 1996: 1507). After analyses of water-soluble salt according to Teutonico, with the exception of carbonate, the samples were dried at 60 °C and weighed before and after analysing for carbonates to calculate the percentage of the residue as well as the amount dissolving (Appendix 14). Water was added again to the samples to analyse for carbonates using concentrated hydrochloric acid, then samples were washed until pH 7 and dried at the same temperature. Physical analysis was also carried out to determine what the salts were. The metal ions were determined, using ICP and X-ray diffraction to define the different crystalline phase.

The chemicals and reagents used for the water-soluble salt were:

•-For sulphates: hydrochloric acid and barium chloride (10% solution) which was used to measure sulphates; a white precipitate indicates the presence of a sulphate.

•-For chlorides: to a solution of a chloride, diluted nitric acid and solution of silver nitrate was added, one drop of nitric followed with one of silver nitrate. A white precipitate indicates the presence of a chloride. Note should be made of the white precipitates as chloride is much deeper than sulphate, which means it contains more salt, the limit of detection being about 5 ppm.

•-For nitrites and nitrates: acetic acid and a mixture of Griess-Ilosvay's reagent in equal amounts were used, to measure nitrites. Sulfamic acid is used to destroy the nitrites in the case they were present, and then one can measure nitrates using the same chemical and a pinch of zinc powder; the solution turns pink in the presence of nitrates.

2.2.3.2-Measurements of Calcium Carbonate ($CaCO_3$)

The method of measuring calcium carbonate ($CaCO_3$) has been described by Jedrzejewska. She mentioned that three basic values of a mortar can be measured; the proportion of carbonates, the proportion of substances soluble in acid, without evolution of carbon dioxide, and the proportion of sand (Jedrzejewska 1960: 132). The third application of the method was not used in my work for two reasons: firstly, because different materials were analysed and in some instances it was difficult to separate the sand, so the third value was of the whole residue; secondly because the standard methods were applied, so that it was very useful to give full details not only about the percentage of the soluble lime, aggregate and silt, but also the particle size of the mixture of each mesh, which can be necessary for reconstruction of the mortar using the same grain sizes.

The method was used not only to measure the calcium carbonate of a mortar, but also for limestone, where it was noticed that some samples were very highly carbonated, as the mortar was a mixture of powdered limestone and some other ingredient. I thought that it was necessary to measure the original limestone used in the mixture, which could give a clue in distinguishing those mortars from others in terms of their geological nature. The same methods of measuring calcium carbonate, as reported by Jedrzejewska (1960), were used with some modifications of both the experimental methods, as can be seen in Figure 6 compared with photograph 2, showing the system used for measuring calcium carbonate ($CaCO_3$). The weight of samples used was mostly about 0.35g with some exceptions from Piddington (see Appendix 15). A standard calcium sample should weigh between 0.35 g and a maximum 0.39 g of pure calcium carbonate.

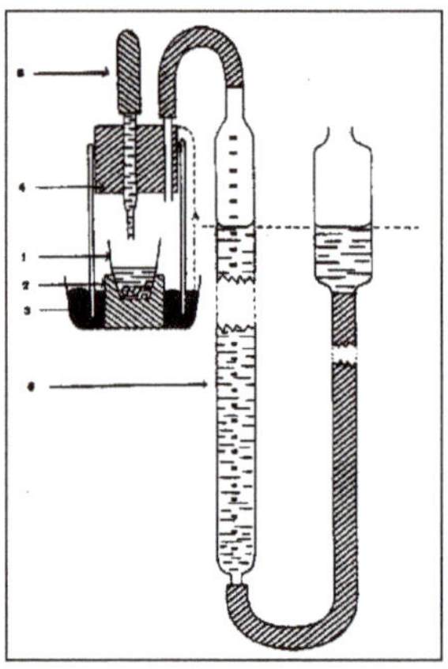

Figure 6, shows the system of the Jedrzjewska method F; (1) Crucible with sample. (2) Cork. (3) Mercury. (4) Rubber stopper. (5) Dropper pipette with con. HCl (6) Gas burette (100 ml) (from Jedrzjewska 1960).

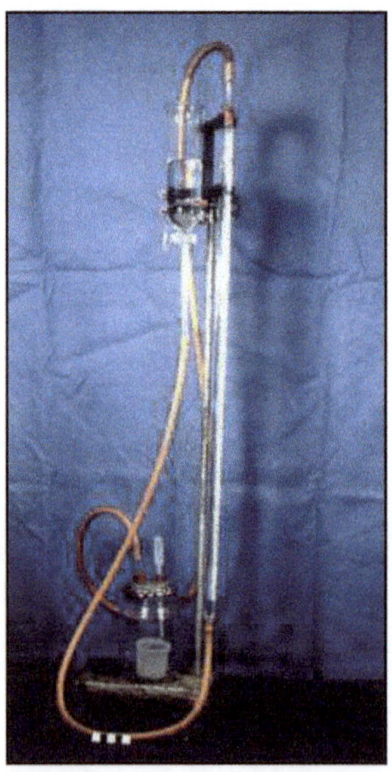

Photograph 2, shows the system used in the present study for measuring the carbonate content. Scale 1-cm division.

2.3-Physical Methods

Different techniques were used to look at the surface structure and its elements, the internal structure, the minerals, the metals ions and the organic materials.

2.3.1-X-Ray Diffraction and X-Ray Powder Diffraction (X-RD & X-RPD)

These tests were used to determine the minerals and their chemical composition. X-ray diffraction was conducted using tiny samples of pigments in the X-RD (Philips PW 1009/80 and Debije Scherrer powder cameras PW 1024), in which the position of the diffraction lines and their intensities were recorded on photographic film. The result involved conversion of the characteristic ring patterns which were recorded on the film into -d- spacing and comparison of the patterns with those of known standards. The process is described by Tite (1972: 285, 288). It is quite different from the spectra machine, which required at least 2-3 mg, that amount being very difficult to collect in some cases. X-RD is ideal for the analysis of only a single crystal of the minerals. Analyses using X-ray powder diffraction were carried on Philips model PW 1729 X-ray generator with a PW 1716 diffractometer and PW 1050/25 detector, operator 40 kV, from 4-64 °2 θ.

2.3.2-Induction Coupled Plasma Emission Spectroscopy (ICP)

This method uses the ionization of gas by radio frequency excitation to produce a plasma into which a solution of the sample is sprayed. The excited atoms produce a linear calibration of emission with concentration over very large concentration ranges (10 ppm to over 10 per cent), and require a small amount of sample, about 0.03 g, for a complete analysis (Kempe & Templeman 1983: 43). The method was used to look at the metal ions of solutions (after measurements of water-soluble salt) and powdered samples, in order to determine the metal ions of different materials (mortar/plaster, limestone as well as salt). Analyses were carried out using a simultaneous spectrometer. The powdered samples were treated in a special manner before ICP (see Appendix 3).

2.3.3-Polarized Light Microscope Examination (PLM)

Some samples were thin-sectioned to look at their mineral structure and texture, and characteristics of the materials, whether plaster, or mortar and rock. They allow us to look at the structure before using HCl acid, which would dissolve all the carbonate materials, such as limestone, fossil shells and oolitic ironstone, etc. It is also useful to identify the kind of materials for possible re-construction of the wall plaster or mortar and to provide us with information about the original materials used. This will help us to understand the deterioration and decay processes, which have taken place inside the mixture due to alkaline factors, which may affected the structure of the mortar. Optical microscopy on thin sections of the mortar samples gave further information about the binder, the type and shape of the aggregates, as well as the internal structures.

Thin section samples were impregnated with a mixture of MY 753 Araldite epoxy resin mixed with HY 951 in the proportion 10:1 and cured at 60 °C for 2-3 hour. The same machine was used for cross sectioning but, care had to be taken when polishing in order to get the right finish. The section should be 30 μm thick, giving quartz grains a grey colour in transmitted light with crossed Polars at 40-50X (see Tite 1972: 215, Jones 1987 and Hutchison 1974). Cover slips were put onto the slices using Euparal for the identification of the minerals and for photography with polarised light.

2.3.4-Scanning Electron Microscopy and Energy Dispersive Spectrometry (SEM & EDS)

This instrument was used for the examination and analysis of the micro-structural characteristics of solid objects (see Goldstein *et al.* 1994 and Tite 1972: 242). Examination was carried out on polished cross sections and fresh samples for detailed characterization of the surface structure.

The machine makes it possible to observe and investigate complex paintings and can give a high magnification image on small samples, with the advantage that the method is non-destructive and one could use the samples for further analysis, although only small samples can be put inside the machine.

The preparation of the samples for SEM involved coating them with gold or carbon or a coating of carbon plus gold or palladium, which gives the best result for porous materials. The coating helps to make the samples electrically conductive and is required to obtain a clear image of insulating materials (Welton 1984: 1-3). EDS analysis was used to determine the elements present, providing more information about the surface structure and its composition. This helps in cases where the surface is covered with layers, for example fragments of wall paintings from Piddington (see Photograph 46) which we need to understand before cleaning, to ensure the use of suitable materials for conservation. It should be remembered that the original surface could be covered with different layers of varied composition. We may find different elements as a result, in one level or another. Scanning electron probe microanalysis showed variation in the microstructure of the surface deterioration. That helped to determine the form of the structural problems, e. g., micro-organisms and micro-cracks where present on the surface, as shown on different samples, not only from different periods, but also different sites such as Alexandria and Piddington (photographs 44 C & D and 46 A & B).

SEM & EDS analysis took place in the University Medical Sciences Unit using a DS 130 with PGT series 111 device and coating unit E 5150. Finishing analyses were carried out in the Geology Department by the author using SEM (Hitachi S-520 & EDS link AN 10,000) and coating unit (Turbo sputter coater K 5675). Unfortunately the machine was unable to provide the quantitative analysis of the element content.

It can be concluded that various chemical and physical methods can be used to distinguish materials' composition, structures and their minerals as well. Mostly non-destructive methods are used in the field of conservation, because they require a small amount of sample, which is commonly available. It was very interesting carrying out these methods of analyses of different mortars/plasters as well as other materials, from different periods and sites. As a result of the research I obtained a more general picture of the different structure of mortars throughout the work, giving more information on the technology of the manufacture of the plaster/mortar used and on the relationships between their composition and place of use. Most of the methods that were involved were non-destructive, as this was an important point to consider in my research. The only exception was in the use of standard mortar analysis methods, where it was necessary to use larger amounts for an accurate results for re-plastering/re-pointing and to have a clear idea of the materials used as filler in ancient and classical times.

Chapter 5

An Introduction to the Sites of Saqqara, Dahshur, Luxor, Alexandria and Piddington

1-Introduction

The materials used for the manufacture of plasters and mortars varied according to the main sources of materials available. For that reason different sites were selected to collect samples, as the aim was to study not only different periods of plasters and mortars, but also different areas. The samples came from four sites in Egypt (map 1), dated to different periods. Saqqara, Dahshur between 2920-2134 BC, Luxor 1550-712 BC, and Alexandria dated to the Macedonian to Roman period 332 BC-395 AD. Fragments of wall paintings were also studied from Piddington Roman villa in Britain (map 2) dated to the 4th century AD. I visited all of these sites and collected samples.

It was useful to carry out this study on the new sites excavated in Saqqara and Dahshur, dated to different periods from the early dynastic period to the Old Kingdom. This because it may add new information on the technology of the manufacture of the plasters and mortars or the walls, as they have not investigated scientifically before and could have received different preparation for the paintings.

Luxor was the second site involved in my research from the New Kingdom (1550-712 BC). The tombs were dated to different periods between the 18th-26th dynasties. From each site two tombs and in some instances one tomb were selected for study. It was extremely interesting to carry out analysis on different materials, areas and periods, to investigate the composition and structure of the plasters/mortars used for the preparation of walls in those periods. The aim was to find out to what extent the materials were similar or different from those found in the Old kingdom, according to not only the geological sources of the materials, but also the developments of the manufacture of cultural materials.

Alexandria is one of the most interesting case studies for several reasons. Firstly, the wall paintings in Alexandria is dated to the Græco-Roman period by means of the different techniques used to execute the wall paintings, because the materials used, the methods of preparation and the techniques of wall painting used in Græco-Roman times differed from those used in dynastic times. The second reason is that this area had never been studied before and it was my personal interest to carry out some work on Græco-Roman wall paintings.

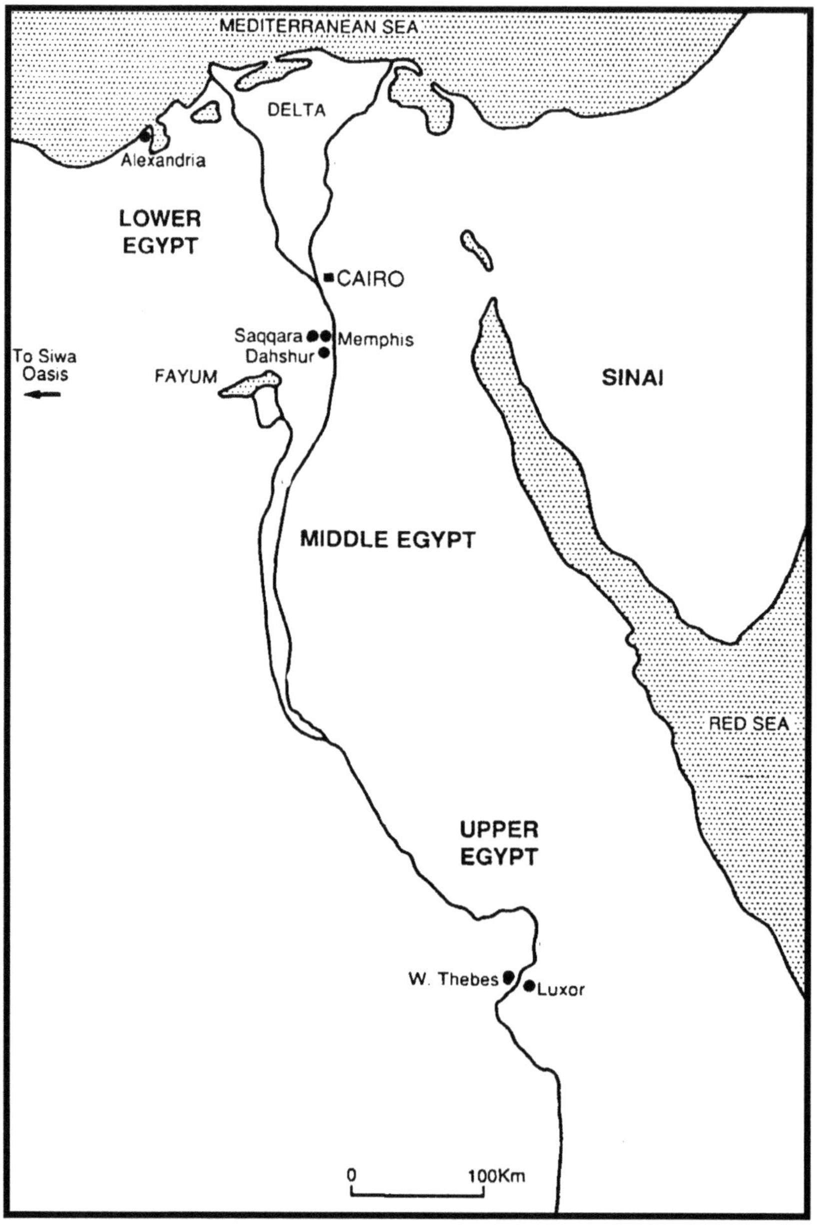

Map 1, shows the sites in Egypt involved in the present research at Saqqara and Dahshur, Luxor and Alexandria (after Dodson 1991).

For comparison, Romano-British wall paintings from Piddington excavated between 1996-97 were examined for the composition of plasters and mortars, as examples of a different period as well as a different environment. It was very interesting to carry out examination and analyses of some fragments of wall paintings, which showed different structures of plasters/mortars. In addition, the decay of excavated materials that have been exposed for a long time in different environments, results in the alteration and deterioration of the plaster and mortar, which varied according to the weathering. Some conservation cleaning and consolidation was carried out (see chapter 7).

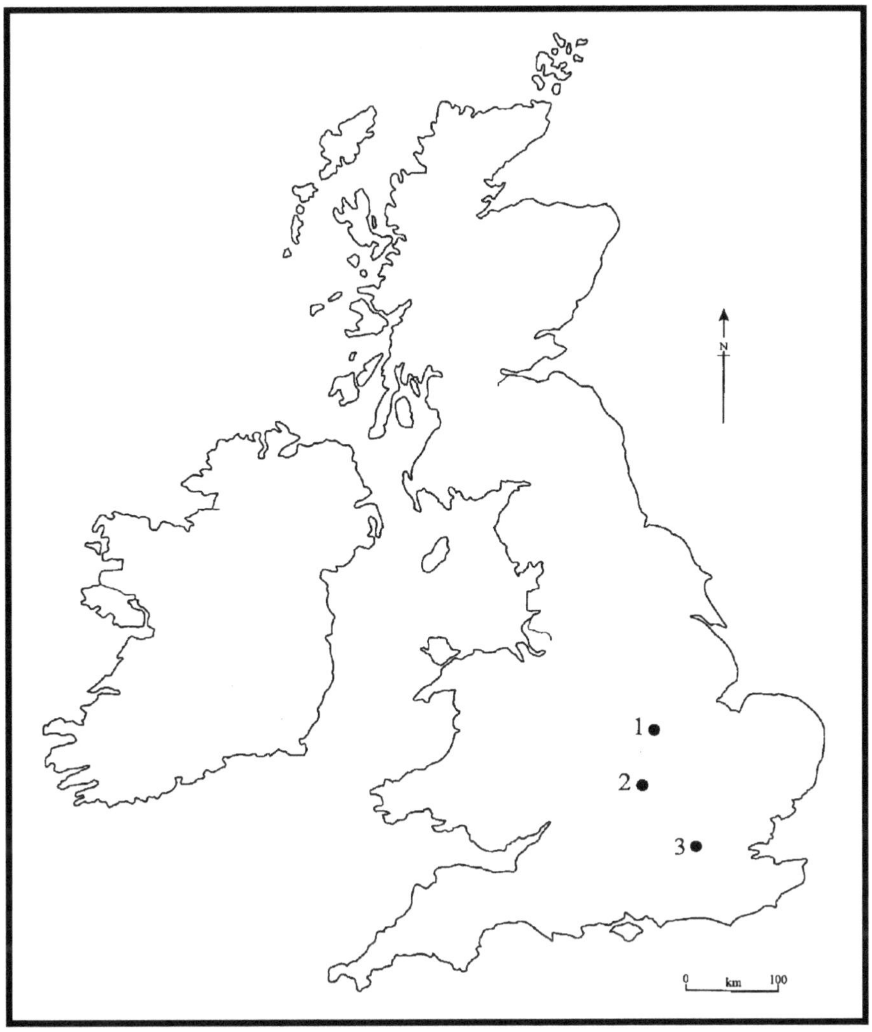

Map 2, general map of Britain

●*1 Leicester*
●*2 Piddington*
●*3 London*

2-Archaeological and Historical Background

In 1994-97, new excavations were carried out by Egyptian archaeologists in Saqqara and Dahshur. The necropolis next to the capital covers an area some eight kilometres long and between 500 and 1,500 metres wide, extending from the Abusir pyramids in the north to Dahshur in the south. Memphis was the capital city of the Old Kingdom of Egypt.

The new sites in **Saqqara** and **Dahshur** (lower Egypt) where samples were collected are located in several areas; the excavations were mostly of tombs/mastabas dated to different periods from the early dynastic period to the Old Kingdom (I-VI dynasty 2920-2134 BC) (see Appendix 4.1). Unfortunately, there is no map to indicate the location of the sites in Saqqara and Dahshur.

The sites from **Luxor** involved in this research were; Sheikh Abd El-Qurna (private tombs No 56 and No 55), Khokha (tomb 192) and Asasif (tomb 414), Drah Abu El-Naga tombs 255 and 13, the Valley of the Kings, tombs 15, 14 and 11and Deir El-Medina tomb 359 (see Appendix 4.2). Luxor is a big site for the Egyptian temples and cemeteries located in the Upper Egypt.

In Alexandria two different sites were selected: Mustafa Pasha and Anfushi (map 4). The tombs which belong to the area of Mustafa Pasha are to the east of Alexandria, beyond the area of Ibrahimya and Sidi Gaber (about 1,400 metres to the north east of Sidi Gaber), and are dated to the late third century and first half of the second century BC. The site was excavated in 1933-34, while Anfushi Bay, on the west side of Ras-at-Tin, dates to the first half of the third century BC, and was excavated in 1901. Adriani (1936), Noshy (1937) and Brown (1957) carried out some studies on the most important Alexandrian sites. Their studies of the tombs of Mustafa Pasha and Anfushi, as well as others in Alexandria, were compared with different examples from the Greek World.

I have already outlined the historical background to the sites in Alexandria, the architecture as well as the styles of decoration, giving particular emphasis to the style applied in Mustafa Pasha and Anfushi. In addition, I have pointed out some of the arguments and discussion on the style used in Alexandria and other areas in the Greek world (see Appendix 4.3).

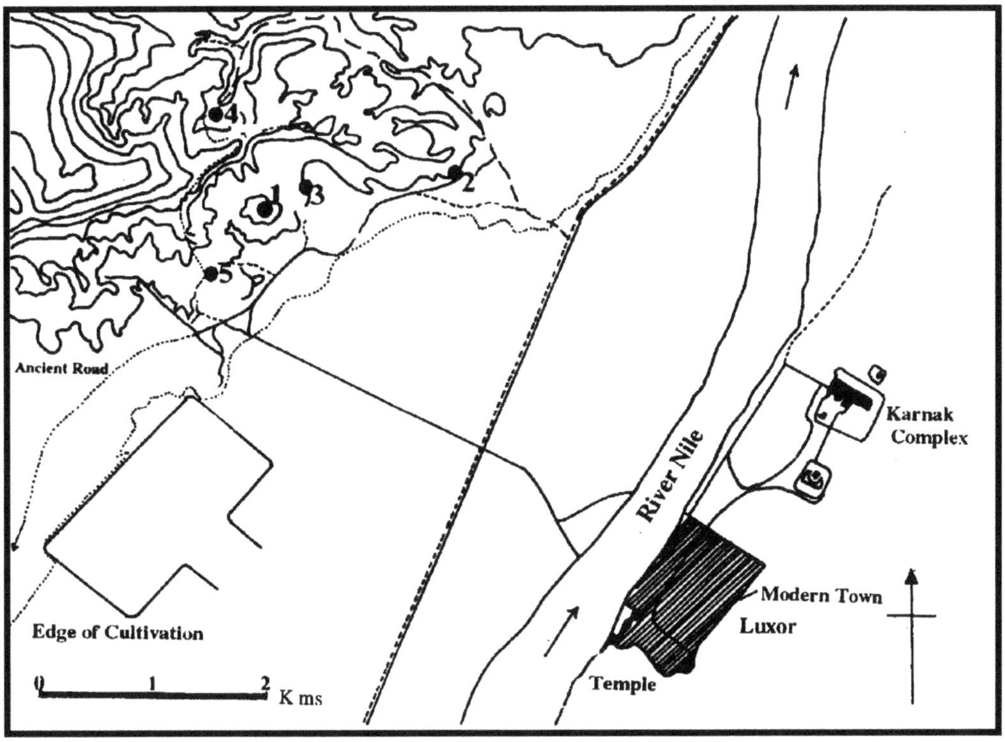

Map 3, showing the general archaeological sites of Luxor, where I took samples, and their geographical locations (after Ruffle 1977).
•*1* Sheikh *Abd El-Qurna* •*2 Drah Abu El-Naga* •*3 Asasif*
• *4 The Valley of the Kings* •*5 Deir El-Medina*

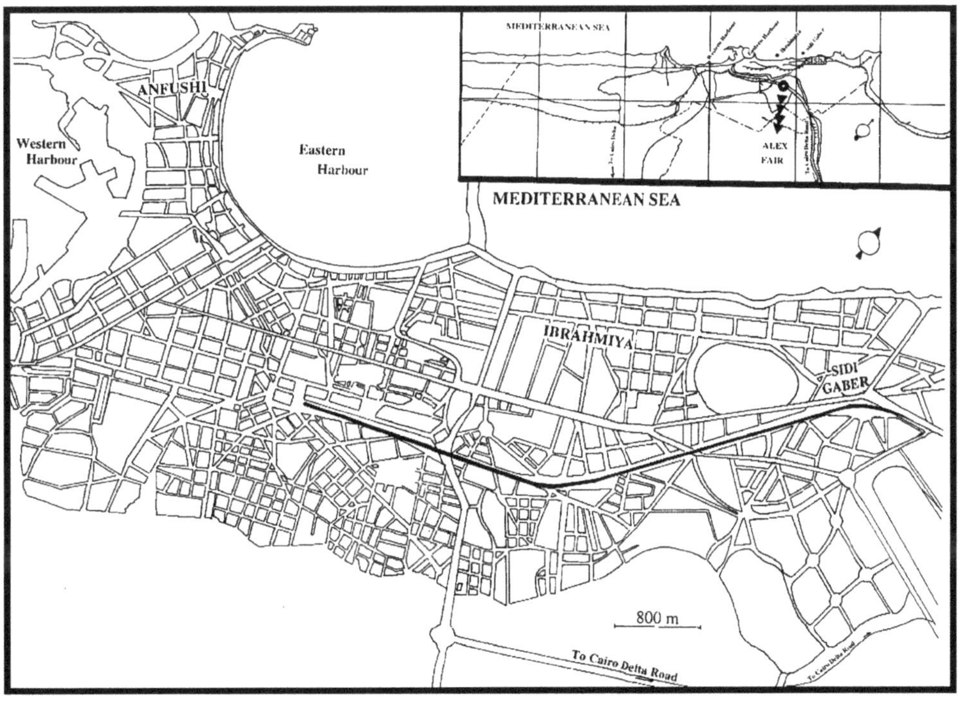

Map 4, Shows the sites of Mustafa Pasha beyond the area of Ibrahimya and Sidi Gaber on the East and Anfushi on the West Side of Alexandria.

Piddington Roman villa is about 35 miles south east of Leicester (map 5), dated from the first to the fourth centuries AD. The villa was rediscovered in 1781. Two Roman villas, 70 yards apart were excavated by the Upper Nene Archaeological Society between 1971 and 1977. These formed parts of an abortive Roman settlement around 175 AD, re-occupied in the fourth century (see Appendix 4.4).

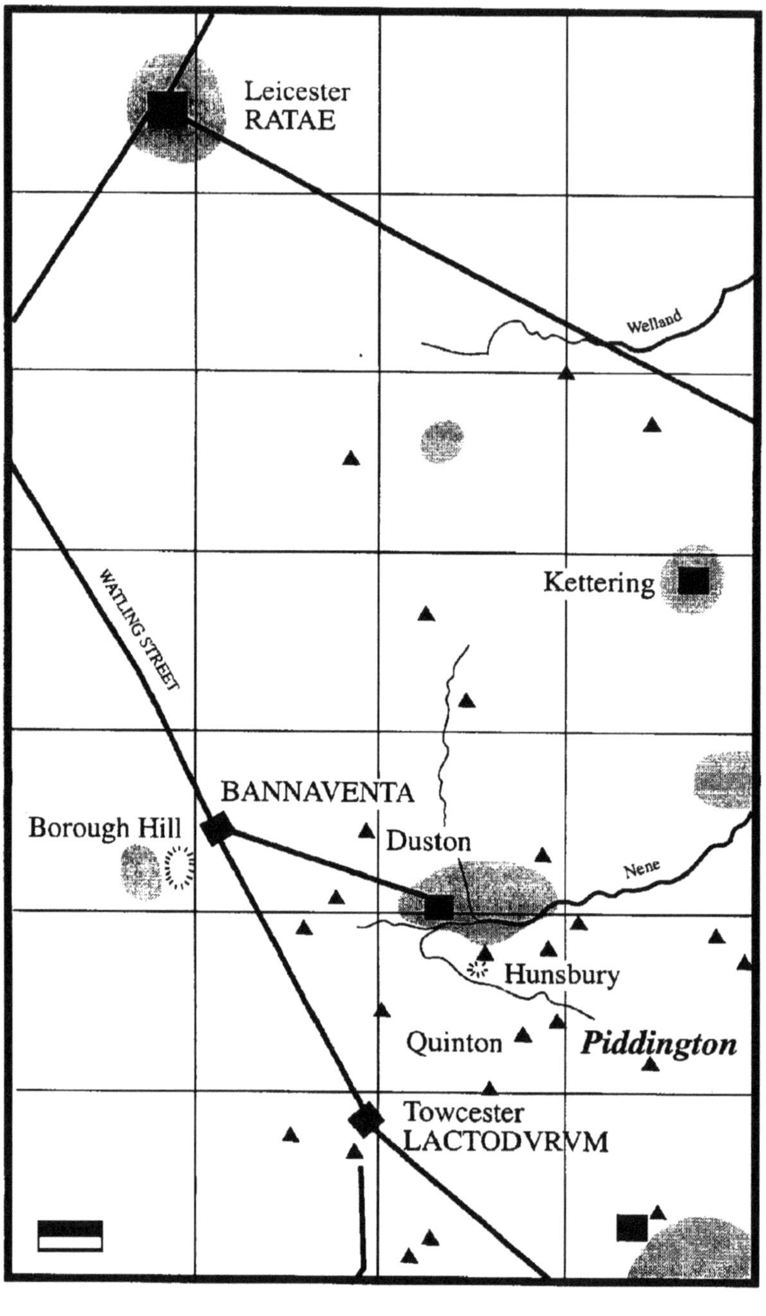

1 cm = 3 miles

Map 5, shows the location of the Piddington Roman villa in Britain: Squares indicate the major late Iron Age and or Roman towns. Triangles indicate the villas (after Rollo 1994).

3-Investigation at the Sites of the Deterioration and Decay of Wall Painting

•-Saqqara and Dahshur

Heavy rain in 1994 in Saqqara had a bad effect on the excavation sites. In the uncovered areas, it was noticed that the plasters/mortars had separated from the walls, collapsing as well as dissolving and losing binding power. As a result the plasters/mortars were soft and became very powdery. The same was noticed in Dahshur, which affected not only the plasters, but also the building materials used. The mud bricks were soft as well. Consequently, great care had to be taken when collecting samples to avoid any contamination of materials.

•-Luxor

In Sheikh Abd El-Qurna in Luxor (T. 56 and T. 55), salt had covered some areas with the appearance of fossil structures (photograph 12 B) and looked very hard and brownish. On some other areas it was very soft and grew as fibres or needle shapes, resulting from recent formations. It was easy to see that the layer of salt on some areas had many holes, as if it had dissolved somehow. The fibres of the needles were generally short, although some were rounded or long in some areas (photographs 47) (see also Photographs 75-76, Part II).

The site of Drah Abu El-Naga was undergoing restoration and re-plastering. The tombs were small and differed in their architectural design from those in Sheikh Abd El-Qurna and Deir El-Medina (see Appendix 4.3 plans 2 A, B, D & E and 3 D). In T. 13, the painting of the tomb, was unfinished and the drawing was in red. Unfinished paintings were found in many tombs in Luxor and it was possible, in some instances, to collect samples. Most of the tombs have been restored and are protected by glass. The opportunity was taken to investigate the processes of restoration and the materials used for re-plastering. It was noted that the old plaster of restoration, which was full of micro-cracks, required re-plastering. A question was asked regarding the stability of the new mixture of the plaster/mortar which had been used. This led to the collection of some samples of the new mixture for analysis, in order to distinguish the new compositions and to compare them with those used in antiquity and in previous restorations.

•-Alexandrian Sites

The sites of Mustafa Pasha and Anfushi in Alexandria are near the seashore, a few metres above sea level. The excavation site slopes gently towards the seashore. Some other sites in Alexandria are away from the shore, such as Kom El-Shuqafa, Amod El-Sawary and Kom El-Dikka. The sites of Mustafa Pasha and Anfushi face the seashore and proximity to the sea causes many problems to arise. Salt and underground water have had a bad effective not only on the tombs, but also on all the buildings which face the sea, due to the sea spray of the salts. The Anfushi site is nearer to the sea than Mustafa Pasha by a few metres.

Examination of earlier record revealed that the tombs of Mustafa Pasha had been previously re-constructed in 1935, because most of the tombs had been destroyed. Adriani mentioned that it was essential to restore them for two reasons; first to conserve and protect the heritage, and second, to complete and reconstruct the missing parts as they were, without any change. Cement materials were used for the restoration and the reconstructed parts, could be clearly distinguished. He stated that in some cases they used small blocks of limestone to build up the walls, which stood on the rock, as if they were covering rocks with limestone. He also reported that both the rocks and the walls of limestone were plastered with a layer of *stucco* (Adriani 1936: 12, 13). Salt crystallisation and micro-organisms have been found to cover some areas in Mustafa Pasha tomb 2, as shown on Photograph 3.

The situation in Anfushi was worse than in Mustafa Pasha. It was noticed that all the walls were covered with very thick layers of salt (photographs 4 -5). The amount of salts inside the wall render dissolved them and made the stone more powdery, shown by the loss of the lime binder from the mortar, as noticed in tomb 3 (photograph 6 A). The problems of salts, underground water and humidity occur not only in Alexandria, but also in different parts of Egypt, as has been mentioned before (see Chapter 3, 5). Micro-organisms were also found at the Anfushi site (photograph 5 B). During the examination of some salts, I found a layer of micro-organism on the salt, which covered the painted layer.

Restoration began in August 1997 and has continued nearly every summer since, to remove salts from the walls. However, these have been accumulating again, indicating a continuing problem in this area. The fact that the amount of salts inside the walls is rising again so quickly suggests that the salts might be from underground water or sea spray deposited onto the walls. It should also be noted that the Anfushi and Mustafa Pasha tombs were covered in protective glass, especially in the courtyard. This has led to a lack of air circulation and rise in temperature. The crystallisation and hydration of the various salts takes place under certain conditions and could be from different sources. It has to be considered that the building stone, mortar, soil, materials used for conservation and repair, the polluted atmosphere, decaying salts, organisms (see Chapter 3, 1, 3.1 & 3.2), as well as some other unknown factors could all be sources of the salts. This led to investigation to determine the main sources of the phenomenon. Different salts from different areas were collected for analysis and identification.

•-Piddington

Fragments were collected from the excavation stores. Of the samples found in position during excavation at Piddington, it was not clear if they were found lying face up or down. This information was needed to understand the causes of alteration and deterioration. Improved methods of excavation frequently reveal painting attached either to surviving walls or fallen from the walls as the villa decayed and lying in layers on the ground. The lifting and re-construction of such painting is no light task and new techniques have had to be invented to solve this problem. A good strategy was used for excavation, conservation, re-construction as well as analysis of wall painting from Settefinestre (Tuscany) by De Vos *et al.* (1982), which might have usefully and applied in the excavation of the fragments from Piddington Roman villa in Britain.

The decay of the fragments of wall paintings from Piddington is represented on the upper surface as a layer of calcite film, found not only on the fragments cleaned but also on the fragment surface as shown in Photographs 50 A & C, 51 A, 53 and 55 A. The calcite film which was covered the painted layers on some fragments to be a kind of the deterioration and decay of wall painting fragments. If the fragments were deposited face up, it might be that these layers of calcite film were produced by something dissolving above them. This film proved to be very hard during cleaning. Watkinson noted that wall paintings are often found face down in the ground. Water-soluble salts, frost and biological agents can all damage the plasters. Moisture content in wall painting should be maintained at the same level as when discovered. It is wise to keep the painting damp by spraying it with water and covering it with black polythene sheet, to reduce the action of micro-organisms and the effect of direct sunlight which fades some pigments (Watkinson 1972: 63).

Investigation of the fragments from Piddington showed different causes of alteration, the samples being covered with different layers and shapes of calcareous deposits and different thickness of calcite. Some were covered with pure white layers and others with very deep yellow, ancient dust deposit, which was very soft and easy to remove (photographs 51 A & 53). Another example had complicated deterioration and decay, not only on the surface, which included a different sort of calcite, but also on the back of the mortar (photograph 56 B). A very thick and hard deep yellow calcite film was found to be the main alteration factor on the wall painting surfaces. The surface of the painted layer was covered with a white calcite film and in some areas with a very deep brown layer of dust deposit, which was very hard and thick on some areas. Biological growth also had a bad affect on both the painted layers and the rendering. In addition, some black specks were found on the paintings, and the colour of mortar degraded to brown, which was caused by fungus. The damp environment had effected the structure of the mortar and made it softer. These layers were analysed to identify the composition. Investigative cleaning also took place, to look at the causes of alteration and deterioration, and to clean the surface of some fragments.

The examination and analysis of wall painting from all sites was carried out to discover what had happened to the materials, what they were made from and how, and to look at the structure, composition and different alterations which related to the position in their geological aspect. The geological location has a very important role in the alteration decay and deterioration of mural painting, as it affects the structure of the mixture of mortar and plaster, which affects the surface of the painted layer, due to the porosity of the materials. The alteration and discoloration of pigments could differ according to weathering and the chemical reactions (see Chapter 3, 3.3.3).

4-The Strategy of Taking Samples and the Nature of Samples

It was very interesting not only to study the plaster/mortar, but also to note and investigate the whole situation, such as salt crystallisation to identify the salts. The smallest samples possible were taken, that gave the required information, but from existing damaged areas, such as edges. Each sample taken should be suitable for the intended analysis. Some instructions for taking samples of ancient mortars for comparative analysis are given by Jedrzejewska (1982: 324-327). The samples in general were taken carefully and under special supervision, using scalpels and dental tools. During the sampling, accurate documentation was made of the areas sampled, as stated in Appendix 5, and some photographs were also taken. In some cases, samples were collected from loose areas and sometimes from fallen ones, but with great care to avoid any contamination, especially of the plaster layers, which were very fragile. All samples were kept in sealable plastic bags, in order to protect them and to avoid any dehydration. The samples collected from Alexandria were in tiny amounts.

5-Samples Areas and Locations of the Fragments

•-Saqqara and Dahshur

Samples were collected from several areas from the new excavations. In addition to a few samples from the remains of the tomb of Herneith, to the north of the archaeological offices (houses of antiquities), dated to the earliest period of the reign of King Udimu in the First Dynasty. The tomb was the burial place of Queen Her-Nit and was excavated by Emery in 1938. The work was published with full details on the architecture and the contents (Emery 1958: 73-97). The tomb was built of mud brick, covered with two layers; the first of mud plaster and the second, another layer possibly of a clay plaster, shown in Photograph 7 B. Descriptions of all samples from the Old Kingdom are given in Appendix 5.1-3. The aim of collecting these samples was to identify the composition and structure and compare the two plasters/mortars used in the same period of the first dynasty.

Unfortunately, there was no published plan of the new excavation sites to indicate the location of the mastabas/tombs and sampling areas, as excavations have not finished yet.

•-Luxor

The procedure for collecting samples from Luxor was a little different, as a result of a survey carried out to investigate the sampled areas. It was a difficult task to find sampling areas for two main reasons. The first was that Luxor is a large area having different sites with a great number of monuments and most of the tombs were closed for maintenance. The second reason was that most of the tombs were restored and covered in glass for protection. On the one hand, samples should be clean and uncontaminated to assure accurate and correct results. However, as some of the tombs were undergoing restoration and re-plastering, (Drah Abu El-Naga and some tombs at the Valley of the Kings), it was a good opportunity to collect samples of both old and new plasters, and to distinguish their composition.

There were also some tombs which needed investigation because it was noticed that the deterioration and decay had been taken place. It was interesting to include tombs such as T. 192 & T. 414. Samples collected from the sites mentioned above were mainly plasters/mortars, used to distinguish not only the different tombs, but also the different areas. These can vary in the composition of the rendering according to the availability of the materials and the geological area. A few samples of limestone and mud brick were collected as well, to investigate the supporting structure of the rendering, as it could be responsible for the deterioration and decay of the plasters/mortars. Limestone in the form of powder was also used in the mixture of the rendering. This led to identification of the composition of the different samples of limestone from different rock cut tombs. Some samples of salt were also collected for analysis. The descriptions of the samples collected from the Luxor tombs are given in Appendices 5.4-7, starting with the building materials (stone/mud brick), mud and clay plaster/mortar, the finishing layers and the pigments.

•-Alexandria

Mustafa Pasha and Anfushi were two more sites where samples were collected. Samples were taken from the site of Mustafa Pasha tombs 1 and 2 (see Appendix 4.3.1, plan 4 A & B). There are big differences between these tombs and the paintings, as described by Brown (1957) and Adriani (1936). The tombs in the Mustafa Pasha site are very poor, and most of the paintings have been lost, except for some walls with paintings in red, yellow, Egyptian blue, and a picture on the lintel, which is covered in glass. However, the architecture survives without any damage, as these tombs are rock cut and re-constructed, as mentioned above. In contrast, the Anfushi tombs (see Appendix 4 plans 5 A & B) where samples were collected, are considered the best, although most of the wall paintings have gone, with the exception of rooms 1 and 2, the rooms in which paintings survive on all wall surfaces. It was very interesting to note that in tomb 2 there were two different decoration styles. It was also possible to see the layers of plaster that were used (photograph 11, Appendix 4, Part II).

Samples were taken from different rooms (see Appendices 5.9-11), but in tiny amounts, making it difficult to carry out chemical analysis. Samples varied between plaster/mortar, pigment and salt. Limestone from the site was sampled, to look at the structure of the original materials which were used to make mortar for painting. The same practice was carried out at Anfushi, looking for differences between the structure of plaster and mortar, its composition and painting, as well as the application techniques. A few samples were collected from the floor of tomb 3, which were found to belong to that tomb. The practice of taking samples was carried out not only because of the two different sites, but also because the architectural style and the decoration schemes differed, which could give varied analyses.

•-Location of the Fragments from Piddington.

These were taken from the excavated areas, belonging to the different parts of the villa, (plan 6) recorded as rooms; 25, 26 and 46 (the south wing) by the archaeologist Roy Friendship-Taylor. Other fragments were from room 21, and also from rooms 53, 56 and 57, as shown on Plan 6. Description of fragments is included in Appendix 5.13-15. Analyses have been carried out on samples from rooms 21, 26, 53, 56 and room 57.

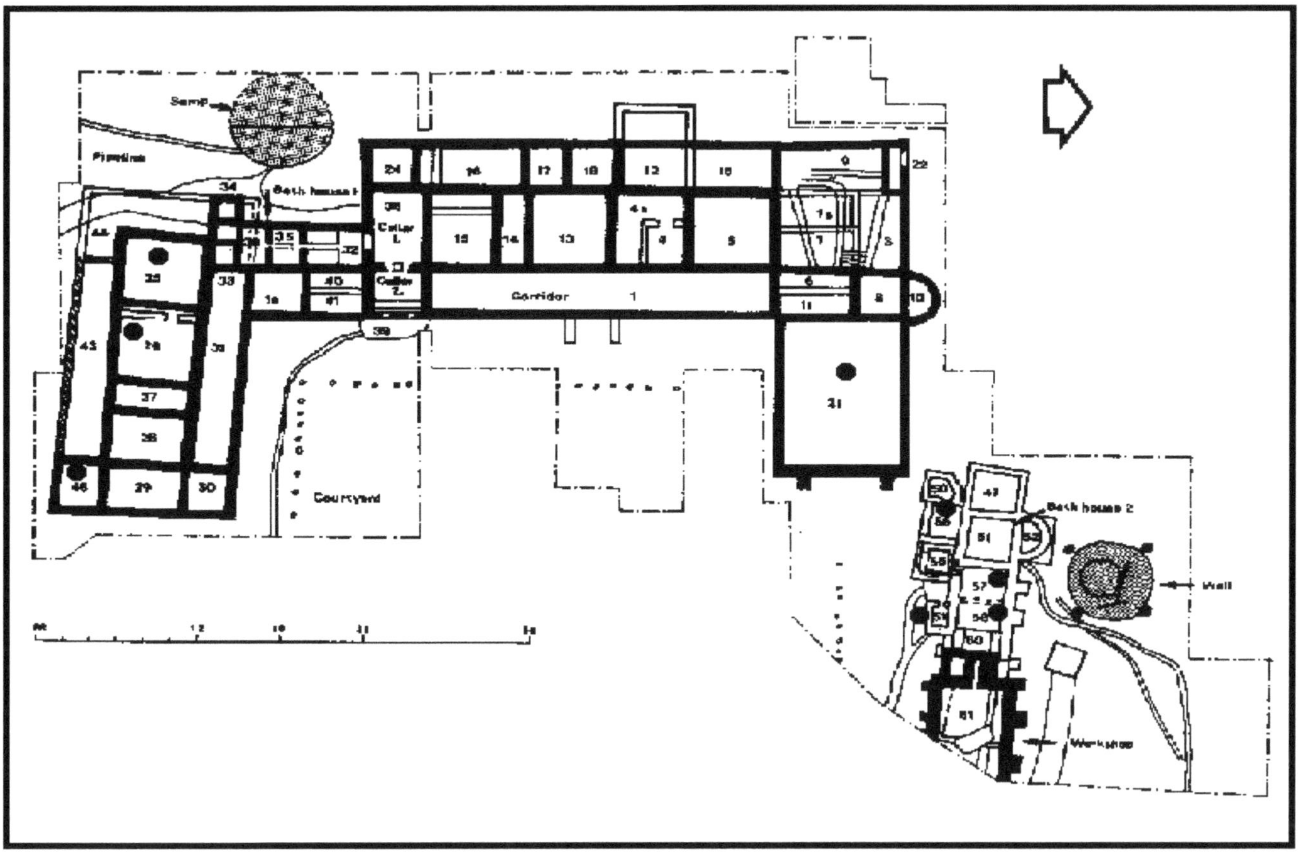

Plan 6 showing the location of wall painting fragments from Piddington Romanvilla (after Current Archaeology no 146)

Chapter 6

1-Preliminary Examination

The strategy of the examination and analysis of wall paintings in this research was as for the analysis of plasters/mortars (figure 5). The methods involved were divided into: primary examination, which included, as a first stage, recording and photographs, microscope (10X) and examination of cross sections, including; broken section structures, and polished sections for measurement, and the use of raking light. In the second phase of examination, chemical and physical methods were used in the examination and analysis of different samples from Saqqara, Dahshur, Luxor, Alexandria (Abd El Salam 1999a & 1999b) and Piddington.

Various plasters/mortars were used in ancient Egypt. Lucas (1989: 79) classified two kinds, coarse plaster and better plaster. In addition, gypsum and/or a mixture of gypsum and powdered limestone covered the mud/clay plasters. Plaster/mortar was to cover the building materials of mud brick or stone, as stated in Chapter 2 (3.2).

My interest was to follow the same classification of the plasters/mortars to distinguish as well as clarify the two kinds by the examination and identification of their additives. I have named them "mud plaster" (coarse clay plaster, used as a first layer, based on Nile alluvium photograph 7 A), and "clay plaster" (better clay plaster, used as a second layer or in some cases as a first layer, based on powdered limestone (*hib*), photograph 7 B and C). The third or fourth layer that covered the mud/clay plaster/mortar is named the finishing plaster layer (as photograph 7 B, top layer) whatever the mixture used. The same terminology was used at Saqqara, Dahshur and Luxor. However, lime plaster/mortar was used in Alexandria and Piddington. The aim of the study was to distinguish the binder used for the two kinds of plasters/mortars and to define their aggregates and additives.

The results of this research are arranged not site by site, but by the analytical techniques, so that the differences between the sites results can be compared more easily.

1.1-Microscope Examination of Samples

1.1.1-Saqqara

The tombs/mastabas in Saqqara and Dahshur where samples were collected, were built of rendered mud brick. Broken sections of different samples of mud bricks from the sites of Saqqara have shown that the mixtures contained mud, quartzite, quartz sand, chopped straw and small pieces of limestone and charcoal. A piece of bone was found inside the mixture of the mud brick from the Idy tomb, but it was difficult to identify such a small piece. A bigger piece of bone was found in the mixture of samples from the tomb of Ka-Nefr, (photographs 26 B) but it was very difficult to distinguish whether it was from a human or an animal. The mud bricks were soft and the straw had degraded leaving only impressions.

Examination of samples (see Appendix 5.1) dated to the **first dynasty (E E)** has shown that mud plaster/mortar which covered mud brick was coated with a finishing layer (?) of variable thickness, showing traces of red and yellow on the surface (photographs 7 A & B). The mud plaster/mortar was a mixture of mud, straw, quartz sand, mica and small lumps of chalk/lime. On one sample the straw had decayed leaving an impression.

The samples from the new excavation, dated to the late **first dynasty (N E)**, showed that the plasters/mortar used was a clay plaster (better plaster). It was noted that a very thin layer of the finishing plaster covered the clay plaster/mortar and one sample had two layers. The cruder under-layer was darker than the second one, and it seemed to be a mixture of quartz sand (red and smoky) with mud as a matrix, in addition to charcoal and a few pieces of lime/chalk. A large bubble was found within the plaster of sample (S.143/1), as shown on Photograph 7 C.

Samples from **Dahshur**, dated to the 4th dynasty, varied giving an outline of the preparation method applied in the Mastaba. Two different samples of limestone from the mastaba and the site of Dahshur have shown that the limestone was very fine and consisted of many layers in a very deep yellow colour, being very similar to the local deposits. It was very soft and broke easily. Some spots of salt were found on the surface and a very thin layer of salt occurred between the layers. This kind of limestone was found in the mixture of the mud brick. The limestone sample collected from the mastaba/tomb was different in some areas, being very white, fine and solid. This kind of stone was used on the building, as mentioned above.

The mud brick was covered with two layers of clay plaster/mortar (better plaster) (see Appendix 5.2-3), of about 1.5-2.5 cm. The plaster/mortar was covered with a very thin finishing layer (?) (photograph 8), identified by spot tests as gypsum plaster in white to yellowish/brown, looking like encrusted material with some black spots on the surface. The thickness of the gypsum plaster was between 0.5-0.6 mm. The surface structures were very crumbly on some pieces (photograph 8 B).

The examination of samples collected from the tomb of **Ka-Nefr** at Saqqara (see Appendix 5.2-3), dated to the 5th dynasty, show differences in the preparation of walls. The samples show that about four layers had been applied to cover the mud brick. The mud plasters/mortars (coarse plaster) of the first layer show that mud was used was as a matrix. The mixture consisted of: charcoal, quartzite, straw, and small pieces of limestone, as previously. The most interesting thing was that the mixture contained traces of white fossil mollusc fragments, and a few ceramic pieces (burnt brick).

The clay plaster/mortar (better plaster) of the second layer was between 7-10 mm thick, covering the mud layer. The same mixture of clay plaster/mortar was also found, but some crystals of a deposit of gypsum were visible inside the mixture, filling the pores. The third layer covering the clay plaster/mortar looked like the same mixture as the second layer, but was about 0.1-0.2 mm thick with very fine particle sizes, as shown on Photograph 9. X-ray diffraction was carried out to identify the composition of the layer (see X-RPD, Appendix 16.8). Another sample from the same tomb showed different methods of walls preparation. It was found that the sample contained many layers of gypsum plaster (identified by spot test), in different thicknesses, covering the mud plaster/mortar (photograph 10).

The fourth layer investigated was the paint layer in bluish/black (photograph 9). There was evidence to suggest that *tempera* technique (distemper) was used for the painting. During the taking of the samples it was noticed that the painted layers had peeled off and separated from the surface; this resulted from the very drying conditions in Saqqara, which cause flaking of the pigment particles, and because the binding media could no longer hold them in place. Although the execution of the painting was poor, it was noted that the painted surface was smooth on some areas, showing some degree of silk.

Examination of a few samples collected from the **Idy** tomb (see Appendix 5.3), dated to the late 6th dynasty, has shown that the same mixture of both clay and mud plaster/mortar was used, as well as the mud bricks. Some seeds in a pod were found inside the mixture of the mud plaster/mortar (see photograph 26 A) and the mud brick. The seeds came from the organic materials used to give strength to the mixture, such as straw.

It was noticed that the straw used with the mixture of the mud/clay plaster/mortar was longer than in the previous sample. In another, a piece of bone was found in the mixture of the mud brick. It was difficult to identify such a small piece of bone (photograph 26 B), but it was possible to pick out the seeds for identification (table 2). All the plasters/mortars were very soft and fragile, except that dated to the first dynasty (E E), which alone was protected from the environment.

1.1.2-Luxor

At Luxor in the rock cut tombs, the different methods of preparation of the walls were due to the varying quality of the stone. Several samples of stone from all the sites were examined, for the identification of the structure of the limestone. The samples are described in Appendix 5.4-7. Samples of Userhēt T. 56 and Ramose T.55 from **Sheikh Abd El-Qurna** have shown that the stone is homogeneous in structure with a very fine grain size and yellowish/brown in colour. It was noted that salt formation had taken place to produce layers of gypsum crystals, which broke down the stone into many layers by mechanical stress. This made the stone easy to break up. The structure of the stone showed microfossil shells in different shapes and also some black spots on the surface, indicating the formation of manganese dioxide or deterioration and decay (insect deposit). Iron oxide covered some areas and bands of iron were also noticed (photograph 11 A).

The examination of the sample of mud brick from T. 55 showed that the mixture contained very long straw, quartz sand, lumps of gypsum and chalk/lime as well as brick and mica. The mud brick was hard and solid. Spot test was carried out which identified phosphates, suggesting that animal dung was added to give more strength to the building materials.

The examination of a sample of clay plaster/mortar from T. 55 showed that the plaster was very fine, consisting of powdered limestone, chalk/lime, fine chopped straw and quartz sand. This kind of plaster was found to cover the rock where the mud plaster (coarse plaster) was not applied. The finishing layer of plaster collected from T. 55, coated the layer of the mud plaster on the column, (photograph 12 A), and consisted of very fine grain quartz sand, chalk/lime, brick and straw to a thickness of 0.5 mm. A very thin layer of whitewash was painted on top. The plaster was pinkish, which could be natural or changed due to deterioration.

Most of the tombs have been restored. Only one sample of mud plaster/mortar collected from T. 56 showed the same composition as previously. The surface was covered in a whitish layer.

The archaeological survey of Kheruef T. 192, **Khokha**, indicated that it was quarried into the irregular surface of the desert rock and was different from other tombs of the 18th dynasty, which were quarried into the hillside. Two samples of stone showed a very fine-grained and homogeneous structure (see Appendix 5.5).

Samples of mud bricks were found in the tomb, covering the bedrock, as mentioned in the Epigraphic survey. The remains of a mud-brick structure rests on the debris filling the upper end of the ramp and covers the bedrock to depth of about one metre. The west wall of this structure is bonded into the mud-brick walls that rest on the bedrock on either side of the ramp passage (*Epigraphic Survey:* 4). The examination of the samples shows that the mud brick contained straw, limestone (of the same structure as the rock), quartzite, quartz sand, pieces of lime/chalk and burnt mud. This mixture was similar to that used during the Old Kingdom. The mortar used between the bricks, as filler, was basd on powdered limestone. Quartz, charcoal and straw were also visible.

The clay plaster (better plaster) which covered the rock looked different in some instances, being pinkish to brownish and slightly darker on top. Smoke or a high temperature had changed the plasters, affecting not only the painted surface, but also the paint layer, blackening the whitish layer found underneath the painting. It had also changed the clay plaster, resulting in a dark and reddish band in the middle of the sample. However, another sample from a different area showed the original colour of the clay plaster/mortar away from the heat source. The clay plaster/mortar was in thicknesses varying between 0.7-30 mm.

The finishing layer, which covered the clay plaster, was of different thicknesses averaging about 0.2 mm. Micro-chemical tests confirmed that the plaster was gypsum containing different minerals as impurities. It was noted that the plaster layer had become blackish compared with the pinkish ones found on a small fragment. As mentioned in Chapter 1, the gypsum found was of different colours and quality. The pinkish ones could be original, unless they had been changed for some reason, as happened in the tomb of Tutankhamun, as mentioned by Lucas (1989) (see Chapter 2, 4.3).

The investigation of small fragments of painted plaster has shown that the technique used was *tempera*, (distemper) resulting in crazed surfaces and even cracks (photograph 13). The surface was painted in red and Egyptian blue (confirmed by spot test) on a very thin whitewash of gypsum/lime or a mixture of both, which turned blackish. Another small fragment showed that the surface was painted in a cream and red panel on the same whitewash, which covered the layer of gypsum plaster.

The archaeological evidence indicated that the ceiling of the passage has been so damaged by smoke and the extrusion of salt, as to obliterate the painting, if there ever was any. In addition, the artefacts found in the subterranean corridors and rooms were the charred remains of reed torches and the bottoms of broken jars slightly blackened, indicating their use as lamps (*Epigraphic Survey*: 5, 9). This may have been from the occupation during the early Christian era, with the result that the wall in some other tombs were covered with a sooty black that had a furred appearance (Jaeschke 1995: 147). From that evidence it can be assumed that the changes to the plasters were as a result of smoke or the heat from a fire.

The examination of samples from **Drah Abu El-Naga,** Roy T. 255 and Shuroy T. 13 (see Appendix 5.6) has shown that more or less the same scheme of plasters was applied. As the tombs were cut into rock, the first layer of mud plaster/mortar (coarse plaster) covering the rock was followed with another layer of clay plaster/mortar (better plaster) as shown on Photograph 14 A & B. The finishing plaster was covered with a very thin layer of whitewash as a ground for the painting and it was not clear wherther if it was gypsum/lime.

The investigation of samples of limestone from the tombs showed that the stone consisted of many layers of very fine structure, similar to samples from Sheikh Abd El-Qurna. Mechanical decay had taken place, as shown by the cracks found inside the stone. Micro- fossil and shell fragments were observed as well, indicating the high percentage of carbonate. Two samples of mud plaster/mortar (coarse plaster) from T. 13 and T. 255 have shown that the plaster was hard and solid, with insect deposits.

The examination of the clay plaster/mortar (better plaster) from the ceiling of T. 13 (photograph 14) gave a thickness of 14-17 mm. However, another sample from the wall showed a thickness of about 15-25 mm. Traces of insect attack and faeces were found through out the plaster, making it very powdery, and were visible in a sample from T. 255. The presence of the organic materials used, such as straw, was the cause of the deterioration and decay. Although the sample was very powdery, micro-tests identified phosphate, due to the effect of the insects.

Investigation of a small fragment showed that the surface was darkened with the covering of a dust layer, as shown in Photograph 15 A. Scraping off the layer revealed that a plastic material covered the plaster. The layer dissolved in acetone, and when heated, the charred material indicated the use of polyvinyl acetate (PVA), as a result of previous restoration. After application, this became insoluble and helped to form a layer on the plaster. Mechanical cleaning was carried out to remove the layer of PVA, which came off easily (photograph 15 A & B) to expose the original plaster, confirmed by a spot test as gypsum plaster. The finishing plaster used was as a mixture of calcium carbonate and gypsum confirmed by micro-chemical tests. The thickness of the plaster was between 0.2-0.6 mm. The surface was painted in different colours, as shown on a loose fragment, indicating that the technique used was *tempera* The Egyptian blue was greenish, due to the percentage of iron contained in the sand used in its manufacture.

As most of the tombs in the **Valley of the Kings** have been restored, very few samples of clay plaster/mortar were collected (see Appendix 5.6-7). It was useful to examine and compare the plaster/mortar used in that valley with those from the other sites. The examination of samples of limestone from the Valley of the Kings showed a very fine grain size, being hard and solid. Micro-organisms were noted as well and confirmed by SEM (photographs 41 A & 16 B, photographs 71-72, part II). The preparation of the walls was different from the previous tombs, due to the good quality of the rock where the tombs were cut.

The plaster/mortar used in the Valley was the clay plaster/mortar (better plaster) covering the rock. Examination of three samples from the tombs (T. 14, Queen Tausert, T. 15, Sethos II and T. 11, Ramesses III) has shown that the plaster was based on powdered limestone, and straw of very fine particle size was visible as well. The clay plaster was covered with a very thin layer of whitewash, which received the paintings.

The tombs of **Deir El-Medina** were different not only in architectural design compared to those of other sites, but also in the preparation of the walls and decoration. Samples were collected from Inherhau T. 359 (Appendix 5.7), which showed that the rock was covered with mud plaster/mortar (coarse plaster) and then coated with a layer of whitewash, where I found traces of it on the surface of the plaster. The samples of stone from Deir El-Medina were different; the stone was brownish to reddish and hard. A very thin layer of iron oxide covered some areas and layers of gypsum in different thickness were formed between the stone.

The examination of two samples of mud plasters has shown that the mixture contained straw, quartz sand, charcoal and small pieces of rock. A very tiny piece of painted plaster showed that the thickness of the finishing plaster was between 0.2-0.3 mm, and the mixture contained different materials such as charcoal, sand and very fine grain lime/chalk. Micro-chemical tests indicated that the mixture of the layer used was gypsum containing different minerals as impurities. The paint layer looked very smooth, indicating that the technique used was *tempera.*

Examination of sample from Ankh-Hour T. 414 **Asasif,** showed that the limestone contained gypsum, in different thicknesses, as a natural formation of the rock, similar to those at Deir El-Medina, but mostly grey. It was found that a very dark brownish/blackish layer of corrosion covered the rock, as shown on Photograph 16 A. The identification of this layer showed iron, suggesting that the layer was iron oxide. This deterioration was probably due to the presence of water. Geological identification shows that the stone was full of fossil organisms such as foraminifera and that the formation of gypsum crystals was as a result of the evaporation of water. The mortar filler filling the voids consisted of crushed and powdered limestone.

The plaster used in T. 414 consisted of two layers: The first was a mud plaster/mortar covering the irregular surface of the rock with mortar filler and then the second layer of the clay plaster coated the mud one, as shown in Photograph 16 A. It was noted that there was no painting in the tomb as if the work had stopped. The plaster was of different thicknesses between 0.5-25 mm, due to the irregular surface of the rock.

The plaster/mortar exhibited bio-deterioration due to fungal growth, which was noted throughout the sample. Analysis and examination under the microscope showed two different stages of germination of the fungus, indicating that the growth was recent. The growth of such organisms occurs under certain conditions of wet or moist environment, as mentioned in Chapter 3 (3.2).

This could be due to floods in the Luxor area since the 1980s. The examination of the mud plaster/mortar from different areas (see Appendix 5.7), showed that the mixture contained straw, quartzite, quartz sand, broken stone and mica. The plaster became powdery, due to the presence of organic material used in the mixture of the plaster/mortar. However, in some instances, the plaster was hard and one sample was covered with a brownish/reddish layer (photograph 16 B). Micro-chemical tests showed the presence of phosphate, perhaps from the faeces of insects. Some samples were not decayed.

The layer of the clay plaster that covered the mud (photograph 16 C), was slightly finer and without straw. Micro-chemical tests showed that the layer contained calcium carbonate. There was no gypsum in the mixture, which was similar to those that found in the Old Kingdom samples (see photograph 6 B). The siimilarities between T. 414 plaster and Old Kingdom interesting because it's possible there was archaism of technique as well as artistic stlye in this period or the Northern influence is possible because the cultural centre had shifted to the North.

1.1.3-Alexandrian sites

Examination of samples from the **Mustafa Pasha tombs** (Appendix 5.8) has shown that the plaster/mortar used was a lime plaster/mortar, a mixture of lime/chalk/marble (identified by spot test) containing fossil shell, with a very fine grain size quartz and some other minerals (green and black). Two different mixtures were noticed; the first layer, in some instances, was darker than the second, due to the use of black sand in the mixture, whilst the second was very fine and lighter (photograph 20 E). The plasters/mortars were mainly of one or two layers whilst some others were of three of different thicknesses, as illustrated in Appendix 6.2. The painted surface was generally slightly rough, although polished surfaces were seen on some samples. The plasters/mortars were hard and dry, but salt was identified inside the structure. The painted layer was covered with a very thin layer of calcite film, indicating *buon fresco* technique and/or *fresco secco*, (see Chapter 2, 4.8.2 and 4.8.3).

Coloured *intonaco* was applied to tomb 1, in red as well as yellow. Acetic acid was used to dissolve the *intonaco* layer and to separate the colour (photograph 20 B), which was a mixture with lime/chalk and sand. The identification of the residue has shown that the mixture contained red and transparent quartz, very fine sand, byroxenens, amphipoda, as well as green olivine. In addition, the splashed technique used different colours, such as red, dark pink and brown on a very light pink (photographs 20 D). Both those techniques were investigated on wall painting fragments from Piddington, dated to the 4th century AD. In tomb 2, the imitation of marble was applied, as shown on Photographs 2 A & B (see Appendix 2, 4.3.2). (It is worth mentioning that some examples from Cyprus were decorated not only in the same style of the imitation of marble, but also applied in the same position, as shown on Photograph 2 A, which was common in the Hellenistic period) (Kakoulli 1998: Pers. Comm.).

Investigation of samples from **Anfushi** tomb 1 (Appendix 5.9) has shown that two different kinds of plasters/mortars of different thicknesses were used. The first was lime/chalk mixed with sand, the grain being slightly different from that used in Mustafa Pasha. Another sample was of two layers of very fine plaster. Some other samples without any paint, revealed that the surface was very rough, indicating the looseness of the lime binder (photograph 17 A). The grain size appeared clearly and could be identified. The samples were hard. The second kind of mortar found was a mixture of burnt brick, quartz, chalk/lime and shell fragments (photographs 21). The plaster/mortar used in Anfushi tombs 2 and 3 was more or less the same mixture and consisted of lime/chalk, fossil shell, as well as other shell fragments and was very fine. As the tomb had been re-plastered probably in Græco-Roman period, the thicknesses of the plasters were about 0.2 mm in both the layers. *Gesso* plaster was also identified. Although these samples were few and tiny, they were varied and gave an idea of the techniques used.

The technique of wall painting used in the Anfushi tombs was true *fresco (buon fresco)* made on wet plaster. This was noticed on the walls, as shown on Photograph 3 A, as well as on 5 A, although the lime binder had dissolved. The colours found on the plasters were yellow, red, black, brown, Egyptian blue and green/grey. The painted layer was covered with a calcite film, whilst some others were covered with a layer of salt, in addition to the crust layer, with dust found to cover the painted layer. The surface was covered with a very thin layer of modern adhesive, which dissolved in acetone.

As mentioned above, salt crystallization, micro-organisms as well as underground water were the main problems in the Anfushi tombs. It was found that different deterioration and decay had affected the plaster/mortar. The first thing noticed was that different kinds of salt were found on the wall. Some areas were very soft and pure, whilst others were in the form of a yellowish and brownish crust. On some samples a very soft and powdery whitish layer had covered the surface (photograph 18 A, the white layer on the top and on the cross-section of the same sample photograph 21 A). Another example was in the form of ancient dust layer, slightly hard in yellowish/brownish colour with a thickness of about 0.1-0.2 mm (photograph 18 B). On one example it was found that the calcite consisted of many layers of different kinds of films, as shown on Photograph 18 C. Some parts of the walls at ground level were very powdery and soft, resulting from salt crystallization.

1.1.4-Piddington

Wall painting fragments from Piddington showed different structures. Microscope (10X) examination was used to look at the mortar, *intonaco* layer and painting, as well as to identify and measure the thickness of the layers present. The plasters/mortars have shown different structures, the mixture being mainly brick. Examination showed that the mortar contained lime as a binder with, burnt brick, ironstone, sandstone, flint, burnt flint, burnt chalk, fossil shell, charcoal, quartz sand and quartzite as well as some mica. Organic material (straw) was also found in the mixture. It was noticed that some fragments did not have any brick in their structure.

Most of the fragments had two layers but some had three (see Appendix 5.13-15). The thicknesses of the first layers were of about 27 mm, with some up to 40 mm, whereas some second layers were about 38 mm. Where present, the third layer was thinner, about 7-10 mm, and was mainly sand. Not all the fragments were the same thickness, because they came from different places or rooms, which related to the preparation of the walls and painting. Some fragments were of red mortar. The first layers were of pink mortar pointing to the use of crushed brick. Different samples were taken for chemical analysis to identify the structure and composition. The mortar was covered with an *intonaco* layer, which was very thin on some samples but mainly about 0.1-3 mm. The *intonaco* layers were painted in different colours with a paint thickness about 0.1-0.1.5 mm. The colour of mortar degraded to brown with, in addition, black specks on the surface. On some fragments, it was noticed that the mortar layer was about 40-80 mm thick. These fragments were from window or door mouldings and floors. The mixture of the mortar was quite different because of the size of brick fragments, which were much larger compared with the other fragments.

The colours were applied in various schemes (photographs 52, 53 C & 55 A). Some were panelled and some splashed with different colours. Samples were taken for analysis to look at the composition of the colours. Some fragments had a very smooth surface, and the *intonaco* layer was of the same colour as that painted on the surface. This indicated the style used in Roman times. The colour was mixed with *intonaco* and spread on the wall and polished to get a smooth surface. It was noticed that some fragments showed pecking of the surface for the application of another layer. This second layer was taken off as a trial to see the first layer (see Chapter 7, 3.2). Where they used more than one layer, we find each one was prepared for new decoration by pecking of the surface for the next layer (photographs 56 B).

1.2-Examination of Polished Cross-sections

Broken and cross-sections revealed the material structure and allowed determination of the characteristics of the materials used, such as the shape of the gravel, sand etc., before using chemical analysis. Polished cross-sections gave an approximate measurement of the layers applied, which are given in Appendix 6**,** and also indicated the composition of some samples, as shown on Photographs 19-23.

As the materials from Saqqara and Dahshur were very soft, it was difficult to carry out sampling or cross-sections. However, it was possible to prepare polished sections from Luxor, Alexandria and Piddington. Several samples were prepared to complete the understanding of the stratigraphy of the wall painting structures, which showed differences, in some instances, due to the preparation of walls.

The examination of different samples from Luxor showed that the thickness of the painted layer of sample T. 192, painted in Egyptian blue and red, was between 0.1-0.3 mm. The finishing layer, coated with a darkened whitish layer of gypsum plaster (confirmed by spot test), had a thickness varying between 0.3-0.5 mm. Heat or fire had affected the plaster, and the differences in the two sections that may have changed from the original was clear as shown in Photograph 19 B and the darker ones as shown on Photograph 19 A. The clay plaster/mortar was of thicknesses varying from 7-13 mm. The mixture of the clay plaster contained ferruginous fossils, as shown in Photograph 19 B. Samples of T. 13 looked different from the previous one, as the finishing layer and clay plaster appeared clearly, as shown in Photograph 19 C. Two samples of mud plaster from T. 359 and T. 414 were prepared (photograph 19 D & E), where it was found that the mud plaster was coated with thin layer of clay plaster. Small cracks were noted as shown in Photographs 19 C & E. Measurements of the layers of wall painting are given in Appendix 6 1.

The cross-sections from the Alexandrian tombs showing much binder of lime/chalk with fossils, fossiliferous limestone and shell fragments. The plaster/mortar used was lime, full of shell fragments and microfossils such as foraminifera and coccoliths, as shown on a polished cross-section using SEM (photograph 20 C), in addition to other organisms. Red quartz, ironstone and quartzite were found with some other minerals (green olivine). Investigation of the polished sections allowed for the measurement of both the layers, of the plaster/mortar, *intonaco*, as well as the painted layers, most the plaster were very thin ranging from 0.1-0.5, as stated in Appendix 6.2-3.

On some sections it was possible to see the different kinds of plaster used in the Alexandrian tombs. One sample of plaster/mortar showed the same structure as that found in Mustafa Pasha, consisting of two layers; the first layer was a mixture of black sand, and the second, white sand mixed with lime/chalk (photograph 20 A and 21 B). The coloured *intonaco* layer showed the same mixture as the plaster (photographs 20 A and B).

Another example not only indicated the mixture of the plaster/mortar used, as shown on Photograph 18 A, but also showed the different layers of calcite film. The surface was covered with a powdery layer and a hard crystal layer underneath, which covered the surface, and had a thickness of about 0.1-0.5 mm (photograph 21 A). The section also revealed that the crystals grew inside the mortar between the cracks, as well as around the bricks. This was due to the transformation of salt from the wall and its migration towards the rendering, resulting from the porosity of the mortar, which allowed for the solution and re-crystallization of the salt (see Chapter 3, 2.1.5).

The difference that was noticed between sections from Mustafa Pasha and Anfushi related to the sand used in the mixture of the plaster/mortar. This led to the investigation of sand from both sites. Examination of the sand was carried out to identify its structure. The Mustafa Pasha sand had large grains and was mostly quartzite, but also contained different shapes of shell fragments, red and smoky quartz, and some sea plants such as calcareous skeletons (of a green colour), formed from colorine algae (red algae).

Sand from the Anfushi beach has shown that the grains are both fine and medium in size, rounded oolitic (egg shape), containing shell sand of different shapes as well as many kinds of shell fragments. The sand from the Anfushi was well sorted.

Investigation of polished sections from **Piddington** has shown that some samples have the same structure. The samples were divided into groups. The first group of samples (S.1, S.2, S.3 & S.5), (photographs 22 A & D) had mortar consisting of; brick, ironstone, flint, quartzite, quartz sand, and chalk or lime, whilst the second group of samples (S.4 & S.10) (photographs 22 B & C), had mortar containing no brick, but ironstone, quartzite, quartz sand, flint, and chalk or lime of different structures.

The mixture in sample S.6 had a lot of quartzite and charcoal, whereas, in samples S.8 and S.9, the mortar colours were quite different, being nearly red, suggesting that they had been burnt. The third group of samples, S.9 and S.11, was completely different. Sample S.9 had a very fine mortar, but the *intonaco* layer was quite thick. There was no brick in the mixture of the mortar, as in group one, but there was some straw. Sample S.11, consisted of two different layers of mortar and a very thick *intonaco* layer, which was covered with a calcite film on the painted surface, as shown on Photograph 22 D. In addition, the mortar colour of the first layer was pink and the second yellow, the same mixture as in group two, but with a lot of brick in first layer and a lot of quartz sand in the second layer. SEM was performed on a polished section to reveal the calcite layer that covered the surface, as shown on Photographs 23, which also indicated the mixture of the mortar. The mixture of the mortar used varied, with different thickness of *intonaco* (see the measurements of the plasters/mortars in Appendix 6.4).

1.3-Summary

There are various lifting techniques used, which preserve evidence to help us to determine the nature of the materials. It is very important to record and describe the samples, as well as the thickness of the paint layer, the *intonaco* and the layers of mortar, as sometimes the fragments may be of different thickness and layers. The colours and the schemes of the painted motifs can help us to understand the technique used. Cronyn gives a good description for lifting materials on site (1995). The visual examination is the the fist stage and the chemical analysis follows immediately. Care must be taken during examination and recording of the materials which are taken from the site. Small samples must be kept under good conditions for later analysis and samples should be treated very soon after excavation.

Preliminary examination, as the first stage of the research, gave a good picture of the stratigraphy and distinguished the different kinds of plasters/mortars used. Although differences were noted in the preparation of the walls, chemical and physical analysis was carried out to identify the composition and structures of the plasters/mortars in order to differentiate the mixture of the plasters/mortars and the additives used for the preparation of the walls.

2-Chemical Methods: Summary of the Results and Discussion

In the analysis of samples from Saqqara, Dahshur, Luxor, Alexandria and Piddington the following methods were used: chemical spot tests, standard methods and analysis of water-soluble salts. Sequence of analysis began with mud brick, mud and clay plasters/mortars and then the finishing plaster of Saqqara, Dahshur and Luxor, then lime plasters/mortars from the Alexandrian sites and Piddington. Measurement of calcium carbonate was not used for samples from the Alexandrian sites, because they were calcareous, containing fossil shell and shell fragments, giving very high percentages of calcium carbonate. The samples dissolved completely in diluted HCl. Micro-chemical analysis was used for identification of the pigments found on the plasters, particularly those from Piddington.

2.1-Micro-Chemical Analysis (Spot tests) was carried out on samples to distinguish the character of the materials used in the mixture of the plaster/mortar (see Appendix 7). Tests were carried out on some samples, using HCl acid and barium chloride to identify gypsum. The identification of gypsum by spot tests has shown that the gypsum was impure, containing other minerals such as charcoal, chalk/lime, and very fine-grained quartz sand as a natural formation. In some instances, the HCl solution was pinkish and gave a strong white precipitate, whilst others were yellowish, indicating iron.

Spot tests were also carried out on samples of clay plaster/mortar, revealed very high levels of carbonation. Phosphates were identified using nitric acid and ammonium molybdate followed by a drop of ascorbic acid, which gives a blue solution in the presence of phosphate, as shown in some samples from Luxor. Also, as an alternative, it is possible to use sodium hydroxide in testing for nitrates, with the evolution of ammonia. Iron was identified using HCl and ammonium thiocyanate giving a red solution.

The samples from the Alexandrian sites were tiny and micro-chemical tests were carried out to identify the materials used. Hydrochloric acid indicated that the plaster used was lime plaster/mortar mixed with lime/chalk. Gypsum plaster was also found in both Mustafa Pasha tomb 1 and Anfushi tomb 2, in different structures. The first sample was on the doorframe, painted yellow. The second was found to cover only the arch in room 2 (burial chamber); the colour was nearly yellowish and contained small pieces of lime/chalk. It was not found elsewhere, suggesting that it was to preserve the architecture detail or may have been added later. It was necessary to identify both the mixtures, so microanalysis was carried out using warm HCl followed by barium chloride to confirm that, in the first sample, the plaster used was gypsum, resulting in a white precipitate. In the second sample, HCl was used showing a little effervescence and then the sample was heated up, followed by sodium hydroxide which indicated that the plaster used was a *gesso,* a mixture of gypsum and some organic material as the sample broke down.

2.2-Standard Analysis (SA)

Chemical analysis was carried out on most of the different samples; mud bricks and mud as well as clay plasters/mortars from Saqqara, Dahshur and Luxor (see Appendices 7.1-5), where it was necessary to identify the composition of the supporting structure for the rendering used. The mortar used to treat the irregular surfaces and/or fill the voids to produce a suitable surface for painting, as reported in Chapter 2 (4.3.2), was also analysed.

Analysis was carried out on several samples of mud brick in order to determine the nature and quantities of additives. This was done to understand to what extent the compositions were similar to or different from the mud plaster/mortar, as they looked similar in appearance. British standard methods recommend the use of about one-kilogram of solid mortar or concrete and at least 100 g fine plaster/mortar (B.S. 4551). However, it was very difficult to apply that standard on some samples because, firstly, it is destructive and the fragments from Piddington had to be returned to the museum and, secondly, most of the fragments were too small. If 100 g were taken nothing would be left. Therefore, samples from different places were taken. At least three similar samples from structures, for example the wall core, face or bonding layers. Problems would occur where walls had been repaired, repainted or re-plastered. The original weight of the samples of plasters/mortars was varied, and in only a few instances it was possible to have the amounts as the British standard specified (B.S. 4551), while some were even less than 10 g, giving less accurate results.

Tests were carried out on two different samples from different fragments taken from Piddington, with acetic acid and hydrochloric acid. HCl could be used, when a sand and aggregate mortar was used to support the structure carrying the painted layers. Samples were treated with HCl as stated in Chapter 4 (2.2.2), and showed very high carbonation, especially the clay plasters/mortars. Some samples were only washed in 500-ml distilled water, as there was no need to dissolve them in acid. It was interesting to carry out such trials and to compare the results from the same kind of samples, using different methods for the identification of the compositions. It was very difficult to deal with such materials in sieving and grading, because the clay/mud became hard after drying. Great care had been taken to avoid such problems by washing off the clay/mud and separating it from the other aggregates. In the analysis of the materials, the classification of the plasters/mortars (mud/clay) took place to identify each separately, and to find their similarities and differences. This was not only because different plasters/mortars might be used, but also because the same kind of plaster/mortar can vary according to different sites.

2.2.1-The Results of the Mud Brick Analysis

The results show that the composition and structures are slightly varied and the percentages of silt and sand found are very different. The composition of the mud bricks shows that the main additives used were; chopped straw, quartz sand (varied in colour in some instances according to the source, north/south), charcoal, feldspar as well as green olivine, in addition to mica (both muscovite and biotite). Other materials were identified in the mixture of some samples, such as ceramic (burnt brick), chert, ironstone, sandstone and cemented material as well, indicating the geological areas where the local materials were available. It was interesting to identify some charcoal as olive wood, in one sample from Ka-Nefr tomb, dated to the 5th dynasty. The percentages of the sediment types are varied, gravel from 1-7%, and sand 16-67%, while mud/silt was 33-82%. However, the acid soluble amount was very low, 2-10%. The result of the analysis of samples from Luxor showed more or less the same composition. However, the percentages of sediment types were different, sand being higher in sample T. 55, and silt about 91.86% in sample T. 192, indicating that the mud brick had finer particles, showing similarities to those from Dahshur (see Appendix 8.8).

The results of the analyses of the original weights, residues, as well as the acid/water used are given in Appendices 8.1 & 8.4. The percentages of sieving weight are illustrated in Appendices 8.2 & 8.5. The percentages of sediment types as well as their residue identifications are reported in Appendices 8.3 & 8.6. Graphs representing the frequency distributions of the particle size of the gravel, sand and silt used in the mixture, are in Appendices 8.7-8.

2.2.2-Mortar Filler/Joining Mortar

Two samples of mortar filler were analysed from Luxor; T. 192, which was used between the bricks, and another from T. 414, used for the treatment of the surface. The first was treated with HCl whilst the second only washed. The composition of the mortar in each sample was slightly different; the first contained quartzite, brick and mica, while ferruginous fossil and seeds were visible in sample T. 414. The identification of samples is given in Appendix 8.11. Graphs 1 A and B, show different distributions of the aggregate particle size in both samples, due to the different method of analysis, showing the different percentages of the sediment types of gravel, sand and silt.

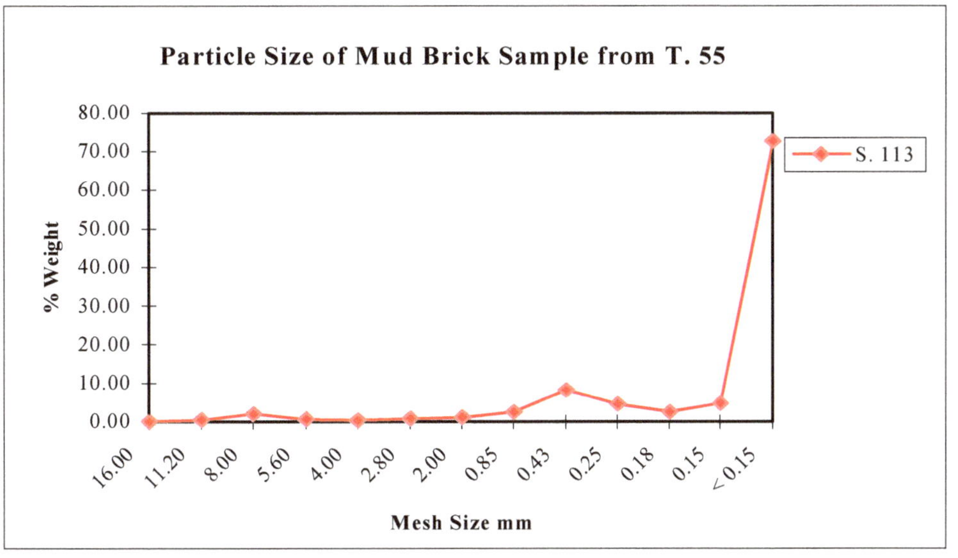

1 A

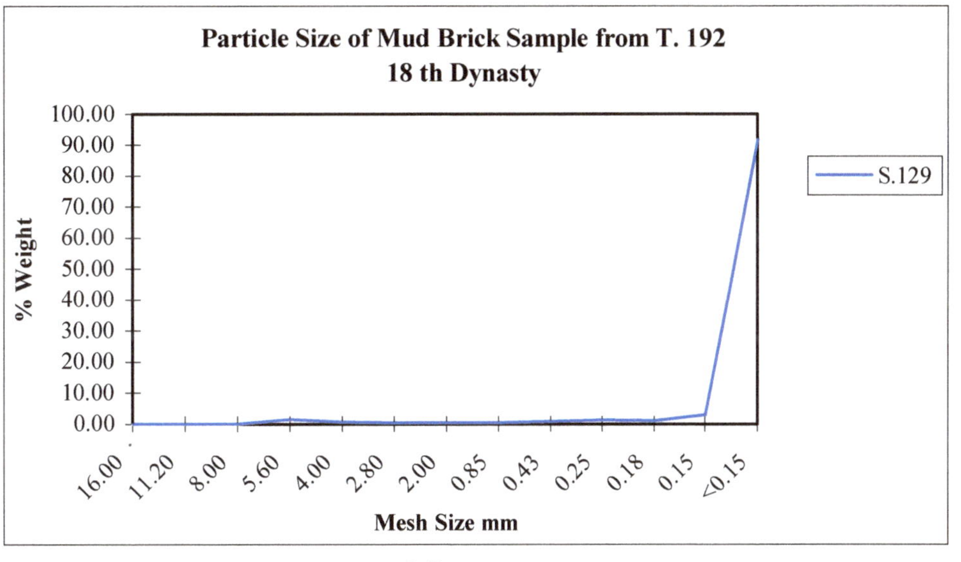

1 B

Graphs 1 A and B, show the distribution of the aggregate particle sizes in mud brick, from T. 55 and T. 192. The differences shown on the graphs are due to the percentage of sand being higher in sample T. 55. The percentage of silt is about 91.86%, indicating that the mud brick had finer particles, showing similarities to those from Dahshur (see Graph 20 D, Appendix 8.7).

2.2.3-Mud Plasters/Mortars

As the mud plaster/mortar (coarse plaster) was the first layer to cover the mud brick/stone, analyses were carried out on different samples from Saqqara, Dahshur and Luxor from different tombs and areas; to identify and distinguish the composition of the mixture used (see Appendices 7.1-5). The qualitative and quantitative analyses gave a full picture of the mixture, composition and percentages of sediment type used as additives (photographs 24). The results of the identification show a variety of sediment types that were mixed with the mud (Nile alluvium) such as, quartz sand of various colours, quartzite, chert, straw, brick, ironstone, feldspar, charcoal, sandstone, and mica. Samples from Luxor showed ferruginous fossil and shell fragments. The same crystals as well as some seeds, were found in the mixture contained in the samples from Ka-Nefr tomb dated to the 5th dynasty, (shown on Photographs 25), and were visible in the Luxor samples as well.

The composition of the samples was more or less the same. Differences were noticed in the percentages of sand from Saqqara samples in particular those from Ka-Nefr tomb, which were much higher, in comparison with the Luxor samples (see Appendices 9.3 & 9.6). Whilst the percentages of the gravel were very low in the samples of T. 414. In contrast there was a very high percentage of silt in all the samples. The percentages of the acid soluble components were slightly different, due to the proportions of limestone contained in the samples. Graphs 2 A & B represent the sediment types of the gravel, sand and silt and show the differences due to the different mixtures used.

It is also possible to determine the percentages of each mesh size of the mixture used (see Appendices 9.2 & 9.5). This information can be required in the case of the restoration of mud plaster/mortar, to ensure that the correct or even the nearest recipe can be applied.

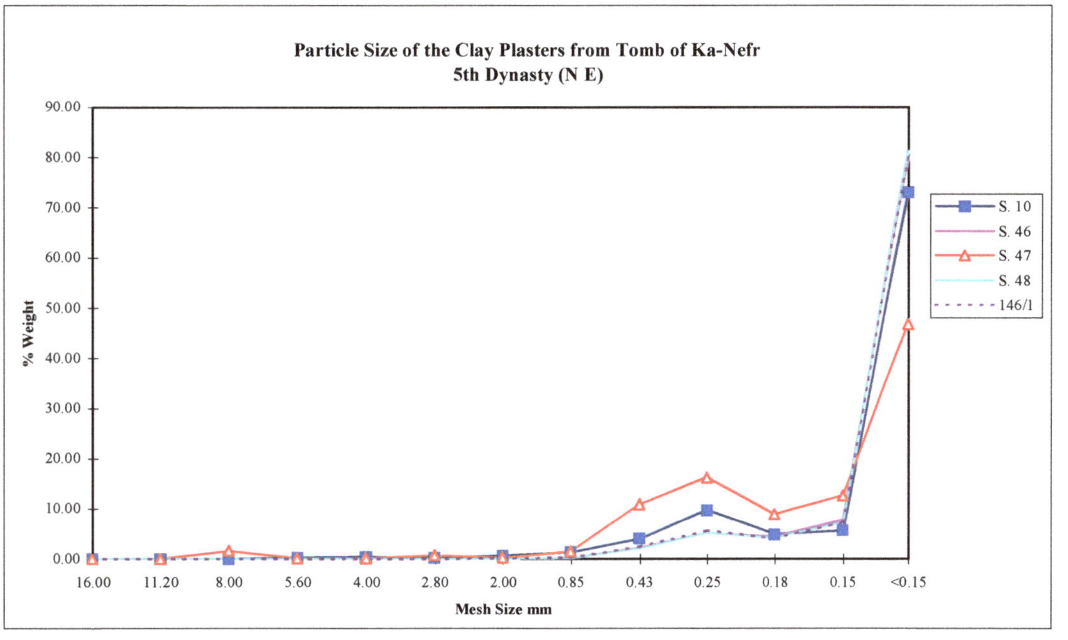

2 A

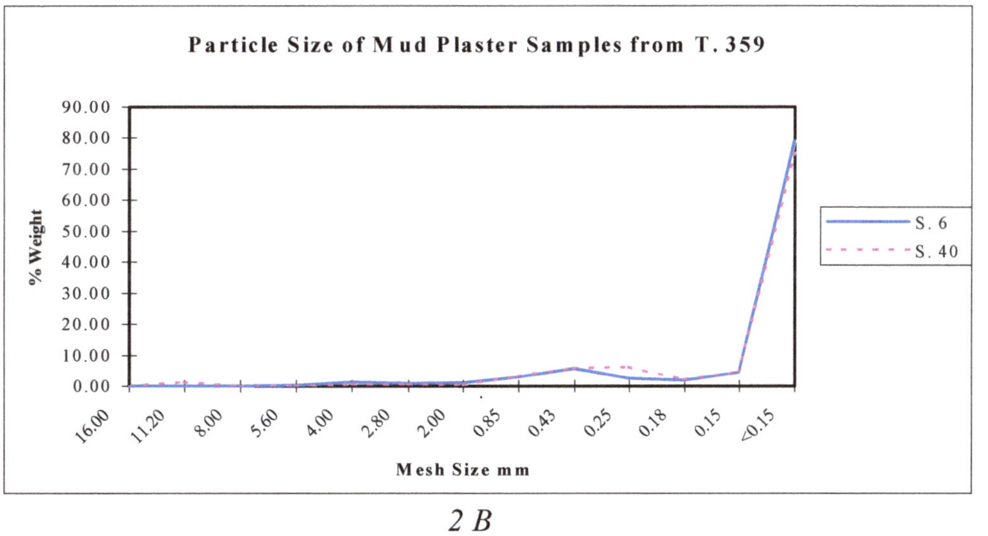

2 B

Graphs 2 A & B show the differences between the distribution of the sediment types in mud plaster samples from Saqqara and Luxor. Note the differences between both sand and silt found in the mixture of the samples.

2.2.4-Clay Plasters/Mortars

To distinguish the clay plasters and to define the similarities or differences of the clay plaster from the first layer covering the rocks, several samples of clay plaster/mortar were analysed from different tombs, areas and periods from Saqqara, Dahshur and Luxor. Clay plasters/mortars, in some instances, were found to be the second and/or the first layer, which covered the mud plasters/mortars (coarse plaster) or mud brick/stone. Chemical analysis was carried out on various samples. It was essential to analyse several samples from the same areas in order to obtain an accurate outline on the composition of the clay plasters/mortars used.

Most of the samples were treated with HCl, they showed a very high carbonate content due to the use of powdered limestone, whilst a few were only washed with distilled water, which made possible to identify the composition of washed samples, as HCl would dissolve the carbonate compounds in the samples (S.134/1W and S.111) (see Appendix 10.11).

The results of the analysis of the clay plasters/mortars show variation, not only in their composition, but also in the percentages of the sediment type. The composition of the clay plasters/mortars and their additives showed similarity to those of the mud plasters, but were clay/silt based. The mixture contained gypsum crystals as well as cemented material. In some samples, the quartz sand in the mixture showed sand lasting, suggesting a desert source, as an environmental effect. Ferruginous fossils were also found in some samples from Luxor. The results showed variation in the percentages of the sediment types used in Saqqara and Dahshur. Gravel was 1-7%, with an exception in sample S.57 being 25%, sand was between 18-27%, with a few up to 50%, and silt showed very high percentages, being 46-81%, except one sample which was slightly low at about 26%. The acid soluble values varied in all samples, as stated in Appendices 10.4 & 5.

The highest percentages of gravel were found in the Luxor samples; sample S.34 being 33.35% and S.86 being 38.74%. The proportions of the mixtures used in the preparation of the clay plaster/mortar varied not only between the tombs but also between different areas and periods. The differences shown between the same samples from tomb T. 192 are reported in Appendix 10.8-9. The acid soluble values were found to be very high, indicating high percentages of carbonate, being 47-84%, due to the use of powdered limestone. Two samples showed low acid soluble percentages, 14-17%. The silt in all the samples was very high, at up to 50%. The results of the identifications of the samples are given in Appendices 10.4-5, 10.9 & 10.10. The original weights, solvents used and residue are given in Appendices 10.1 & 10.6. More details of the chemical analysis and the percentages of the difference mesh sizes are illustrated in Appendices 10.2-3 & 10.7-8. The graphs show the differences in the distribution of the aggregate sizes of the sediment type of the samples, which in some instances are slightly similar as shown in Graph 4 A, B and C (see also the graphs in Appendices 10.11-12).

It was mentioned above that the aim of collecting a few samples of the restoration of the tombs, was to distinguish the plasters/mortars used and to see to what extent they are compatible with the old ones. Different results have been obtained by analysis of a few samples used of re-plastering in the tombs; No. 255, 13 and 14. The plasters consisted of silt/clay, quartz, amorphous silica, ferruginous fossil, charcoal, ironstone and shell fragments. Green olivine was noticed in S.87 as well. The percentages of sediment types were very different. It was interesting to observe that gravel was no longer found in all samples. Different proportions of sand and silt were noticed in the mixture; sand was very low in S.74 about 1.33%, whilst silt was up to 98%. The highest amount of sand was in sample S.79, being up to 85%, however, silt was lower at about 15%. The percentages of sand in sample S.87 were 36% and silt was up at 64%. The percentages of acid solubles found were between 43-54%. The results of analysis are given in Appendices 10.13-15. Graph 4 shows the distribution of aggregate sizes of the re-plastering samples used for restoration and the differences between those samples and the old ones used in the antiquity at the same tombs can be noted.

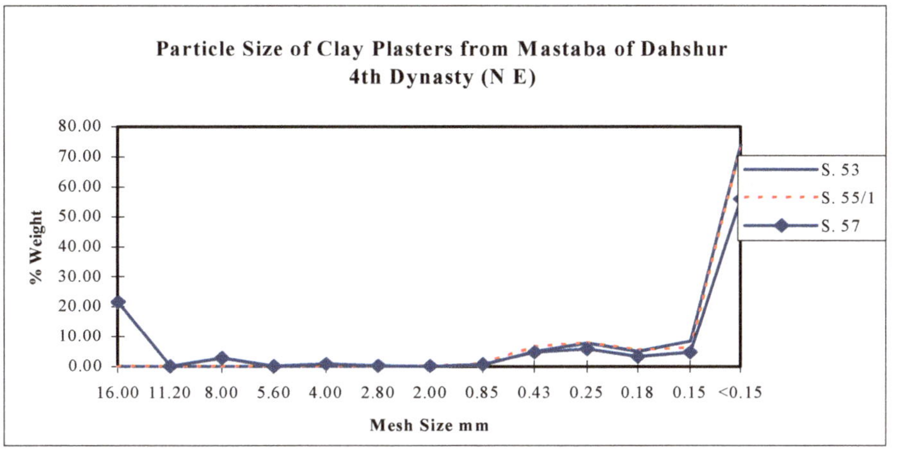

3 A

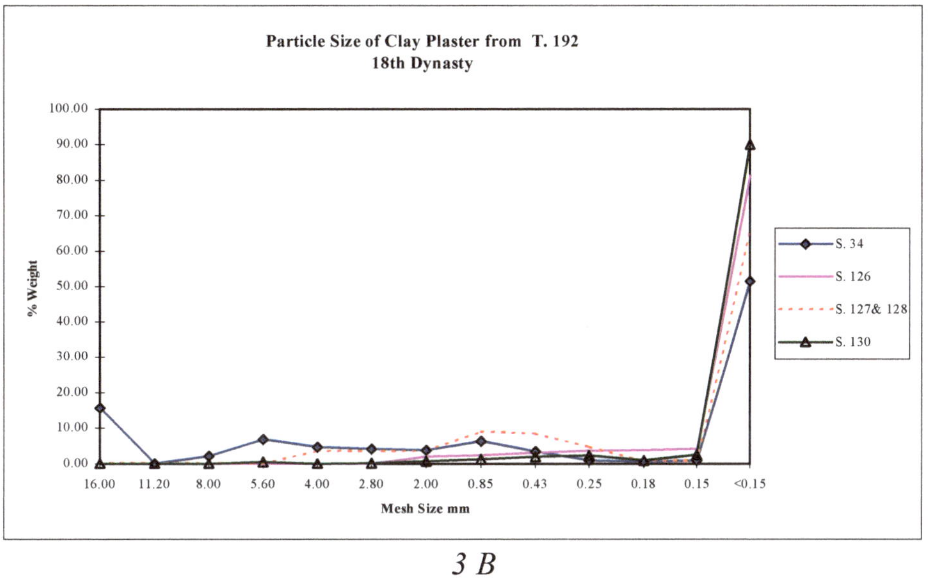

3 B

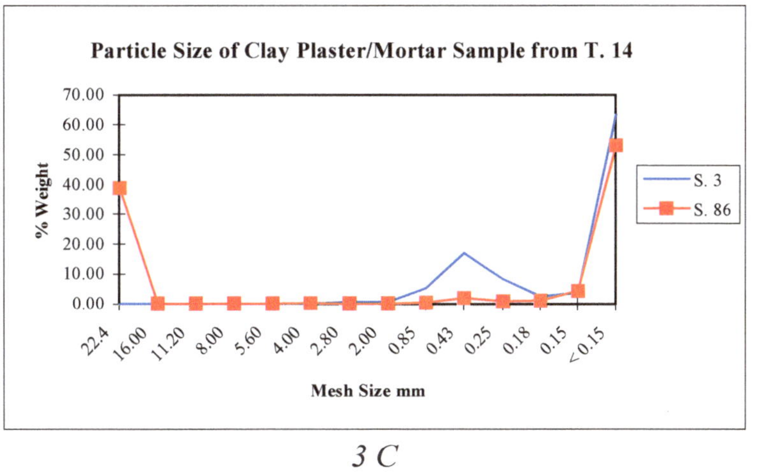

3 C

Graphs 3 A-C, show the distribution of the aggregate particle sizes of the clay plasters/mortars from different periods and areas. Note the similarity between graphs A & B, although the first represented the particle size of samples from Dahshur, while the second was of T. 192 from Luxor. The differences between the samples are visible, although they came from one place, and are due to the sediment types used in the mixture. However, C shows very different percentages of sand and gravel.

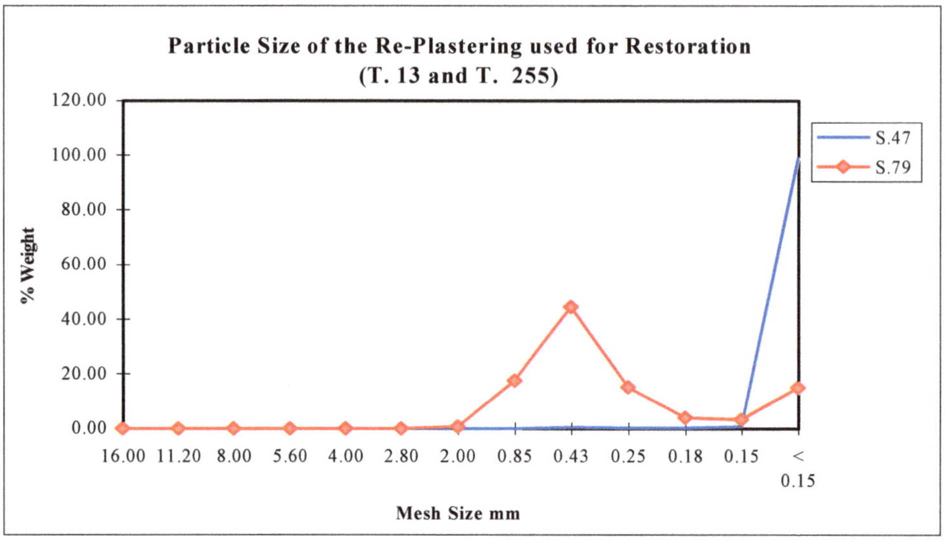

4 A

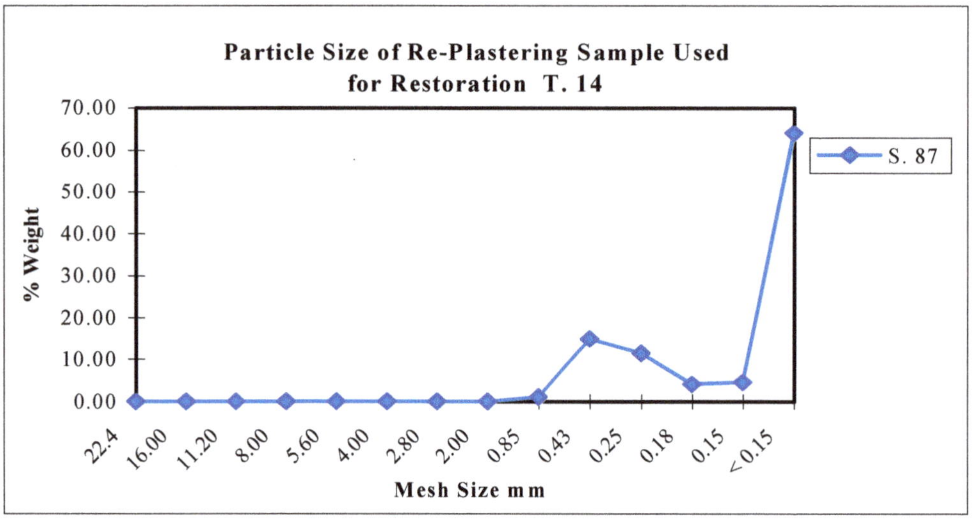

4 B

Graphs 4 A & B, show the distribution of the aggregate particle size of the re-plastering samples used for restoration. Note the differences in sediment types used, such as sand and silt, and the absence of the gravel (compare this with the original samples from the same tombs, Graphs 3 C, and Appendix 10.12, Graphs 26).

2.2.5-The Finishing Plaster

The finishing plaster layer was applied for the painting. This layer was found, in some instances, to be gypsum (indicated by spot test) containing fine aggregates, possibly an impurity of the material used. Another kind of finishing plaster found was a mixture of calcium carbonate (powdered limestone) and some other minerals, as shown by very high carbonation. Since the samples were very small, as well as contaminated with the plaster underneath, it was very difficult to carry out chemical analysis. In order to obtain accurate results, a physical test was used to confirm or complete the microanalysis. Analysis was carried out on a few samples using X-ray powder diffraction to determine the composition of the finishing layer and to distinguish the plasters used in the Old Kingdom. Similarity was noted in the finishing plaster applied in the first dynasty (E E) and the ones form Asasif (T. 414) (see photographs 6 B & 16 C) (see also X-ray identification Appendix. 16.11-13).

2.2.6-Lime Plasters/Mortars

As the samples were small (most were less than 20 g), it was difficult to carry out analysis from the **Alexandrian sites**, but a few samples were large enough to be analyzed (see Appendix 11.1-2). Three samples were analysed; two of limestone from the site of Mustafa Pasha and Anfushi, and one of mortar from Anfushi tomb 1, to identify the composition using different acids. The mortar sample was analysed (photograph 5 A), after cleaning the surface. Then the layer which covered the surface, was analysed. The mortar sample showed a different structure containing quartz, mica and sandstone. Brick and brick dust were found to be present, which was known in Roman times but the rarity of burnt brick building before then. The identification of samples is given in Appendix 11.3. In Graph 5, note the difference of the percentages of gravel, sand and silt.

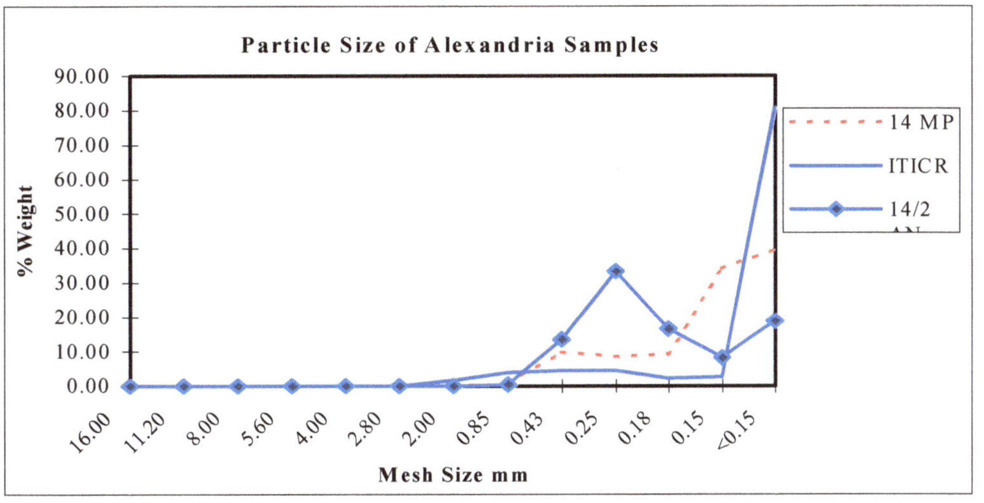

5

Graph 5, shows the distribution of the aggregate particles of different samples from Alexandrian sites.

The fragments of wall paintings from **Piddington** were mainly two layered, although some had three with different colours, as mentioned before. The samples were varied; some from wall plaster, window and door moulding, as well as floors. Some were yellow and the second layers, applied later, were pink. For that reason, twenty-three samples were taken from two layers, as well as different fragments to see the different structures. Analysis was carried out on the layers separately to identify the composition of each and determine the approximate percentages of the mixture. The samples were treated with dilute acid as mentioned above, to separate the carbonated lime from the aggregate of the sand and other ingredients. This made it possible not only to study the grading and other characteristics of the aggregate, but also to determine the approximate proportion of lime used in the original mixture.

The analyses showed that the character of the mortar varied accordingly to the local materials available. The original weights of the plasters/mortars were varied, resulting in different percentages of lime and residue, as well as the percentage of the lime dissolved, as stated in Appendices 12.1 & 12.4. Appendices 12.2 & 12.5 show more details of the results of the sieving with various mesh sizes. Graphs 6 A & B, show the particles sizes of gravel, sand and silt used in the mixture of the plasters/mortars of the first and second layers (see Appendix.7-9). Appendices 12.3 & 12.6 show that the mortars were generally very rich in binder.

2.2.7-Mortar for Window, Floor and Door Mouldings

Analyses were carried out on a few samples of mortar for window, floor and door mouldings to identify the composition and determine the percentages of the sediment types used, in comparison with the plasters/mortars found elsewhere. The results of chemical analysis of the original weights and acid soluble as well as the residue of the samples are illustrated in Appendix 12.9. Appendix 12.10 gives the percentages of each mesh size. The graphs show different distributions according to the aggregate particle size of gravel, sand and silt (graph 6 D) (see also Appendix 12.12). This is evidence from coarser material to make a more hard wearing floor.

The mixture of the plasters/mortars of the first, second and third layers, as well as the moulding mortars of windows, floors and doors, indicate that a lot of different materials were used in the manufacture of the plasters/mortars. The most common additive was brick, in different colours (black & red-brownish), ironstone and sandstone, quartz sand, quartzite, flint (burnt in some samples), charcoal, coal and silt, in addition to silica and mica. Some samples did not contain brick e.g. S.13 A & B, but may have been burnt, because they were reddened. The mortar was lime mortar, which dissolved in HCl. The average of the acid solubles found in the samples was between 32-58%, as illustrated in Appendices 12.10 & 12.11. Sample S.14 A, had the highest percentage of acid soluble components. The percentage of acid soluble lime found was different in some samples, indicating the use of different mixtures of mortar and lime. The proportions of the sediment types of gravel, sand and silt were also different.

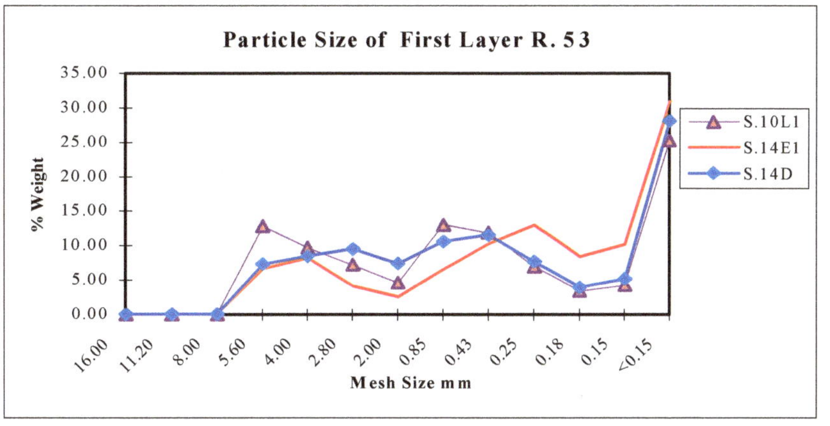

6 A

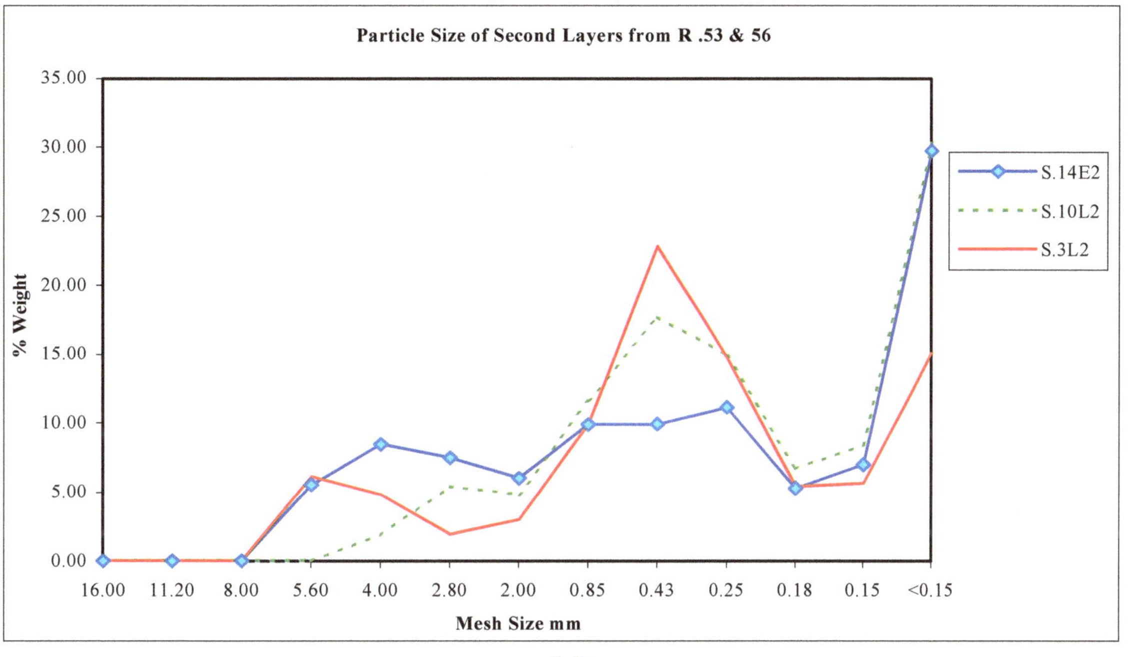

6 B

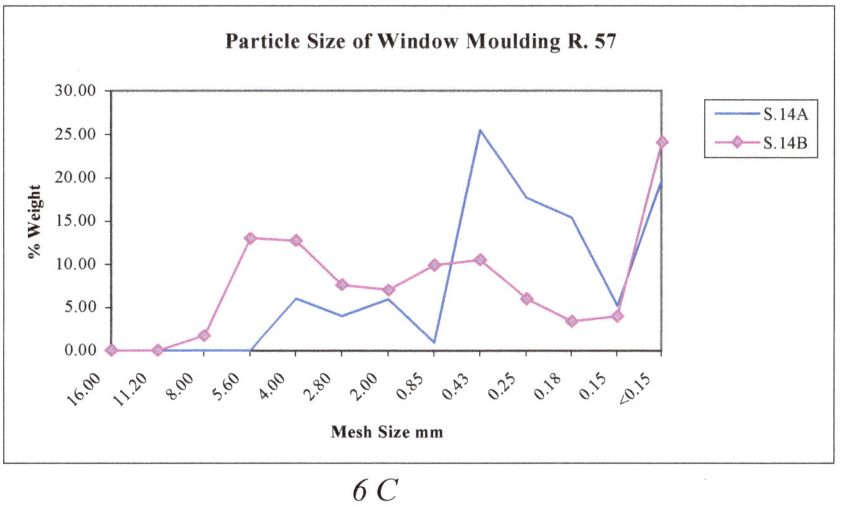

6 C

Graphs 6 A-C, show the different distribution of aggregate particle sizes of lime plaster and mortars from Piddington: A, shows the first layers. B, represented the second layers of the same samples. Graph C, showing window moulding samples, and shown very different due to the proportions of gravel, sand and silt (see also Appendix 12.7-8 & 12).

2.3-Discussion of the Analytical Results

Analyses show variation in the composition of the samples of mud brick, mud and clay plasters/mortars collected from Saqqara, Dahshur and Luxor, indicating differences in the proportions of the sediment types used.

The mud brick, as well as the mud plaster/mortar, showed various additives, as stated in Appendices 8.3 & 8.6, and 9.3 & 9.6. This led to the analysis of one sample of mud from Luxor (Nile mud), in order to determine what the original mixture contained and to distinguish the mixture in the samples used. The sample was washed and the result given in Appendix 13.1-3. It is possible to compare the composition of the sample with those of the mud plasters/mortars. It should be noted that this one sample may not be representative of Nile mud generally.

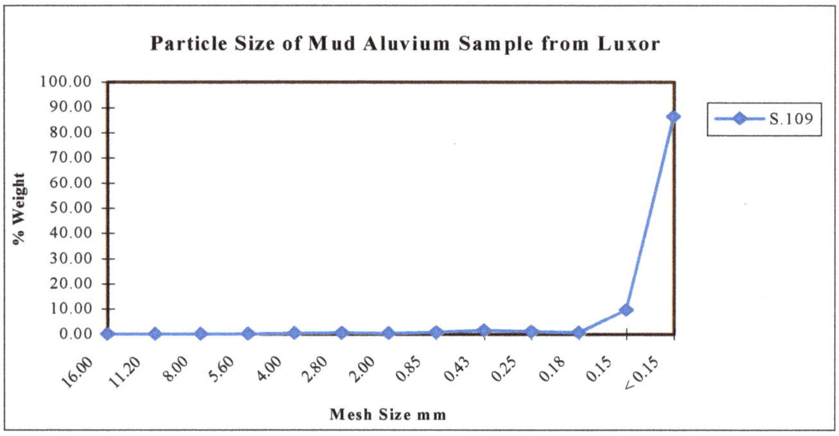

Graph 7, shows the distribution of aggregate particle sizes of a sample of Nile mud from Luxor.

The composition of the washed sample (S.16) used as mortar filler is described in Appendix 8.11. It contained very interesting microfossil shells. Identification of their characters was made using SEM (photographs 25). The presence of such microfossils and shell fragments was noticed during the examination of the different limestone, indicating the high percentages of calcium carbonate.

The identification of the mud plasters/mortars has shown them to be more or less the same as the mud brick. The most interesting thing noticed was the use of long straw in the mixture of mud plaster/mortar, where it contained some seeds as well as pieces of bone, as shown on Photographs 26. As the layer of mud plaster/mortar was the first, the mixture was coarse in order to adhere well to the building material, which is why it is similar to the composition of the mud brick, as stated in Appendices 8.3, 9.3 & 9. 6.

The identification of the plants (table 2) has shown that the plant remains are desiccated, except the *Lolium* seed in S.137 and a seed in sample S.141, which are charred. *Phalaris cf. minor* was mostly found in all the samples. Seeds were found in different samples from Luxor (S.113, S.16, S.6, S.20 and S.121). The identification of the seeds has also shown cereal grains, wheat, and some fragments of chaff sticking to the surface, in addition to *Lolium* and *Phalaris*.

Table 2, Identification of seeds

Samples name	Identification and classification of seeds	
	Cultivated plants	**Wild plants or weeds**
S.137	One *Hordeum vulgare* (hulled, 6-row barley Two, *Lens culinaris* (lentil; seeds) and one *Lens culinaris* (lentil; pod) One *Linum usitatissimum* (flax; capsule fragment)	Two *Lolium temulentum* One *Gramineae* One *Leguminosae* (seeds)
S.139		Two *Leguminosae* (seeds) and (pot fragment) undetermined One *Phalaris cf. minor*
S.144		One *Phalaris cf. Minor*

It was very interesting to find a beetle thorax, with head and bases of the first pair of legs and an elytron (wing case) in sample S.20, pointing to bio-deterioration and decay as noticed in the samples of T. 414, as described above. The loss of organic binders such as straw through being eaten by insects, e.g. (white ants) or organisms such as bacteria, was found in sample T. 414 (S.17) and was partly responsible for it being so fragile, in addition to the dry atmosphere.

The additives used in the mixtures from the Luxor samples were the same as those found in the Old Kingdom, with slight differences depending on the geological source of the limestone. The percentages of acid soluble materials differed due to the powdered limestone in the samples. The only thing noted, which was not found in the samples of the Old Kingdom, was the presence of ferruginous fossils and iron oxide.

Brick/ceramic was also visible in the mixture, which was indicative of relatively low temperature in the kiln. This was indicated by the presence of a shell. Ceramic/burnt brick was found in the mixtures of both mud bricks and mud/clay plasters/mortars. It was noticed that some bricks were brownish/reddish, indicating low temperature firing, evidence that the Egyptians knew the technology to make burnt-bricks, but that it was not employed in Egypt. A possible source of brick material in the recycling of materials from burnt building. Lucas mentioned that burnt brick was used in Mesopotamia and at Mohenjo-daro, in India, at a very early date and was known before the Roman occupation (Lucas 1989: 49, 50). However, there are a number of archaeological finds of fired bricks and floor tiles, such as at Nubian fortresses of the Middle Kingdom, and in graves at Theben of the New Kingdom. Finds increase from the 20th/21st dynasties, both within civilian and religious/burial monuments. They become most common in Ptolemaic and Roman times. Fired buildings are known in the Karnak and Medinet Habu areas from the 21stdynasty. There are also a number of literary sources for the use of brick: an account bill from the times of Ramsses II, a medical text from the 19th dynasty, and another text from 259/8 BC. First sources date from the Middle Kingdom (Verhoeven 1987).

During the identification of clay plaster/mortar, crystals of gypsum, as well as fossil shell and iron oxide, were found on samples from the Mastaba of Dahshur, dated to the 4th dynasty, as well as on other samples. Lucas mentions that animal dung, as well as chopped straw, was added to the mixture to increase both the strength and plasticity of the clay/mud (1989: 49-50). However, the materials found were very soft and fragile, indicating the loss of binding strength, whether or not it had been used, with the exception of those samples collected from the first dynasty (Emery excavation) tomb in Saqqara, as it was still roofed.

It was easy to distinguish between the two different kinds of clay and/or mud, as a dark mud was used for the mixture of the (coarse plaster) and a light one was a mixture of clay and powdered limestone as a clay plaster/mortar "better plaster". A small amount of both, heated to 800°C, showed that the same resultant colour was red, but that the dark mud had more iron. The main source of dark soil was silica, plants and similar organic matter. The light colour clay from (geble) was known as "*hib*", natural mixture of powdered limestone and clay. The result of analyses of the clay plaster/mortar has shown very high percentages of the clay/silt residue (Appendices 10.4-5, 10.9 & 10.10), giving a high ratio of clay as an insoluble residue, indicating calcareous clay plaster/mortar.

Therefore, the clay plaster/mortar contained the same additives as mud plaster in different proportions. As it was a finer mixture to cover the first layer of the mud plaster/mortar, the percentages of the silt/clay were found to be higher than those of the other additives used, except in those of the first dynasty (N E), where sand was found. The straw was used as organic material to bind the aggregates, mixed with clay and/or mud alluvium as a binder or matrix due to its natural capacity for cohesion.

Time is an important factor, which affects the analytical process of dissolving the lime. Long dissolution times lead to loss of the lime in the mortar or plaster and if limestone is used in the composition, this will also dissolve. If the samples contain limestone it should be treated for only a few minutes, but a short time may be not enough to dissolve lime in mortar. It is known that some mortars dissolve more quickly than others. Therefore, the dissolution time is a very important consideration. The mortar should not be crushed before dissolution as this may damage the aggregates.

It was noticed that sample S.1/A, from Anfushi tomb 1, did not dissolve completely in HCl, resulting in the first analysis showing that the percentage of residue was 68.35% compared with soluble 31.65%. Investigation indicated that the sample contained some gypsum as a hard residue. The sample was re-dissolved in HCl and heated up to ensure that any gypsum present was dissolved (see Appendix 13.2). However, as the sample was covered with a white layer (photograph 5 A), and a spot test indicated the presence of gypsum, it can be assumed that the mixture of the mortar contained gypsum as well, produced from a mixture of gypsum and silica in the water. Another point is that that sample was collected from zone A, which probably contains gypsum crystals, as mentioned before in Chapter 3 (see 4.3).

The mixture of the mortar used indicated that Tomb 1 was re-used in Roman times. This was because the mortar mixture used in Roman times was also found at Piddington, but with different geology. In addition the literature confirmed the increasing numbers of rooms in Tomb 1 in Roman times. Although there was no longer any Egyptian painting in tomb 1, as tomb 2 which combined both Egyptian and Greek styles, and gave evidence that the tomb was requisitioned in Roman times (see Appendix 4, the decoration style in Anfushi tombs, 4.3.2.2).

Other examples were noticed from Piddington. After calculating the percentages of each sample; it was noticed in S.7A1 that there was some difference in its structure. The sample had quartzite, oolitic ironstone, iron slag, hæmatite and most interestingly a piece of Egyptian blue, which did not dissolve in HCl. Sample S.10 L2, contained fossil shells and sand stone, in addition to a piece of mortar with mica. Both S.7A1 and S.10 L2 had unique structures which I did not see in other samples. The colour of the mortar was pink in S.7A1 and yellow in S 10 L2. Hydrochloric acid was used to dissolve both samples S.7A1 and S.10 L2 for a second time, as both of these samples did not fully dissolve, probably because the time allowed for dissolution was too short. Appendix 13.4, shows the residue weight and the differences after using hydrochloric acid for a second time. The correct result is given in Appendix 13.4.

2.4-Analysis of Water Soluble Salts

The analysis of water-soluble salts is one of the essential aspects of the examination of works of art, particularly wall painting, as it is known that wall paintings usually contain salts from different sources. It is possible to obtain primary information on the presence of salts using micro-chemical tests, which can be confirmed by physical methods. Analyses were carried out on most of the samples (see Appendix 7), collected from Saqqara, Dahshur, Luxor, Alexandria and Piddington.

As the Alexandrian tombs were covered with salts, it was necessary to define the different crystalline phases present. That led to the analysis of the water-soluble salts in the plaster/mortar. The salts were mostly from zone B, where they were heavily accumulated. Some were also from zone A, according to Arnold's theory that the most important salts were usually in zone B, as stated in Chapter 3 (see page 31).

The amount of the samples was large enough to carry out such analyses. The method of analysis used on the water-soluble salts was the same as mentioned in Chapter 4 (2.2.3.1). After analysing for water-soluble salts (sulphates, chlorides, nitrites and nitrates), the residues were dried at 60 °C and weighed to a constant mass. The weight loss was calculated from the differences between the weights of samples before and after the water extraction procedure (Appendix 14, which include the original weights, the percentages of the residue as well as the soluble components). Then analysis was carried and samples (residue) were hydrated again with the same solution (about 1-2 ml) to analyse for carbonates using concentrated hydrochloric acid. Samples were washed until pH 7 was achieved and dried. The weight loss was calculated from the differences between the samples after analysing for water-soluble salts and the residue after analysing for carbonates. The rest of the solution was filtered into polyethylene bottles for determination of metal ions using ICP. The weight loss of the samples after the water extraction gave an indication of the total amount of salts present (see Appendix 17). The results of analyses of the water-soluble salts are given in Tables 3 A-D.

The results of analyses indicated that the most significant salts were chlorides, nitrates and sulphates, and less commonly nitrites. The result of the analyses for carbonates show variation due to different samples; carbonates were notably present in the clay plaster/mortar samples, which were highly carbonated. The mud plasters/mortars samples did not show very much carbonate, as they contained only small amounts of lime/chalk or limestone.

Analysis of water-soluble salts was carried out on a few samples from Piddington, and no chlorides were detected using the silver nitrate test. Sample S.10 had no sulphate whereas, sample S.4, gave a white precipitate when barium chloride was added to the acid solution, confirming a sulphate. This indicated that it might be from a different wall or that the ground contained sulphates.

The following tables give the results arranged in order of dates from the earliest to latest.

Table 3 A, The Results of Analysis of Water-Soluble Salts from Saqqara

Samples name	Sulphates (SO_4)	Chlorides (Cl)	Nitrites (NO_2)	Nitrates (NO_3)	Carbonates (CO_3)
First dynasty (E E)					
S.104/M	++	++	-	+	-
S.100/M	-	++	-	-	-
S.101/M	-	+-	++	-	+-
S.102/M	-	++	-	-	+-
S.103/M	-	+-	-	-	+
First Dynasty (N E)					
S.143/1/C	++	++	-	+-	++
Dahshur 4th Dynasty (N E)					
S.54/L	+-	++	-	-	-
S.50/M	-	+-	-	+-	-
S.51/M	-	+-	-	+	-
S.55/1/M	-	+-	-	+	++
S.56/M	+-	+++	-	+-	-
S.53/C	+-	++	-	+-	++
S.55/2/C	-	+++	-	++	-
S.55/3/G	+	+-	-	-	-
S.57/C	-	+	-	+	+
S.57/3/G	+-	±	-	+-	-
Ka-Nefr Tomb 5th Dynasty (N E)					
S.9/M	+-	+++	+-	-	-
S.145/M	+-	+-	+-	-	-
S.49/M	+-	++	-	+	+
S.134/3/M	+-	±	+-	-	++
S.144/M	+-	++	-	-	+
S.10/C	+-	+++	-	-	++
S.46/C	+-	++	-	-	++
S.47/C	+	++	+	-	-
S.48/C	+-	++	-	+	++
S.146/1/C	+-	++	++	-	++
Idy Tomb 6th Dynasty (N E)					
S.139/M	-	++	+	-	+-
S.137/M	-	++	-	+	+-
S.138/M	+-	++	++	-	-
S.140/C	++	++	-	+	+-

– = Absent + – = Traces ± = Concentration with slight traces
+ = Present ++ = Notably present
+ ++ = High concentration "more than standard"

C = Clay plaster *G = Gypsum* *L = Limestone* *M = Mud plaster*

Table 3 B, The Results of Analysis of Water-Soluble Salts from Luxor

Sample name	Sulphates (SO_4)	Chlorides (Cl)	Nitrites (NO_2)	Nitrates (NO_3)	Carbonates (CO_3)
T. 56					
S.37/L	-	+	-	-	+++
S.105/L	+-	+	++	-	++
S.107/L	+-	+-	+-	-	+++
S.67/S	-	±	-	-	-
T. 55					
S.113/M	-	-	-	-	++
S.111/C	+	+++	+	-	+
S.110/1/S	+	±	-	-	++
S.112/S	++	±	+-	+-	++
T. 192					
S.27/3/L	+-	+-	-	+-	++
S.124/MF	++	++	-	+	++
S.129/M	+-	+++	-	++	+
S.131/M	+-	++	-	++	+-
S.28/P	++	+	-	++	+
S.29/P	-	+-	-	+-	++
S.31/P	++	+-	+	-	++
S.33/P	-	-	-	+	++
S.33/2/P	+	++	-	+	++
T. 255					
S.83/1/S	-	±	+-	-	+++
S.122/M	+-	+	+	-	+
S.121/C	+	+-	-	-	++
S.120/F	++	+	++	-	+++
T. 13					
S.118/L	-	+-	-	+-	++
S.115/M	+-	++	-	-	+
S.116/C	++	±	+	-	++
S.119/3C	+-	+-	-	-	+++
S.119/P	+-	+-	-	-	-

– = Absent   + – = Traces   ± = Concentration with slight traces
+ = Present   ++ = Notably present
+ ++ = High concentration "more than standard"

C = Clay plaster   *F = Finishing plaster*   *G = Gypsum*
L = Limestone   *M = Mud plaster*   *MF = Mortar filler*
P = Plaster (from fragment)   *S = Salt*

Table 3 C, The Results of Analysis of Water-Soluble Salts from Luxor (continued)

Samples name	Sulphates (SO_4)	Chlorides (Cl)	Nitrites (NO_2)	Nitrates (NO_3)	Carbonates (CO_3)
			T. 14		
S.RT/L	+-	+-	-	+-	+++
S.61/L	+-	+-	-	+	++
S.3/C	++	-	++	-	+++
S.86/C	+	+	-	-	++
S.88/S	+-	±	+	+-	+++
			T. 15		
S.4.C	+-	+-	-	-	++
			T. 11		
S.59/L	-	+-	-	+-	++
S.64/C	+-	+-	++	-	+++
			T. 359		
S.1/L	++	++	+	-	++
S.8/L	+	+++	-	++	++
S.6/M	+	++	-	+-	+++
S.40/M	++	+++	-	-	+++
			T. 414		
S.14/L	+-	++	-	+-	++
S.15/M	+-	+-	+++	?	+++
S.16/M	+	++	-	+-	++
S.18/M	+-	+-	++	?	+++
S.19/M	+-	+	+-	?	+++
S.20/M	+	++	+	?	+++

– = Absent　　+ – = Traces　　± = Concentration with slight traces

+ = Present　　++ = Notably present

+ ++ = High concentration "more than standard"　　?= undetermined

C = Clay plaster　　*L = Limestone*　　*M = Mud plaster*

MF = Mortar filler　　*P = Plaster (from fragment)*　　*S = Salt*

Table 3 D, The Results of Analysis of Water-Soluble Salts from the Alexandrian Sites

Sample name	Sulphates (SO_4)	Chlorides (Cl)	Nitrates (NO_3)	Carbonates (CO_3)
Mustafa Pasha T. 1				
S.3/I/L	-	+	-	
S.4/I/L	+-	-	-	+++
S.5/I/L	-	-	-	
Anfushi T. 1				
S.4/A1/LP	+	++	+	
S.4/A2/LP	+	++	+	
S.4/A3/LP	++	++	+-	+++
S.4/A4/LP	++	++	+	
S.4/A5/LP	++	++	+	
S.5/1/S	++	+-	++	
S.6/S	++	++	+	-
S.7/S	+	++	+-	(Salts)
S.8/S	+	++	+-	
Anfushi T. 2				
S.14/L	++	++	+	+++
S.1/2/P	++	++	++	+++
S.9/S	++	++	+	
S.10/1/S	+-	++	+	
S.10/2/S	+-	++	+-	-
S.10/3/S	+-	++	+	(Salts)
S.10/4/S	+-	++	+-	
S.10/5/S	+-	++	+	
S.12/2/S	++	-	+-	
Anfushi T. 2, 1998				
S.4/S	+	-	-	
S.5/S	++	++	+-	-
S.6/S	++	++	-	(Salts)
S.7/S	++	+	-	
Anfushi T. 3				
S.19/L	+	-	-	
S.22/L	+	+	+	+++
S.23/L	++	+	+	

– = Absent   + – = Traces   ± = Concentration with slight traces
+ = Present   ++ = Notably present
+ ++ = High concentration "more than standard"

L = Limestone   *LP = Lime plaster*   *P = Plaster*   *S = Salt*

As illustrated above, it is obvious that the most common salts present are; chlorides, sulphates and in some instances nitrites as well as nitrates. Chlorides, sulphates and nitrates are commonly found coming from different sources, such as impurities in the materials used, e.g. sand. However, nitrites are not found frequently in wall painting. The presence of nitrites suggested an old burial or recent formation, which may originate from the contamination source. It can be assumed that the presence of nitrates is a result of the oxidation of nitrites. Another possibility is atmospheric pollution from the industries in Helwan to the south east of Cairo and the limestone quarries close to the monument sites; the use of dynamite may help to precipitate chlorides and sulphate, as well as nitrates. Rain also occasionally occurs in Saqqara and Dahshur, which are in desert areas.

Teutonico said that the decomposition of nitrogenous organic materials produces nitrites. Where there is infiltration of sewage or the proximity of old burial sites, nitrites rapidly oxidized into nitrates (Teutonico 1984: 60).

The high presence of carbonates found was due to the stone and the clay plaster/mortar used. As mentioned before, salts could come from different sources. In addition, the wall painting contained salt as impurities in the material used. It is common to find such salts, but the sources may be different. Some samples gave a yellowish colour in testing for nitrates/nitrites instead of pink, as stated in Table 3 C.

Luxor is dry and there was a clear atmosphere without any contamination, so the main source of salts may be underground water and the stone itself. Lucas has analysed several samples of limestone, sandstone, granite and quartzite, which were occurring in Egypt, and those used in the building as well. He found that the salt was mainly sodium chloride, which caused disintegration of the building materials, particularly limestone, indicating that the salt was derived from both the ground and the stone itself (Lucas 1915: 5, 6). The salts have been identified in the tomb of Nefertari where the sodium chloride was of relatively high purity (Preusser 1987: 6). However, recent contamination may have affected the tombs and near by habitations that could have casused environmantel change in a varitey of ways over many centuries

The qualitative analysis of soluble salts provided information about the type of ions present in the samples, as shown in Table 3 D. The presence of sulphates and chlorides was notable, which led to the identification of the crystals to confirm the mineral content. Wall paintings usually contain soluble salts; sulphates, chlorides, nitrates as well as carbonates. Calcium carbonate is the main component of the both plaster/mortar and the limestone. Teutonico mentioned that there are some factors, such as agricultural practice and associated chemical use, which can affect a wall through capillary action. In addition to chlorides, seawater contains small amounts of sulphates, especially of magnesium. Sea spray can deposit sulphates on a surface as mentioned above. Another possibility is that the materials used in the preparation of mortars and plasters can contain small quantities of sulphates as impurities, which can dissolve in water as efflorescence, leading to the presence of sulphates. Nitrates could result from the decomposition of extraneous organic materials or from old burials. Agriculture and industrial pollution can also produce nitrates (Teutonico 1988: 60).

The different methods used to identify the salts were X-ray diffraction (X-RD), induction coupled plasma spectroscopy, and ICP for metals ions, in order to confirm the micro-chemical results.

2.5-Measurement of Calcium Carbonate ($CaCO_3$)

The method used for the analysis of calcite films, and the measurement of $CaCO_3$ was described by Jedrzejewska (1960) and was also recommended for the analysis of mortar, but here it was only used for the measurement of calcite films. Standard methods were used to identify the structure and composition of the mortars. Analyses were carried out on different samples; stone, mud brick, mud and clay plaster/mortar from the sites to determine the percentage of calcium carbonate (see appendix 7). The method used for the measurement of calcium carbonate is stated in Chapter 4 (2.2.3.2).

Analysis was carried out on one sample of limestone from Dahshur, as it was a fundamental part of the composition of the clay plaster/mortar, where it was used in the form of powdered limestone and clay, in addition to the samples of clay and mud plasters/mortars. The results show a very high percentage of calcium carbonate in the Dahshur samples, about 64. 8 % and 40.2 %, whilst the limestone sample from the site of the Idy tomb from Saqqara shows a higher percentage, about 75 %, as stated in Table 4 A. Evidence that the composition of limestone mixture contained very high percentages of calcium carbonate was indicated when dissolving samples in HCl as a vigorous effervescence.

Most of the plasters/mortars from T. 192 show very high percentages of calcium carbonate, as illustrated in Table 4 B. Samples of mud plaster/mortar from T. 414, showed an average percentage between 13-23%. Others varied even in the percentages of the calcium carbonate in the limestone. These showed slightly differences, however, as most of the stone showed high percentages of calcium carbonate due to the microfossils and shell fragments inside the stone. This led to the examination of some samples using SEM and EDS to identify the composition of the stone and compare the "*hib*" used in the mixture of the plasters/mortars.

The results show variation in the calcium carbonate content, depending upon whether the samples were stone or mud/clay plasters/mortars. The results of the measurement of calcium carbonates has shown variation between the percentages of calcium carbonates and the soluble components, as stated in Tables 4 A-C. Full detail of the samples (weights, acid soluble component and the percentages of calcium carbonate), are illustrated in Appendix 15. The following tables give the results in order of date from the earliest to the latest.

Table 4 A, The Results of the Measurement of Calcium Carbonate ($CaCO_3$) from Saqqara

Sample name	Calcium Carbonate %	Acid soluble %	Residue %
First Dynasty (N E)			
S.143/1/C	18.2	25.75	74.25
Dahshur 4th Dynasty			
S.52/L	64.8	97.95	2.05
S.54/L	40.2	51.03	48.97
S.50/M	4.4	11.45	88.54
S.55/1/M	20.9	22.03	77.97
S.56/M	17.6	11.45	88.54
S.53/C	23.1	22.2	77.8
S.57/C	24.2	28.89	71.11
Ka-Nefr Tomb 5th Dynasty (N E)			
S.145/M	11.6	18.78	81.22
S.10/C	35.2	47.46	52.54
S.46/C	38.6	56.43	43.57
S.48/C	55.7	62.89	37.11
S.146/C	47.7	54.86	45.14
Idy Tomb 6th Dynasty (N E)			
S. Stone	75	88.75	11.25
S.140/C	2.3	15.09	84.91

Table 4 B, The Results of the Measurement of Calcium Carbonate ($CaCO_3$) from Luxor

T. 56			
S.36/L	66.67	86.06	13.94
S.105/L	62.07	65.97	34.03
S.106/M	9.54	21.03	78.97
S.66/L	60.11	66.23	33.77
T. 55			
S.113/M	8.49	83.8	16.2
S.111/C	68.60	12.14	87.86
S.76/F	15.12	89.77	10.23
T. 192			
S.27/3/L	71.11	88.17	11.83
S.124/MF	38.89	43.09	56.91
S.27/1/C	40	47.63	52.37
S.27/2/C	64.44	82.31	17.69
S.29/C	86.66	87.43	12.57
S.32/2/P	82.22	86.17	13.83
S.33/2/G	82.22	79.66	20.34
S.125/G	82.22	76.31	23.69
S.126/C	68.89	73.69	26.31
S.31/C	15.38	55.06	44.94

C = Clay plaster *F = Finishing plaster* *G = Gypsum*
L = Limestone *M = Mud plaster* *MF = Mortar filler*

Table 4 C, The Results of the Measurement of Calcium Carbonate ($CaCO_3$) from Luxor (continued)

Sample name	Calcium Carbonate %	Acid Soluble %	Residue %
		T. 255	
S.83/1/L	92.31	95.32	4.68
S.121/C	65.93	75.17	24.83
S.120/F	30.77	90.89	9.11
		T. 13	
S.118/L	95.12	90.71	9.29
S.115/M	14.99	21.34	78.66
S.116/C	42.68	49.91	50.09
S.119/3/C	68.29	74.94	25.06
S.119/P	62.44	81.29	18.71
		T. 14	
S.RT/L	80.23	87.03	12.97
S.61/L	84.88	89.77	10.23
S.3/C	54.65	80.2	19.8
S.86/C	45.35	71.4	29.6
		T. 15	
S.4/C	46.32	57.14	42.86
		T. 11	
S.59/L	80.46	85.94	14.06
S.64/C	60.92	72.09	27.91
		T. 359	
S.1/L	10.84	20	80
S.8/L	12.65	21.54	78.46
S.6/M	8.43	19.49	80.51
S.40/M	13.52	29.74	70.26
		T. 414	
S.14/L	67.39	78.29	21.71
S.15/M	13.04	42.51	57.49
S.16/MF	68.69	83.57	16.43
S.18/M	21.96	26.26	73.74
S.19/M	22.83	27.49	72.51
S.20/M	18.70	26.46	73.54

C = Clay plaster *F = Finishing plaster* *L = Limestone*
M = Mud plaster *MF = Mortar filler* *P = Plaster (from fragment)*

As these are approximate results of the percentages of calcium carbonate contained in the samples, more analyses are required to determine the exact percentages. This would be desirable for re-construction of the ancient plasters, in order to match the formulation of the new plaster to that of the original.

2.6-Analysis of Calcite Films and Measurement of $CaCO_3$

The Piddington fragments were covered with different layers of calcite film, which was found to be very hard during cleaning. Samples were taken from fragments (S.1A, S.1B, S.2, S.4, S.7 and S.10), to identify the calcite film and its crystalline state, or what other sort of calcareous film was present and to measure the calcium carbonate content. It was interesting to look at these layers of calcite film, not only because of the thickness and hardness, but also for the colours and forms which covered most of the fragments (photograph 27). The samples were treated with HCl, resulting in the calcite film effervescing, indicating calcium carbonate or "calcite", in its crystalline state. Table 5 gives the percentages of the calcium carbonate content and, the residue of the percentages dissolving in the samples.

Table 5, Calcium Carbonate Content of Plasters from Piddington

Sample name	Calcium carbonate %	Dissolving samples %	Residue %
S.1 A	91.24	94.91	5.9
S.1 B	87.9	11.2	11.2
S.2 B	83.71	1.2	1.2
S.4	85	6.23	6.23
S.7	85.18	12	12
S.10	70.62	14.93	14.93

2.7-The Results of the Chemical Analysis of the Pigments from Piddington
Micro-chemical tests (spot tests) were carried out on pigments collected from different fragments (Appendix 7.9), these were; grey, three types of pink, yellow, red ochre, dark red, purple, and green. Dilute hydrochloric acid was used to determine the composition of the pigments. All the colours dissolved in HCl, indicating *buon fresco* with the use of lime as a medium. The solution did not turn different colours in HCl, which would have suggested other elements in their composition. The results are given in Table 6.

Table 6, The Identification of the Pigments from Piddington

Fragment name	Colour	Composition
S.10	Green	Not malachite, but green earth
S.15	Grey	A mixture of Egyptian blue and carbon black some traces of Egyptian blue did not dissolve in HCl
S.8	Pink 2	Brick dust (more silica)
S.8	Grey	(Different tone) "lime" and carbon
S.8	Purple	Red ochre and brown
S.8	Yellow	Yellow ochre, with calcite
S.4	Dark red	Natural red ochre
S.6	Pink 1 & 3	The same mixture as S.8, being calcite with red ochre

It is necessary to know the composition of the pigments and their mixtures in the event of the re-painting of missing areas. The results of chemical analysis of the pigments from Piddington indicates the use of these colours in Roman times, as stated in Chapter 2 (see 4.9), and has been identified (Béarat *et al.* 1997). However, it is hoped that the results of Piddington (Abd El Salam 2000) will be published by the Upper Nene Archeological Society in due course.

2.8-Summary
Chemical analysis helped not only to identify the composition of the samples, but also gave an approximation of the materials used. Qualitative and quantitative analyses could be carried out for the re-construction of plasters/mortars and therefore, it is necessary to determine the sediment types used as well as their percentages. Analysis of the soluble salts has identified the most important salts present, and given an initial idea about the salts, before carrying out more detailed physical methods to confirm the results indicated by the use of micro-chemical tests (spot tests).

3-Physical Methods: Results and Discussion

The methods used in the analysis of samples of plasters and mortars, as well as other materials, were X-ray diffraction for both salts and pigments, and X-ray powder diffraction for the mineralogical compounds, ICP for the metal ions, SEM & EDS for investigation of the surface structures and their minerals and PLM to identify the internal structures (Appendix 7). These methods were almost non-destructive and were used for the examination, analysis and the identification of wall paintings. The samples used were very small.

3.1-X-Ray Diffraction and X-Ray Powder Diffraction (X-RD & X-RPD)

X-ray diffraction was used to identify the crystalline compounds present of both salts and pigments, in particular from the Alexandrian sites and to confirm the results of the micro-chemical tests. X-ray powder diffraction was used where necessary to obtain more information on the mineralogical content.

3.1.1-Pigment Identification

X-ray diffraction was carried out on different samples from the Alexandrian tombs, both Mustafa Pasha and Anfushi; to identify their mineralogical structures (see appendices 7.6-8). The equipment and methods employed were described in Chapter 4 (2.3.1). Spot tests using acetic acid were carried out on sample S.5 (*intonaco* layer, photograph 20 B), to wash off the calcite film. X-ray diffraction was used to identify the pigment. The result indicated red ochre (iron oxide, Fe_2O_3). In addition, micro-chemical test was also carried out on the green pigment indicating green earth (glauconite/celadonite). The experimental results of X-ray diffraction are summarized in Table 7.

The analysis of the pigments has shown that the peaks of yellow ochre ($Fe_2O_3.2H_2O$) as well as red ochre (iron oxide Fe_2O_3) from Mustafa Pasha are different from those of Anfushi, suggesting the presence of clay. That led to the comparative analysis for kaolinite, which confirmed its presence, resulting from impurities in the pigments used or present in the natural red oxides used in Mustafa Pasha. Egyptian blue was identified on samples from the Mustafa Pasha tombs and investigated in some samples from Anfushi tomb 2. It was difficult in some instances to measure the weak lines present, due to the small amount of the crystal in the sample. All samples showed the presence of calcium carbonate $CaCO_3$ (calcite) from the plaster. However, sample S.13/2 showed the presence of gypsum on the peak line, which could be from the transformation of calcium carbonate into calcium sulphate. Iron oxide was identified as a black pigment (sample S.21 A T. 3).

Identification of pigments from the Hellenistic walls in Acre showed that the pigments used in this periods were more or less the same as found in Alexandrian tombs, with some consideration of the geological sources where the materials were available (Segal & Porat 1997). As a result of the analyses, it could be seen that the iron oxides pigments were widely used and have been identified in previous studies. These pigments were traditionally used in ancient as well as classical times, as stated above. Egyptian blue, a synthetic pigment, was widely used. It can vary according to the composition of the pigment. It has also been identified (Profi *et al.* 1974, 1977, Filippakis *et al.* 1979, Calamiotou *et al.* 1983, Corbeil *et al.* 1996, Járó 1997 and Rozenberg 1997). However, the differences found in the pigments may relate to the impurities as well as the geological sources of the pigments used, whether natural or artificial (burnt). Burnt yellow was commonly used, as mentioned previously in Chapter 2. However, distinguishing between natural or artificial is difficult by X-ray diffraction. X-ray microdiffractometry can be used to distinguish, as well as classify, very small samples and can differentiate, for example, burnt yellow earth from naturally occurring red earth (Helwig 1997). The analysis of pigments established the presence of binding media. A case study (Kakoulli 1998: Pers. Comm.) has shown that organic materials, such as animal glue and gum Arabic, were used in Greek times and might have been used in Alexandria.

X-Ray diffraction was carried out on some pigments found on samples from Piddington, to confirm the chemical results, but unfortunately this did not give accurate results of what minerals the pigments included. This was because most of the pigments contained calcite, except purple, which contained traces of hæmatite and magnetite, and has been identified; iron and silicates (Bugini & Folli 1997: 125). This indicated that the *fresco*s are likely to be true *frescos*, in which the pigments were placed directly onto the wet support, rather than *fresco secco*, as no organic binding medium was detected in the samples (see Chapter 2, 4.8.2 & 4.8.3). Chemical analysis was therefore used to identify the composition of the pigments in *fresco* rather than X-RD as stated above. It can be concluded that both chemical and physical analyses are required in the identification of the pigment

Table 7, Analytical Data on Pigments and Other Minerals from the Alexandrian Sites

Sample Name	Colour	X-ray diffraction X-RD	Identification	JCPDS File no
		Mustafa Pasha T.1		
S.5	Yellow	yellow ochre, kaolinite calcite	hæmatite	13-534
S.6	Red	red ochre (Fe_2O_3), hæmatite		
S.7	Yellow	yellow ochre, kaolinite calcite	Egyptian blue, calcite	2-512
S.8/1	Blue	Egyptian blue ($CaCuSi_4O_{10}$), calcite	hæmatite, calcite	13-534 6-0221
S.8/2	Red	red ochre (Fe_2O_3), hæmatite, kaolinite, calcite		
		Mustafa Pasha T.2		
S.11/1	Green	unidentified, calcite		
S.11/2	Yellow	yellow ochre, kaolinite, calcite		
S.13/1	Yellow	yellow ochre, kaolinite, calcite		
S.13/2	Red	red ochre (Fe_2O_3), hæmatite, kaolinite, gypsum	hæmatite, gypsum	13-534 6-0221
		Anfushi T.1		
S.5/2	Yellow	calcite		
S.5/3	Brown/ Red	red ochre (Fe_2O_3) Hæmatite, calcite	hæmatite, calcite	13-534
		Anfushi T.2		
S.11	Black	Carbon, calcite	calcite	
S.13	Yellow	yellow ochre $Fe_2O_3.2H_2O$, calcite	yellow ochre, calcite	
S.15/2	Black	Carbon, calcite	calcite	
S.15/4	Red	red ochre (Fe_2O_3) hæmatite, calcite	hæmatite, calcite	13-534
S.16	Yellow	yellow ochre, $Fe_2O_3.2H_2O$, calcite	yellow ochre, calcite	
S.17/2	Reddish	red ochre (Fe_2O_3) hæmatite, calcite	hæmatite, calcite	13-534
		Anfushi T.3		
S.20	Yellow	Calcite	calcite	
S.21	Black	iron oxide (Fe_3O_4), magnetite, calcite	magnetite, calcite	19-629

3.1.2-Identification of Salts and Other Minerals

Analysis was carried out on a few samples using X-ray diffraction of single crystals to confirm the results obtained by spot tests. Two samples of finishing plaster and others from the inside of the mixture of the plaster/mortar were examined. The first was a piece of lime/chalk, the second a deposit of gypsum, which was noticed in several samples from Saqqara. The examination of salts collected from different tombs in Luxor (Appendices 5.4-7) showed different forms of crystals, as stated before. This not only led to the identification of the salts to determine the minerals contained, but also using SEM to identify the form of the crystals and confirm the results of X-ray diffraction (see SEM Photographs 47-49 and also Photographs 68-78 , Part II).

The identification of the finishing plaster, using spot tests confirmed by X-RD, showed the presence of gypsum, (S.146/2) as stated in Table 8. A piece of the lime or chalk found in the mixture of the plaster/mortar was confirmed as calcium carbonate ($CaCO_3$) (S.144). The identification of the salts of a deposit of gypsum was proven to be calcium sulphate ($CaSO_4$) (S.140). EDS identified the elements present, as sulphur (S), calcium (Ca) and silicon (Si).

Analysis of one sample (S.2 photograph 18 C) from Anfushi T. 1, showed a completely different structure, with two different layers of salt crystals and other powered minerals. X-Ray diffraction was used to identify the minerals before and after washing the sample. The result showed sodium chloride (NaCl) was the crystal layer, but with a few peaks on the sample after washing indicating another mineral. The powdery layer was calcite or calcareous mud and it may be that other minerals, which did not show their peaks on the sample after washing, could have dissolved. Another sample (S.12/2) showed different minerals of gypsum, silica and calcium hydroxide (slaked lime). The results of X-ray diffraction are given in Table 8.

In analysing the salts found, a standard of natural salt was used for comparison. Although the salts were collected from different areas and tombs, the main salt found was sodium chloride (NaCl) and was indicated on the samples, as in the standard, with other salts, e.g., S.110/1. This showed a different structure, as indicated before (see photograph 18 B) and salt found underneath the plaster was in the form of needle shapes. Analysis of powdered samples of 110/1, 110/2 and S.112 using ICP indicated that the main ions were Na, Ca, Mg and K, and Al in sample S.110/2 (see Appendix 17.3). Other salts from different areas in the Alexandrian tombs indicated sodium chloride (NaCl) and sodium manganese phosphate ($NaMnPO_4$) as found in sample S.6. Another sample S.3 was calcium sulphate ($CaSO_4$) and calcium carbonate ($CaCO_3$), which covered the surface. However the whitish layers that covered the surface, as shown on Photograph 18 A (S.1/1) and another sample from tomb 2 (S.2) at Alexandrian site of Anfushi, were analysed to obtain an accurate result, as micro-chemical tests had shown the presence of gypsum. X-ray diffraction confirmed the mineralogical compound. The results of the salt analysis of Luxor and Anfushi are illustrated in Table 9.

Table 8, Mineral Identifications

Sample name	X-Ray Diffraction X-RD	X-Ray Powder Diffraction X-RPD	Identification JCPDS/File no
	Saqqara		
S.144	calcium carbonate ($CaCO_3$)		42-27
S.146/2	calcium sulphate hydrate/gypsum ($CaSO_4.2H_2O$)		21-816
S.140	calcium sulphate hydrate/gypsum ($CaSO_4.2H_2O$)		21-816
	Anfushi T.1		
S.1/1	-	calcium sulphate hydrate ($CaSO_4$. $2H_2O$) silicon oxide/ moganite (Si O_2) magnesium sulphate hydrate/ hexahydrate(Mg $SO_4.6H_2O$)	33-0311 38-0311 24-0719
S.2/ C S.2/CL	sodium chloride (NaCl) (before washing) calcium carbonate ($CaCO_3$) calcite with other minerals (clay)		5-0628 5-586
S.2/DS S.2/M	sodium chloride (Na Cl) (after washing) calcite ($CaCO_3$)		5-0628 5-586
	Anfushi T.2		
S.12/2	calcium sulphate hydrate/gypsum ($CaSO_4.2H_2O$) calcium aluminum silicate (Ca Ai $_2$Si $2O_8$) calcium hydroxide/portlandite (Ca $(OH)_2$)		33-0311 20-0020C 4-07331

C= Crystal *CL= Cream layer (C & CL before washing)* *DS= Dry solution* *M= Mud/clay (DS & M after washing*

Table 9, The Identification of Salts

Sample name	X-Ray Diffraction X-RD	Identification JCPDS/File no
	Luxor	
S.67	sodium chloride (NaCl) (same as standard)	5-0628
S.110/1	sodium chloride (NaCl) sodium sulphate ($NaSO_4$) (see Photo. 12 B) calcium carbonate ($CaCO_3$)	5-0628 6-0226 5-586
S.110/2 S.112	sodium chloride (NaCl) with some peak unidentified in S.112	5-0628
	Anfushi T. 2	
S.3	calcium sulphate ($CaSO_4$) calcium carbonate ($CaCO_3$)	6-0226 0226
S.5 S.6 S.9	sodium chloride (NaCl) sodium magnesium phosphate ($Na/MnPO_4$)	5-0628 5-0620
S.7 S.10/1 S.10/2 S.10/3 S.10/4	sodium chloride (NaCl)	5-0628

As previously mentioned and stated in Chapter 3 (3.1.5.1), the source of salts was varied, from different environmental factors, by the chemical reaction of the materials, as well as the reaction of some factors under certain conditions.

In Alexandria, it can be concluded that the deterioration produced results from a reaction between materials. Underground water, humidity and polluting compounds were present in the water, as the site was a burial ground in Græco-Roman times. In addition, sea spray is a factor, as the sites of Alexandria are so close to the sea, especially at Anfushi. The conservation materials used to remove salts (bicarbonate/E.D.T.A), helped to reduce another efflorescence quickly (see Chapter 3, 3.3).

3.1.3-X-Ray Powder Diffraction

Different samples were analysed using X-ray powder diffraction to identify their minerals (Appendix 7). This gave details of the mineral composition and indicated the manufacturing technology of the plasters/mortars used. It was necessary to analyse different samples from different areas and periods to distinguish them and to compare them with the various geological sources, in order to find out to what extent that had affected the composition of the plasters/mortars used in the Old and New Kingdoms.

Analyses were carried out on different samples of stone, mud brick, mortar filler, mud, and clay plaster/mortar, in addition to some samples of the finishing plaster which covered the mud/clay plasters/mortars. These results were used to compare the composition of the layers of the plasters/mortars applied both as rendering and finishing layers, and to compare the similarity and differences of their mineral structures. The identification of the minerals was used to compare with those used in the Old Kingdom, in particular the finishing plasters. The results are given in order of limestone, mud brick, mortar filler, and mud plaster/mortar, then the clay and the finishing plasters (see Appendix 16).

The identification of a sample of **limestone** from Dahshur was carried out to determine the original composition used in the mixture of the clay plaster/mortar. The results show that the limestone consisted mainly of calcium carbonate ($CaCO_3$) with some gypsum ($CaSO_4$) and other minerals as stated in Appendix 16.1. The presence of gypsum, as a part of the composition of the limestone, was found in the mixtures of the clay plasters from Dahshur.

Examination of different samples of stone from the Luxor sites has shown variation in the colour and solidity. In addition, a white layer was found, in some cases, between the layers of the stone, as mentioned above. This led to the identification of the natural minerals contained in different samples of stone, as it was a fundamental part of the making of the plasters. The results obtained from the identification of the stone minerals showed different compounds with calcium carbonate ($CaCO_3$) being the most common found in all samples, as well as silica (SiO_2). Some samples (S.105, S.8 & S.14) contained gypsum ($CaSO_4$). Kaolinite was identified in samples S.8 & S.14, as well as hæmatite in samples S.105 & S.59. The other minerals are listed in Appendices 16.1.

The minerals found in the **mortar fillers** were slightly different from each other except in the presence of calcium carbonate ($CaCO_3$), silica (SiO_2), and calcium sulphate ($CaSO_4$). Sample S.16 was very similar to the stone (S.14) (see Appendix 16.3). Graph 8 represents the mineral compounds of the samples. The results are given in Appendix 16.4. Two samples of mud brick from Luxor showed the same minerals as stated in Appendix 16.5.

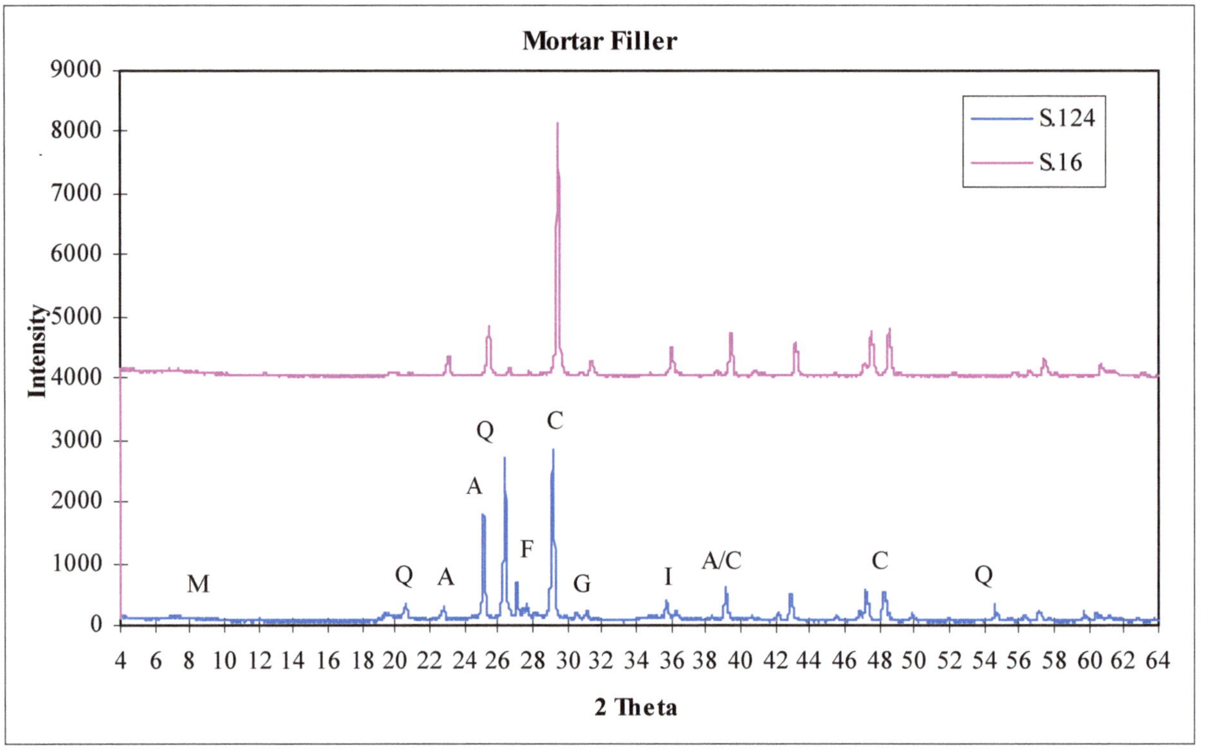

Graph 8, shows the more important mineralogical compounds in samples of the mortar filler.

A = Anhydrite	*C = Calcite*	*F = Feldspar*
I = Illite	*M = Muscovite*	*Q = Quartz*

The results obtained from the identification of **the mud plaster/mortar** from Saqqara showed variations in their mineral composition. It was noted that silica (SiO_2), montmorillonite, albite and muscovite were found in all samples, with some differences in their compounds formula whilst sample S.142/2L had a different composition as stated in Appendix 16.6. The mineralogical compounds in sample S.18 of mud plaster from Luxor T. 414, were very different to those from Saqqara. It was interesting to find hæmatite (Fe_2O_3), as illustrated in Appendix 16.7.

The identification of minerals in the **clay plasters/mortars** from Saqqara and Luxor showed different compositions. The differences were not only due to the manufacturing technology, but also to the sampled areas, as in samples S.46 and S.104. Calcium carbonate ($CaCO_3$) was present due to the use of powdered limestone in the mixture of the plaster/mortar, which, with silica (SiO_2), was found to be the main compound in the composition. Calcium sulphate hydrate ($CaSO_4.2H_2O$) and anhydrite ($CaSO_4$) were also noticed in some samples. Samples S.3 and S.4 show differences with the presence of dolomite in the first and hæmatite in the second. The results are given in Appendices 16.8-10. Graphs 9 and 10 show the distributions of the mineralogical compounds in clay plasters/mortars from both Saqqara and Luxor. Note the differences in graphs.

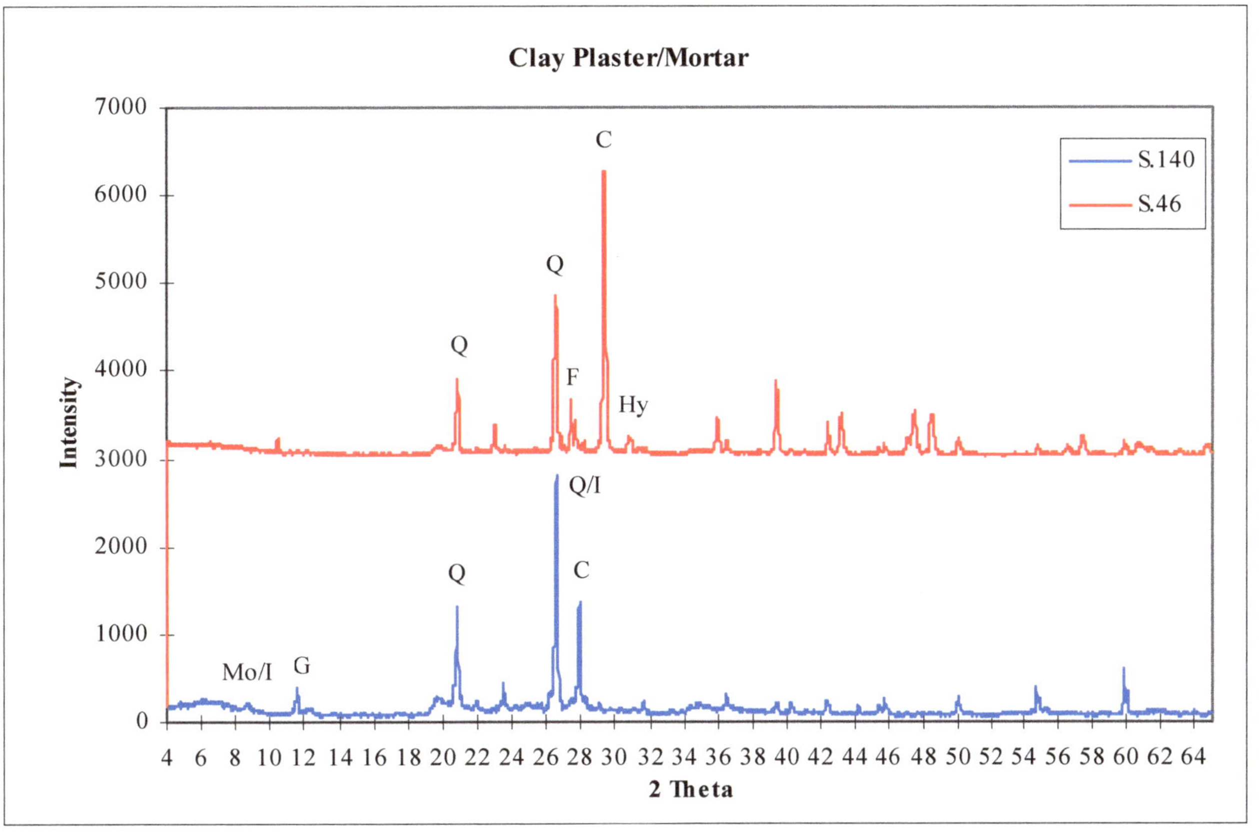

Graph 9 shows the more important mineralogical compounds of the clay plaster/mortars from Saqqara.

C = Calcite	*Hy = Hydromagnesite*	*F = Feldspar*
G = Gypsum	*Mo/I = Montmorillonite/Illite*	*Q/I = Quartz/ Illite*

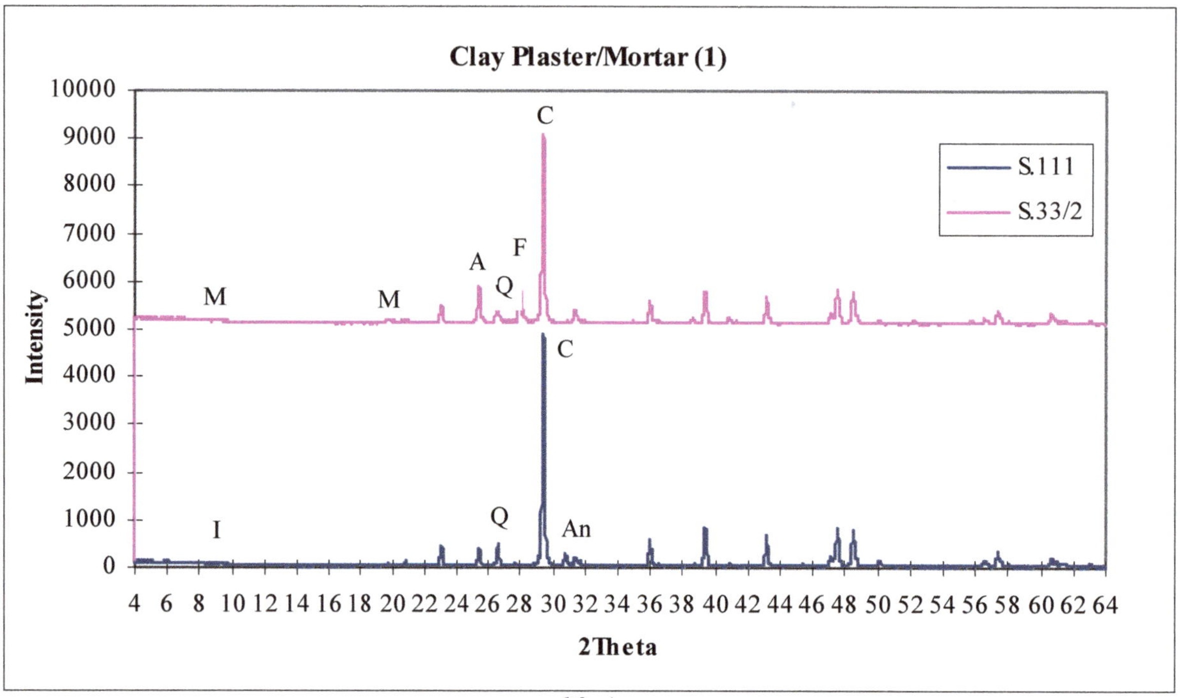

10 A

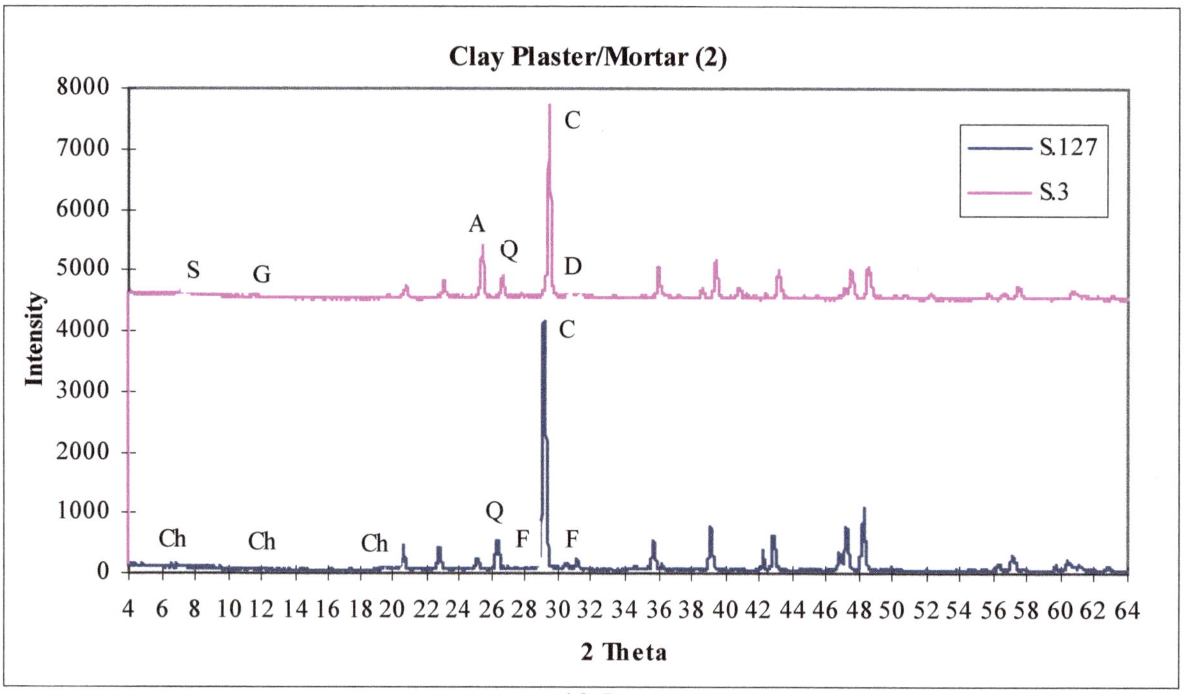

10 B

Graphs 10 A-B, show the more important distribution of the mineralogical of clay plasters/mortars from Luxor.

A = Anhydrite	*An = Anlorite*	*C = Calcite*
Ch = Chlorite	*D = Dolomite*	*F = Feldspar*
G = Gypsum	*I = Illite*	*M = Muscovite*
Q = Quartz	*S = Sepiolite*	

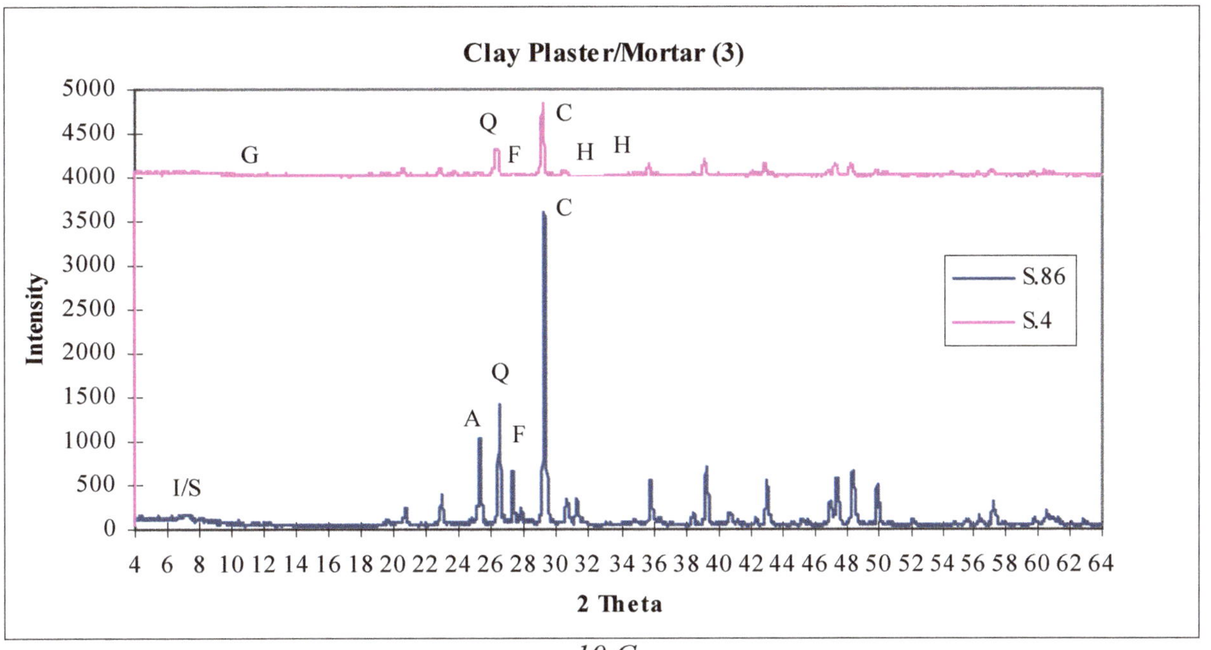

10 C

Graph 10 C, shows the more important mineralogical compounds of different samples of clay plaster/mortar from Luxor.

A = Anhydrite	*C = Calcite*	*F = Feldspar*	
G = Gypsum	*H = Hœmatite*	*Q = Quartz*	*I/S = Illite/Semite*

Several samples of **the finishing plaster** from Saqqara and Dahshur, which covered the mud/clay plaster/mortar varied due to the different sampled areas and periods (see Photographs 1 B (S.102/1), 2 (S/143/1) and 3 (S.55/3). The results obtained show differences between S.55/3 and S.57/3 although from the same sampled area. Sample S.102/1, dated to the first dynasty (Emery excavation), did not show gypsum, the mixture being mainly calcium carbonate ($CaCO_3$), with other minerals, as stated in Appendix 16.11-12. Samples, S.55/3 and S.57/3, contained gypsum ($CaSO_4.2H_2O$), and silica, but the amount of some other minerals were slightly different, although both samples came from the same area. Sample S.134/2 indicated the main component of the plaster layer to be gypsum ($CaSO_4.2H_2O$) and anhydrite ($CaSO_4$), with silica, which was more or less the same as sample S.55/3, the only exception being the presence of potassium sodium aluminum silicate.

The results of the identification of the minerals in samples from Luxor showed that gypsum ($CaCO_4.2H_2O$), anhydrite ($CaSO_4$), calcite ($CaCO_3$) and silica (SiO_2) were the main compounds in the mixture. Sample S.119/2 was different due to the presence of calcium magnesium manganese carbonate, kutonhorite or magnesian. The results of the identification of the composition of samples are reported in Appendix 16.13. Note the differences and/or the similarities between the compositions of the mineralogical compounds found in the mixtures of the plasters/mortars (see graphs 11 and 12).

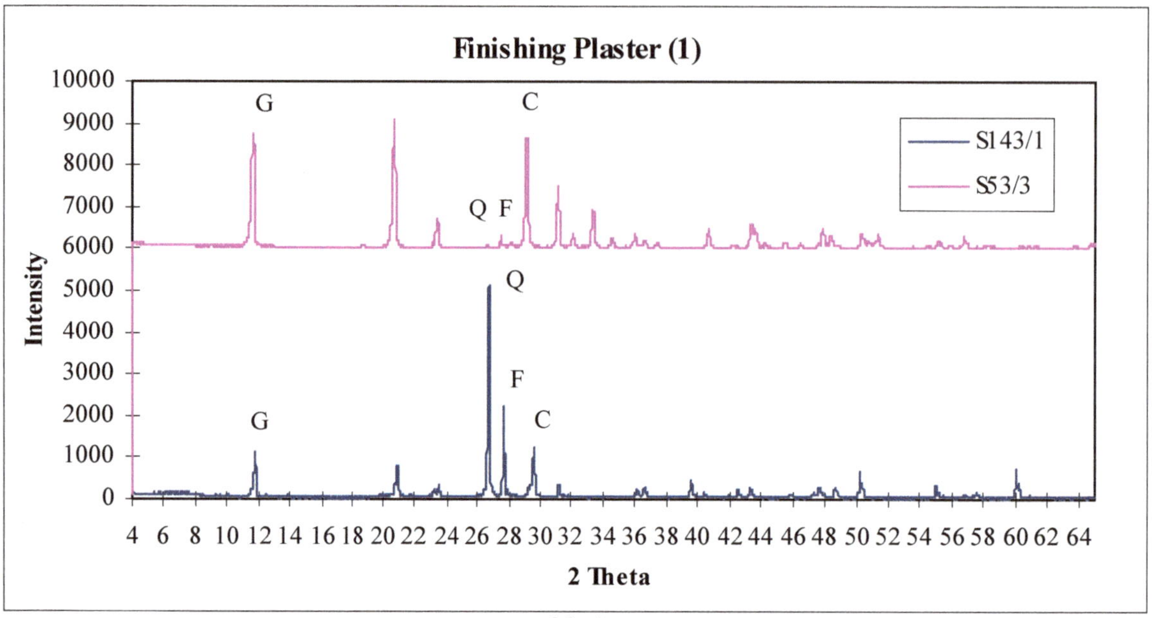

11 A

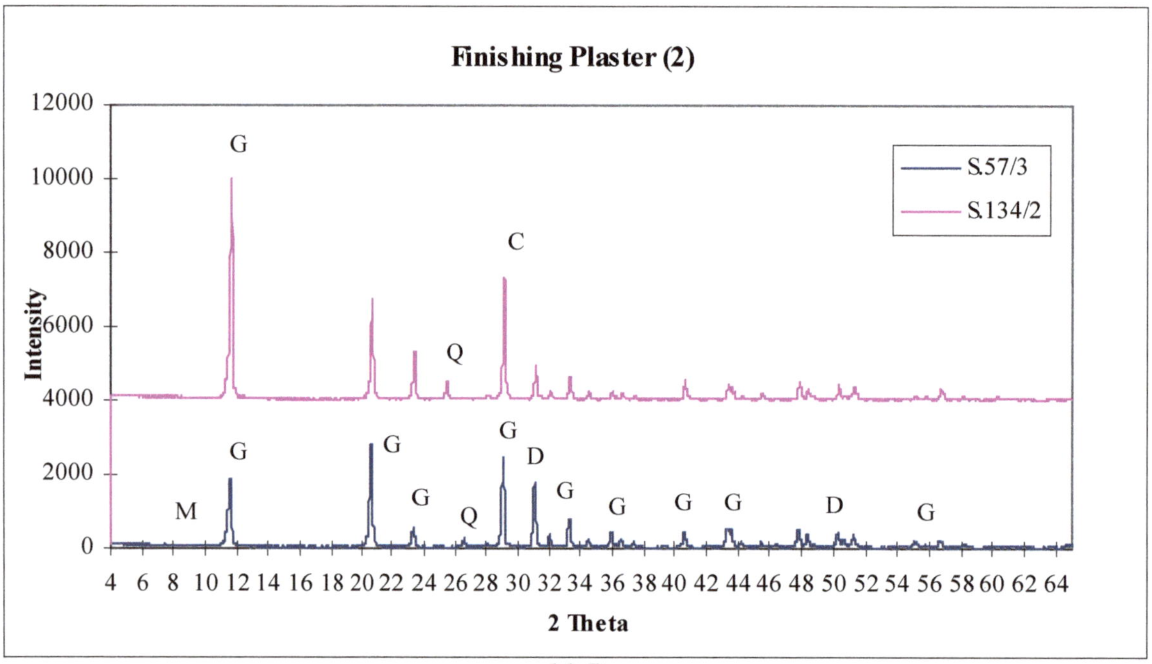

11 B

Graphs 11 A & B, indicate the more important mineralogical compounds found in the finishing plasters from Saqqara and Dahshur, note the differences between the samples.

C = Calcite | *D = Dolomite* | *F = Feldspar*
G = Gypsum | *M = Muscovite* | *Q = Quartz*

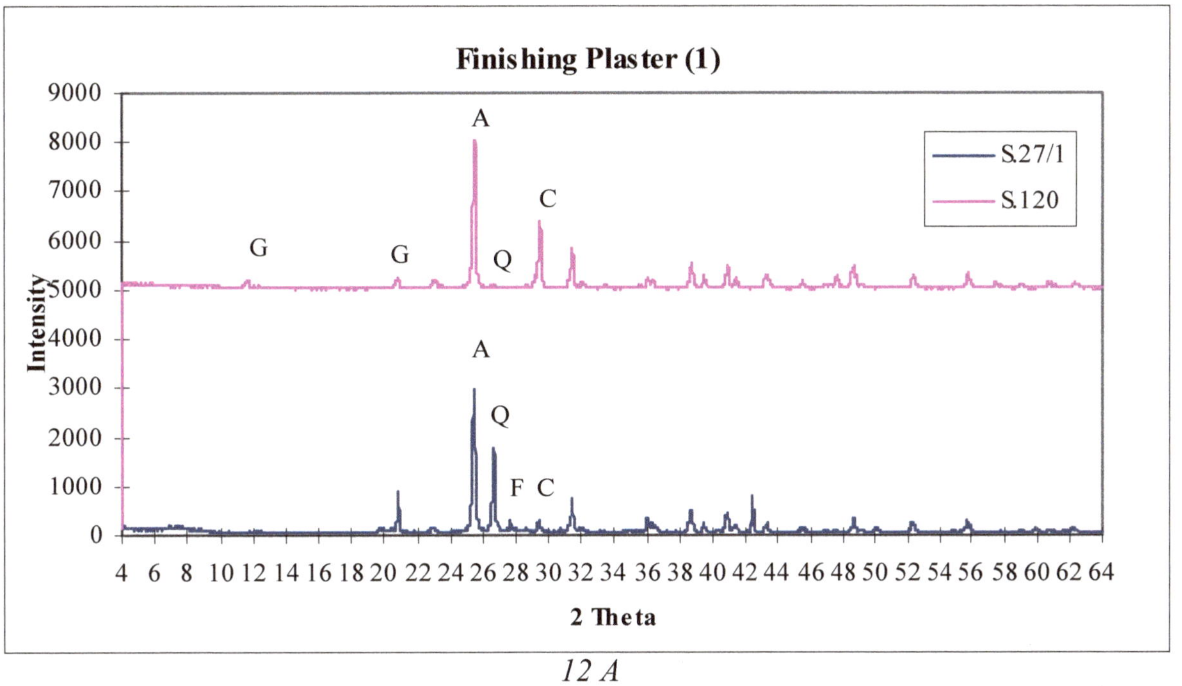

12 A

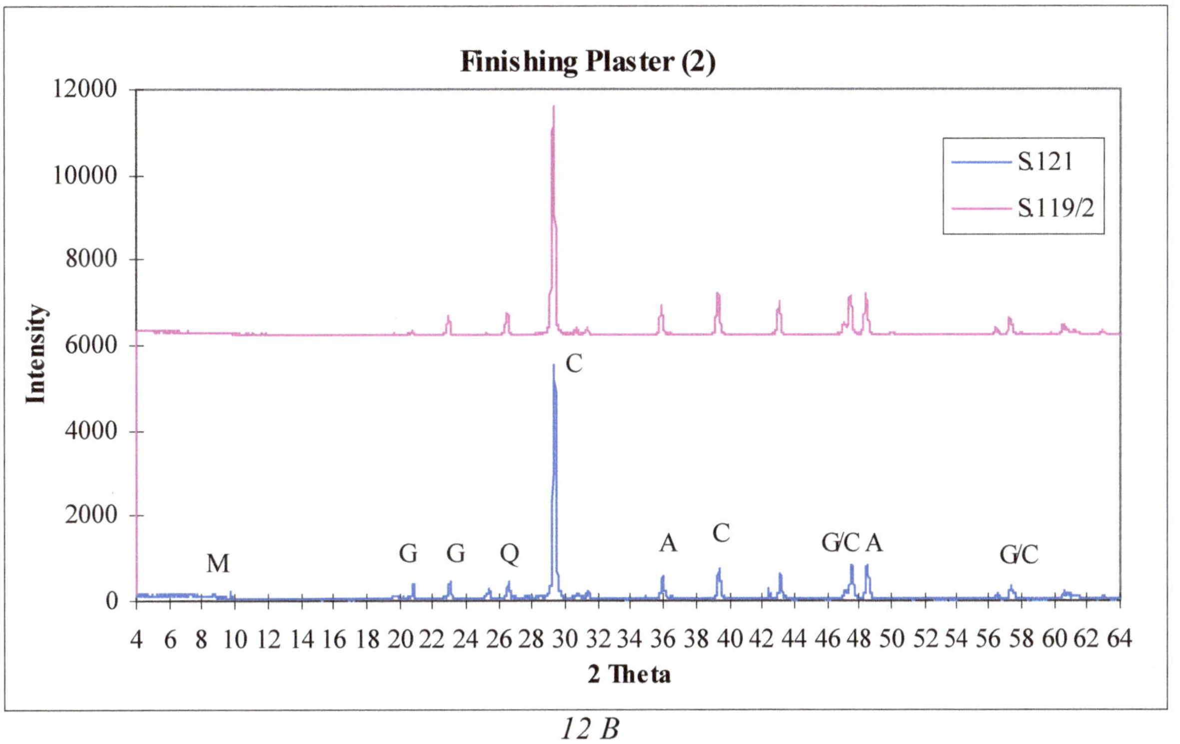

12 B

Graphs 12 A & B, showing the distributions of the more important mineralogical elements in the finishing plasters from Luxor.

A = Anhydrite	*C = Calcite*	*F = Feldspar*
G = Gypsum	*M = Muscovite*	*Q = Quartz*

X-ray diffraction of the samples showed the presence of montmorillonite, albite and kaolinite in addition to other minerals, as stated in Appendix 16. The minerals found in the finishing plaster showed a mixture of calcite and gypsum. There is a possibility that the mixture of the finishing plaster used was a natural formation of gypsum, as mentioned in Chapter 2, that the gypsum used was of different quality for different purpose (see page 14).

This was confirmed by PLM (see photographs 32 and 33). The mixture found may be impure gypsum, which contained calcite, quartz and other minerals.

From the results obtained from the identification of the minerals in the samples from the Old Kingdom, compared with those from Luxor, it can be concluded that there is a similarity in the mineral compositions, but with slight differences in the clay minerals, such as kaolinite, noticed in some samples in Luxor. The presence of hæmatite was due to the geological sources from which the materials were available for making the plasters/mortars. The presence of calcium magnesium carbonate/dolomite ($CaMg\ (CO_3)_2$) in sample S.57/3 (Appendix 16.11) indicated that the temperature used to burn limestone was not enough to obtain a complete de-carbonation of calcite and dolomite. It was noticed that sample S.46, was different to sample S.140. The results obtained show the presence of magnesium carbonate hydroxide hydrate/hydromagnesite ($Mg_4\ (OH)_2\ (CO_3)_3.3H_2O$) in sample S.46, as given in Appendix 16.8.

It is worth mentioning that the increase in the temperature, even if not enough to obtain a complete de-carbonation of calcite and dolomite, could have produced the initial formation of calcium magnesium (CaMg) oxides on the external surface of the stone fragments with the final effect of obtaining a kind of lime-like product once mixed with water. To distinguish lime or lime-like there are two main points: the first is the presence of glauconite in the plaster up to which temperature it remains unchanged; the second point, which has been mentioned, is the presence of magnesium in the binder. Magnesium lime is transformed into calcite and magnesite and hydromagnesite when reacting with carbon dioxide. Mg segregation phenomena should only be present in the case of the precipitation of calcium and magnesium (Ca and Mg) in the use of lime even if unintentional (Borrelli & Laurenzi-Tabasso 1996: 1450). Recent analysis to distinguish dolomitic lime mortar from other lime mortars has been carried out by Bläuer-Böhm and Jägers (1997).

Based on the above, in sample S.46 from Ka-Nefer tomb, the presence of magnesium carbonate hydroxide hydrate/hydromagnesite, it is possible to say that the plaster used was lime mortar, as this is the only sample showing such minerals. X-ray diffraction in all the samples qualitatively determined the mineralogical compounds present. The presence of gypsum ($CaSO_4.2H_2O$) was noted in several samples of limestone, clay plaster and finishing plasters. Also anhydrite ($CaSO_4$) was observed in different samples as well. However, it was most interesting to find both anhydrite and hydrate in clay, mortar joining and finishing plasters, as summarized in Table 9 and illustrated in Appendices 16.1, 3-4, and 16.7-13.

Table 10, The Presence of Hydrate and Anhydrite in Samples by X-RPD

Hydrate/gypsum ($CaSO_4.2H_2O$)	**Anhydrite ($CaSO_4$)**	**Hydrate/gypsum & Anhydrite ($CaSO_4.2H_2O$) & ($CaSO_4$)**
S.52/L	S.8/L	S.124/MJ
S.105/L	S.13/L	S.3/C
S.140/C	S.14/L	S.134/2/F
S.143/1/F	S.16/MJ	S.120/F
S.53/3/F	S.3/C	
S.57/3/F	S.18/M	
	S.33/C	
	S.86/C	
	S.27/1/F	
	S.121/F	
	S.119/2/F	

L = Limestone *M = Mud plaster/mortar* *MJ = Mortar joining*
C = Clay plaster *F = Finishing plaster*

Note in this table hydrate and/or anhydrite was found in limestone, mortar joining, mud and clay plaster, and finishing plaster.

The presence of hydrate ($CaSO_4.2H_2O$) and anhydrite ($CaSO_4$) may be due to two causes. The first was the over-burning of the gypsum rock and second to long-term environmental dehydration. We cannot understand what happened exactly and whether the changes happened during the manufacture of the plaster or after the plaster was applied to the walls. We must also consider that that could be natural formation due to the geological source, as found in some samples of limestone of both hydrate/anhydrite which were used in the manufacture of plasters/mortars (see Appendix 16.1 and 16.3).

The presence of anhydrite together with gypsum had been previously reported. Gypsum plaster ($CaSO_4.2H_2O$) dehydrates in two stages;

$$CaSO_4.2H_2O \rightarrow CaSO_4.\tfrac{1}{2}H_2O+1\tfrac{1}{2}H_2O$$

$$CaSO_4.\tfrac{1}{2}H_2O \rightarrow CaSO_4+ \tfrac{1}{2}H_2O$$

The kinetics of gypsum-anhydrite reaction is a combination of dehydration and a structural transformation of gypsum as monoclinic structure, then to hemihydrate in hexagonal form then to anhydrite in two forms, hexagonal and then orthorhombic.

$$(CaSO_4.2H_2O) \rightarrow (CaSO_4.\tfrac{1}{2}H_2O) \rightarrow CaSO_4 \rightarrow CaSO_4$$

The transformation of gypsum through hemihydrate to anhydrite occurs at surface temperature and pressures in the absence of a liquid phase. The common parameters conducive to the natural dehydration of gypsum are; moderately high temperature and low humidity (Smeaton and Burns 1988: 300, 301, 303).

Egyptian craftsmen had such a good mastery of gypsum technology as to be able to reach different temperatures as required, about 160 °C for rapid setting plaster and between 300-500 °C for a slower setting type. The high temperature corresponds to the formation of insoluble anhydrite with a more difficult hydration, and consequently a slower setting. The craftsmen probably used up to 200 °C, which is sufficient to de-hydrate gypsum as well as facilitate the desegregation of the bulk materials (Borrelli & Laurenzi-Tabasso 1996: 1449, 1450).

However, Torraca gave a different interpretation to the presence of anhydrite, based on the temperature and relative humidity measurement of the air and hydration process which can occur in gypsum and clay plaster, mentioning, for example, the tomb of Nefertari. The gypsum plaster lost cohesion and adhesive strength because of complete dehydration in the local environment (28-29 °C, and 30-40% relative humidity), assuming that the de-hydration of gypsum to anhydrite occurred due to the rather high temperature and low humidity inside the tomb for a long time (Torraca 1970: 171). (See also Smeaton & Burns 1988).

3.2-The Results of Induction Coupled Plasma Spectroscopy

Analyses were carried out on powdered as well as solution samples from the sites of Saqqara, Dahshur, Luxor, Alexandria and Piddington. The methods of analysis for metal ions are discussed in Chapter 4 (2.3.2). The results from the powdered samples show very high amounts of metal ions. It is possible to compare the result of both powdered and dissolved samples to see which ions have dissolved, as a result of extraction. The results are illustrated in Appendix 17.2 and 17.4-5. Aluminum (Al), calcium (Ca), sodium (Na), potassium (K) and in some instances magnesium (Mg), iron (Fe), and strontium (Sr) were the most important elements present in powdered samples. The solution results have indicated that the highest ions were mainly Ca and Na. The results of all analysis are given in Appendix 17. The most significant results were obtained from the Alexandrian sites, were I found high amounts of ions, in particular sodium (Na) and calcium (Ca), in addition to potassium (K) and in some instances magnesium (Mg), (see Appendices 17.10-12). Sodium is much more common than other ions, indicating the source of soluble salt as the soil.

This usually formed in the ground and contains sodium, potassium and calcium nitrate, as reported in Chapter 3 (see 3.1.5.1). The salt present in sea spray is mainly sodium chloride, which affected deposits in the Alexandrian sites.

The result of the analysis of the metal elements contained in the calcite film, collected through cleaning of wall painting fragments from Piddington and also after measuring of the calcium carbonate content, has shown that calcium (Ca), aluminum (Al) magnesium (Mg) and iron (Fe) were the highest elements on all the samples. Titanium (Ti) was much higher on sample S.10 (see Appendix 17.13) and this element was also higher in the Luxor samples, as shown in Appendix 17.4. Note the differences of the distributions of the metal ions of both the powdered and solution samples from Saqqara, Dahshur, Luxor, Alexandrian sites and Piddington in Graphs 13 and 19.

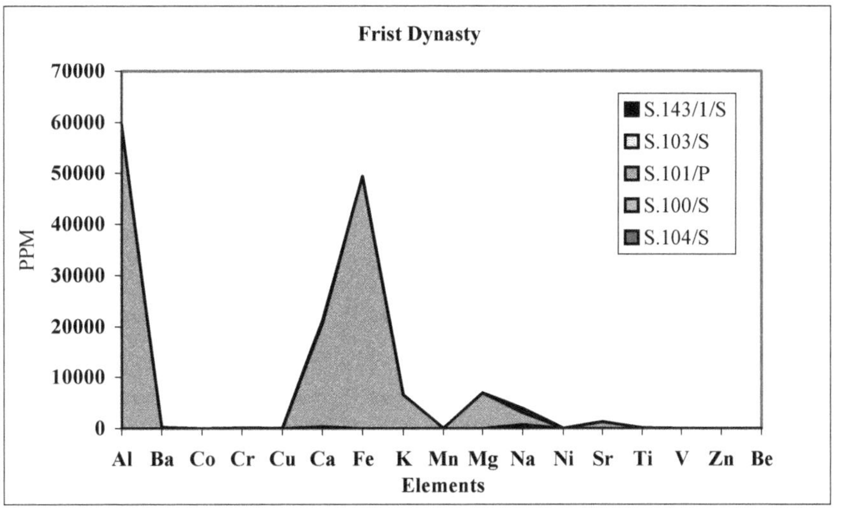

13 A

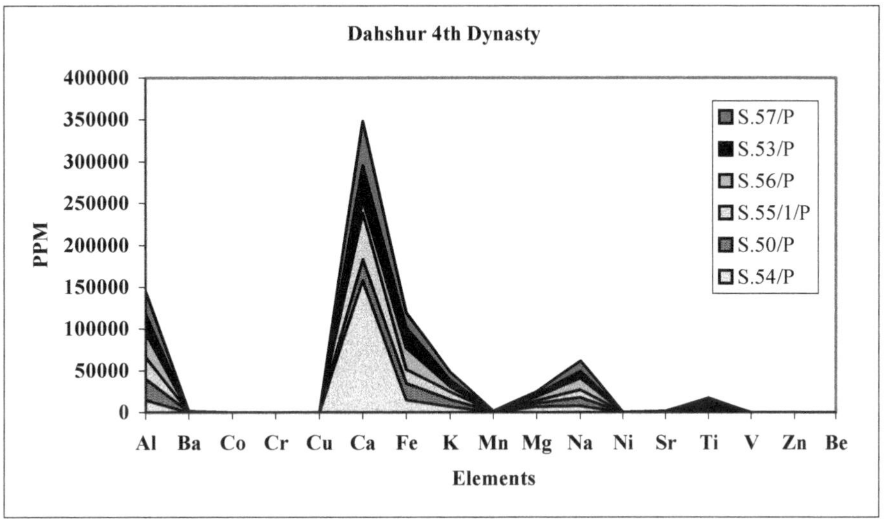

13 B

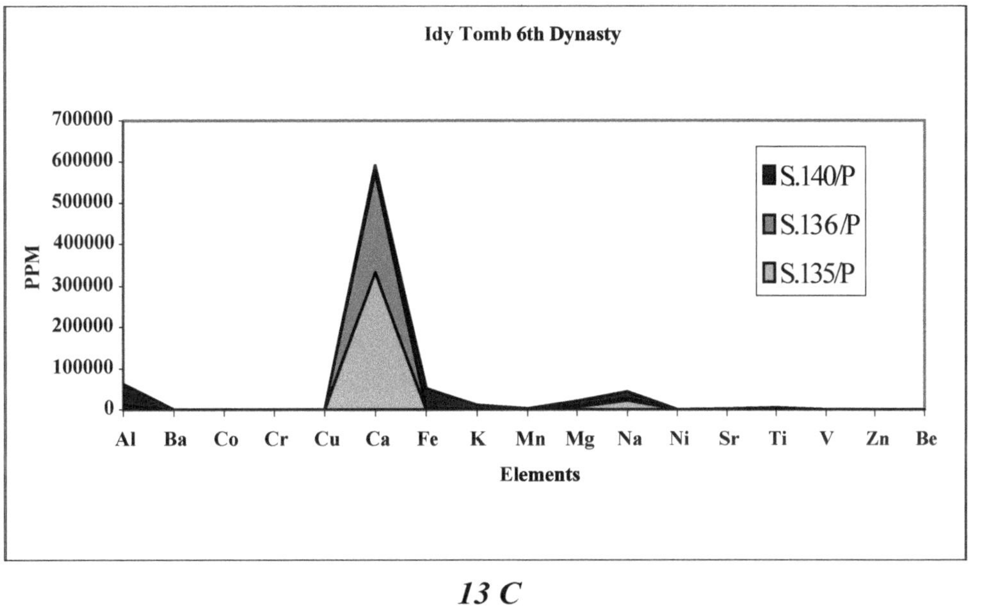

13 C

Graphs 13, A-C show the distribution of the metal ions found in samples from Dahshur and Saqqara. A and B, indicate Ca to be the highest element. On C the highest amounts are calcium (Ca) aluminum (Al) and iron (Fe).

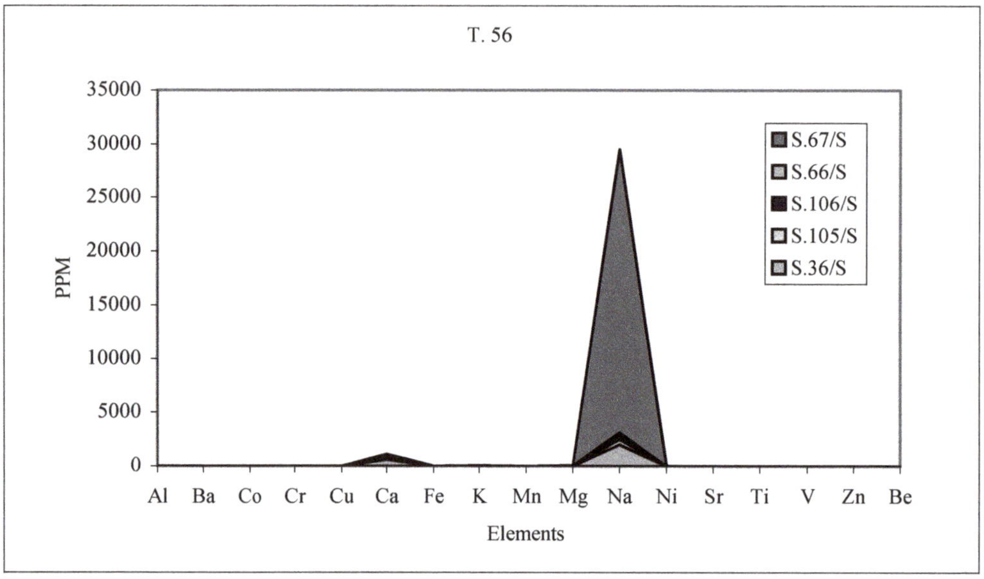

14 A

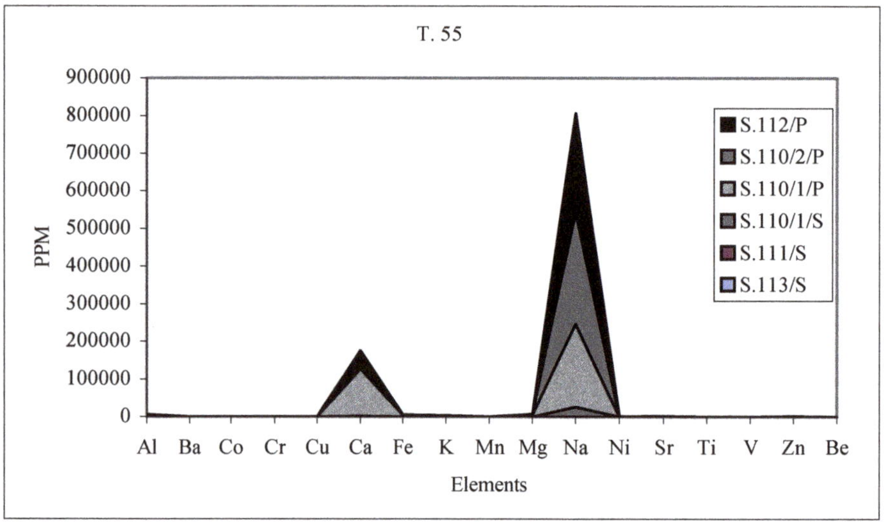

14 B

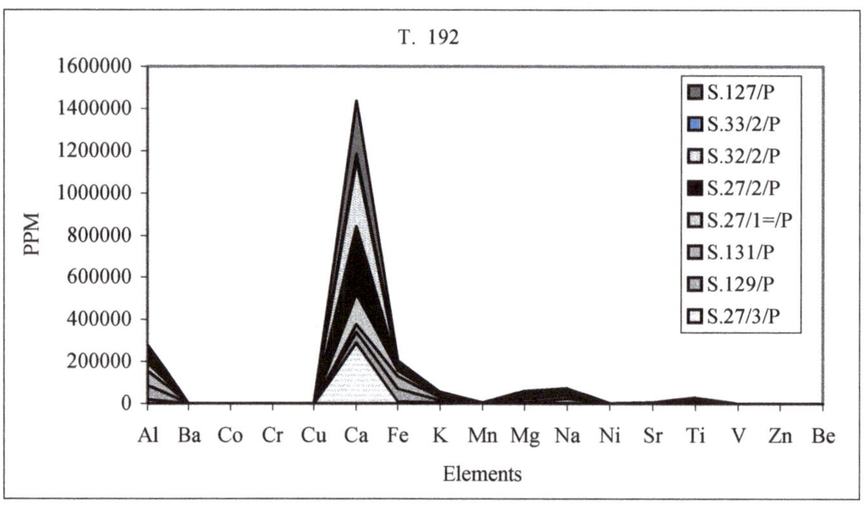

14 C

Graphs 14 A-C, show the metal ions found in different samples from Luxor. A shows sodium (Na). In graph B sodium is present. Graph C show calcium (Ca).

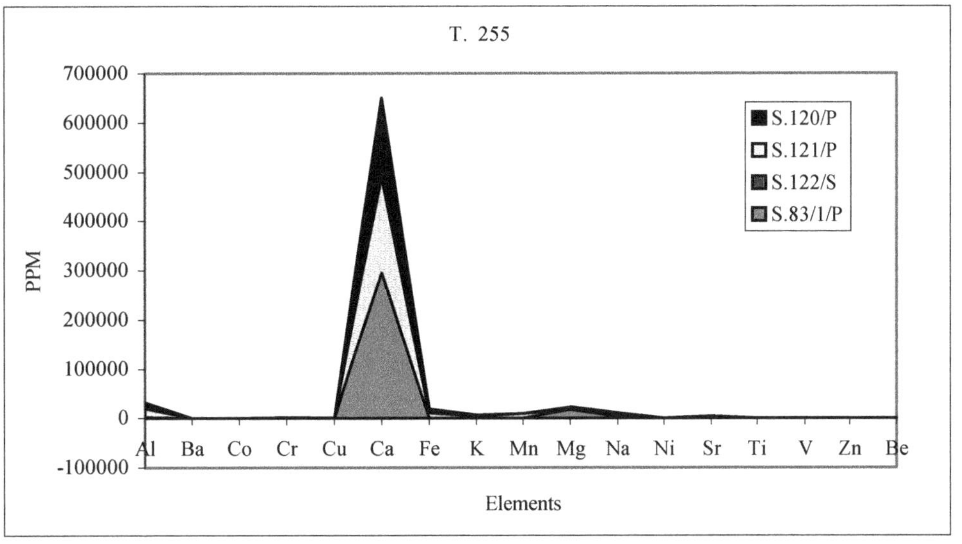

15 A

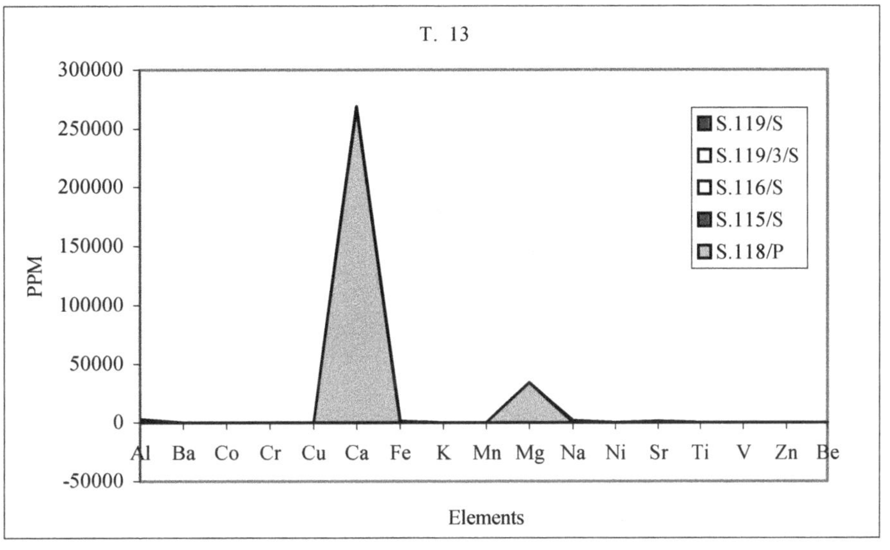

15 B

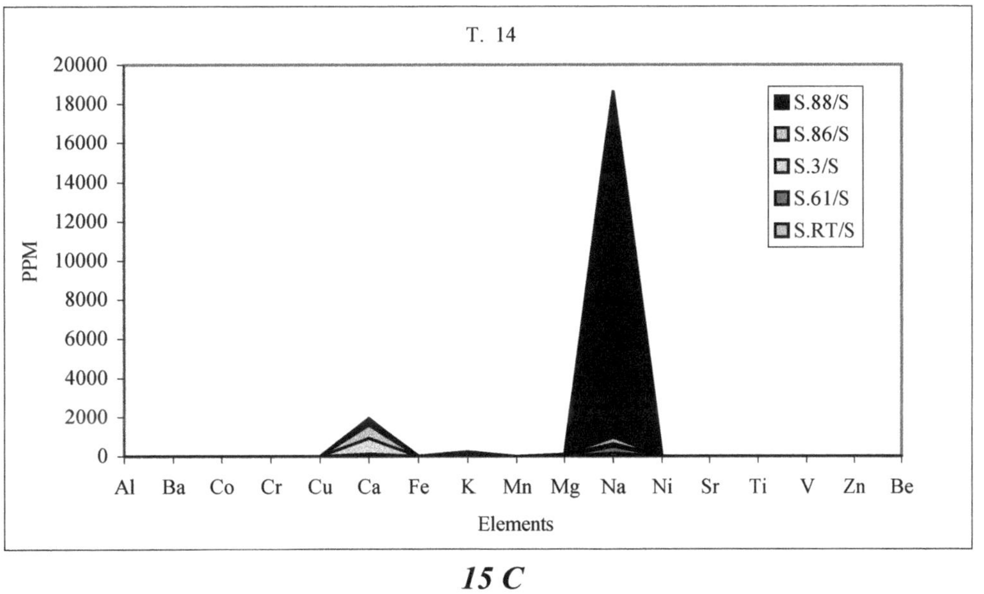

15 C

Graphs 15 A-C, show the metal ions found in different samples from different tombs in Luxor indicating the presence of calcium (Ca) in graph A & B. In graph C the highest element is Sodium.

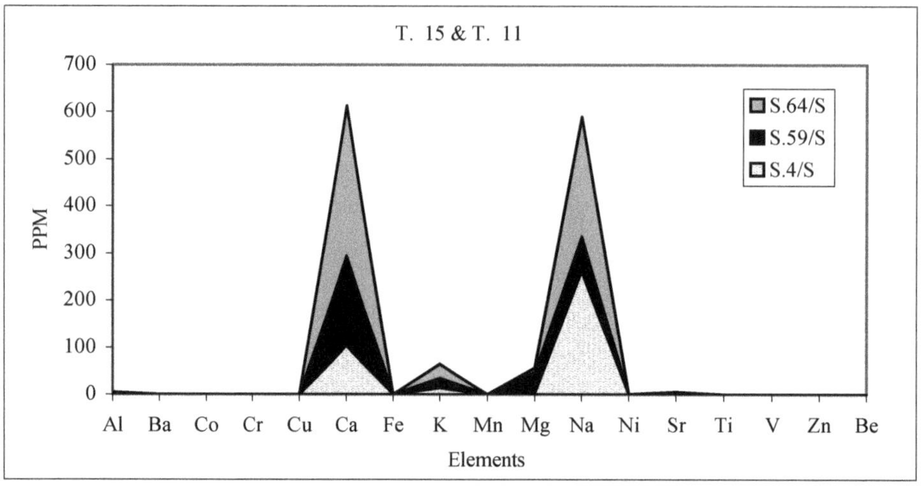

16 A

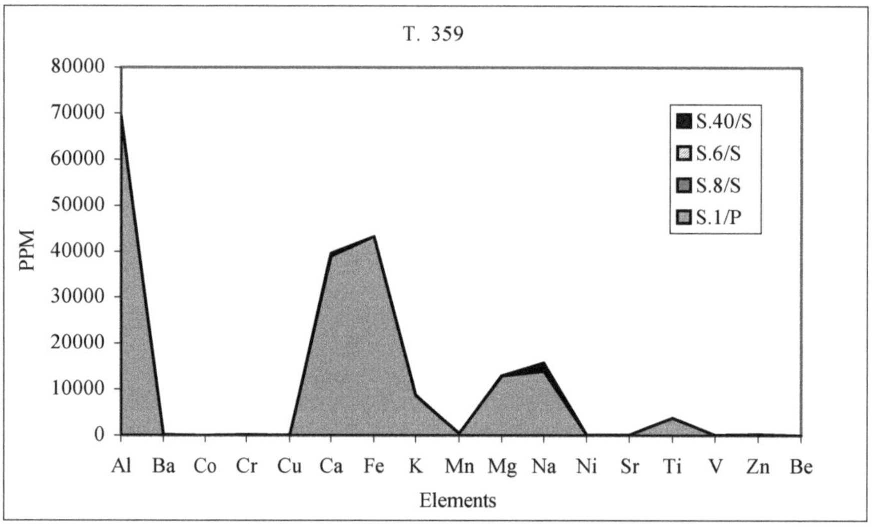

16 B

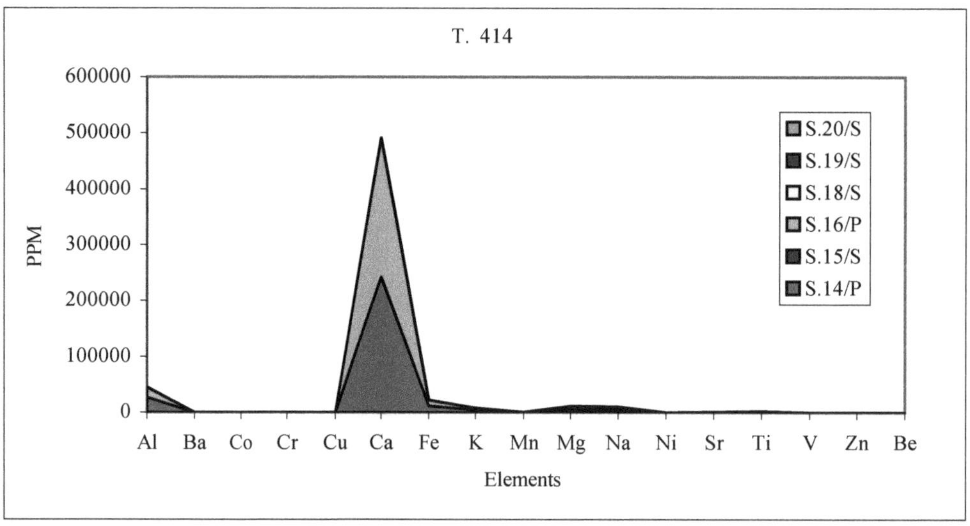

16 C

Graphs 16 A-C, Show the distribution of the metal ions found in samples from Luxor. On Graph A sodium (Na) & calcium (Ca) is the highest ions. Graph B shows iron (Fe) being much higher than calcium, in addition to other elements, such as potassium (K), magnesium (Mg), and sodium (Na). On C shows calcium (Ca).

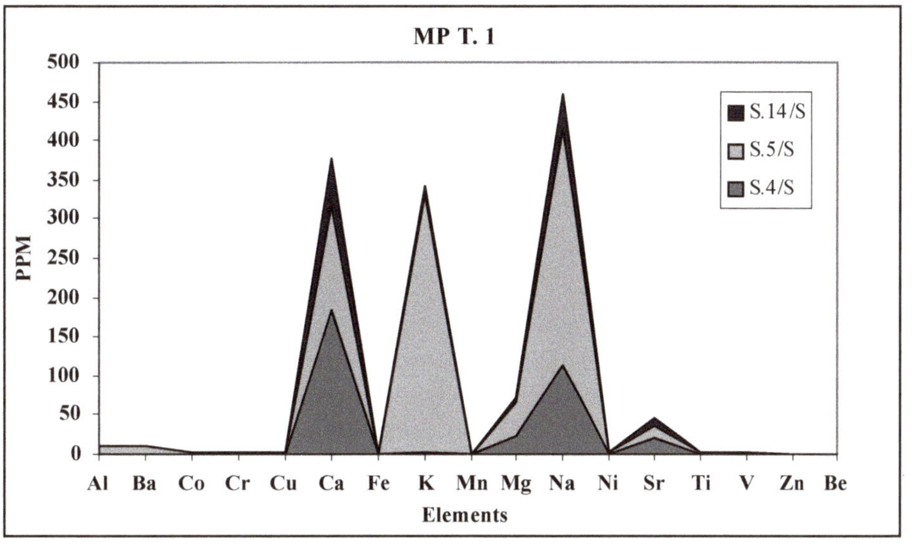

17 A

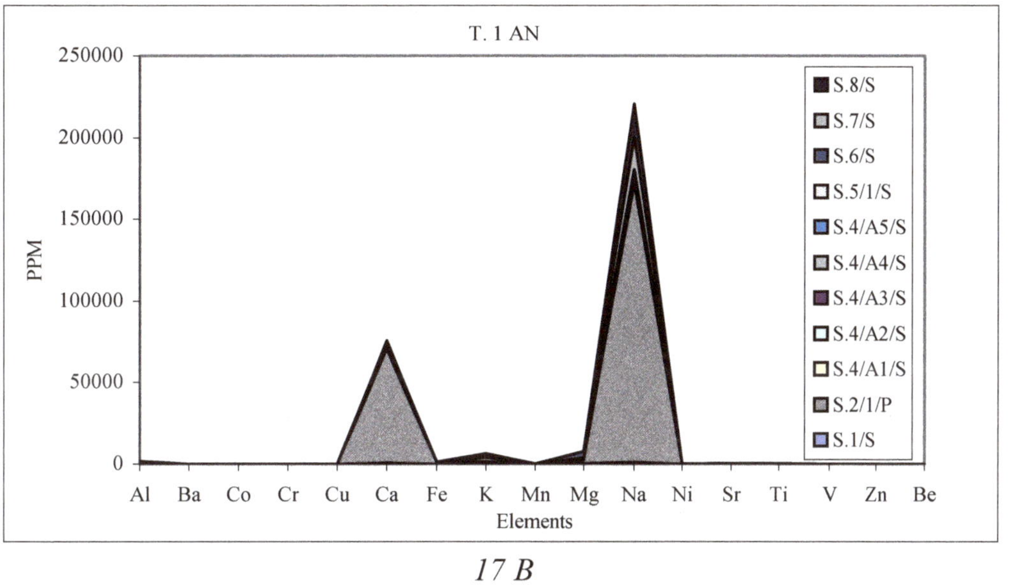

17 B

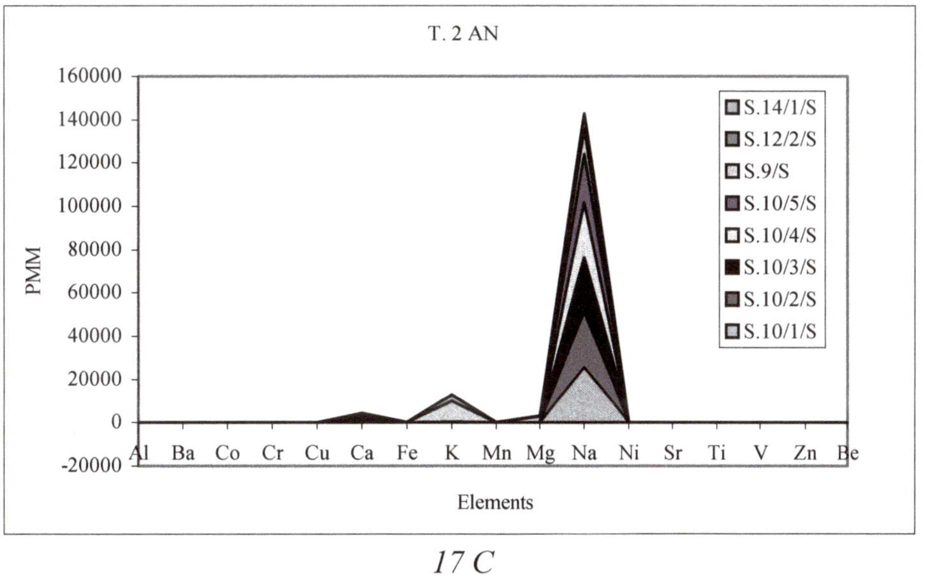

17 C

Graphs 17 A-C, show the distribution of the metal ions found in the samples from the Alexandrian sites. Note the presence of K, Na & Ca, which were most common elements found in the samples. Sodium showed the highest quantity of ions, as given in Appendix 17.9-12.

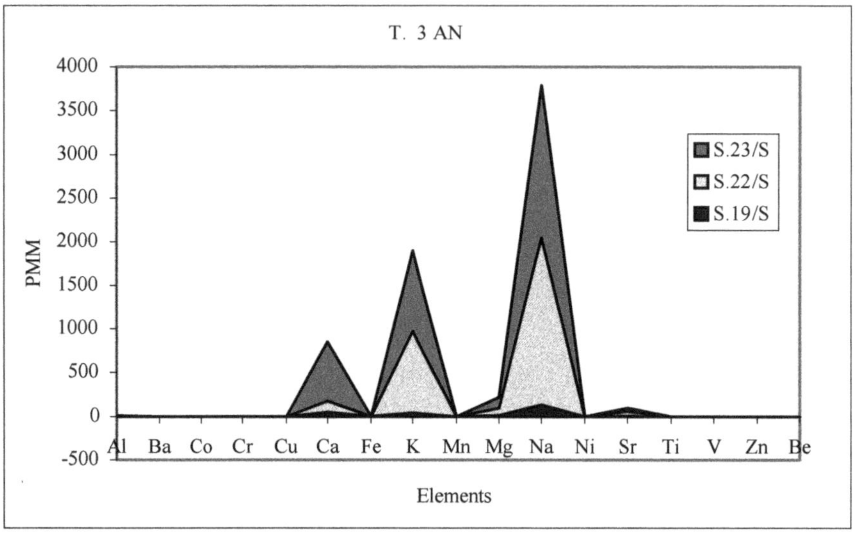

18

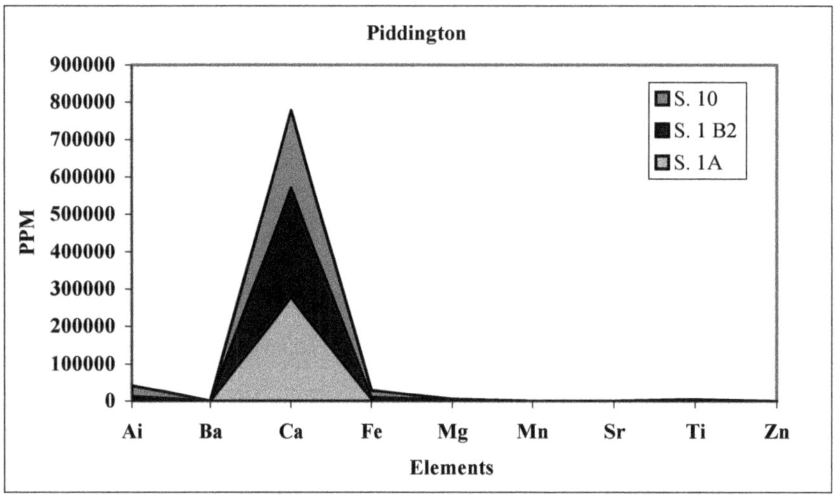

19

Graph 18, show the distribution of the metal ions found in the samples from the Alexandrian sites tomb 3, Note the presence of Na, K, & Ca.

Graph 19, show the distribution of the metal ions found in the samples from Piddington, show the elements of the calcite film which were covered the painted layer on some fragments.

3.3 Polarised Light Microscopy Identifications

Several sections were made to identify the internal structures of a few samples of limestone used in making the plasters/mortars. Sections were of different samples of plasters/mortars from the four sites of Dahshur, Luxor, Alexandria and Piddington, using the methods stated in Chapter 4 (2.3.3). Although it was difficult to carry out examination of samples from Saqqara and Dahshur, as they were very friable, a few samples from Dahshur were suitable.

Microscope examination using PLM has shown that the structures of **limestone** from Dahshur were very similar; it was very angular, uniform and had a very fine grain size (photograph 28 A). Two samples of mud brick showed sand to silt size, quartz sand, ill-sorted organic plant cells and small pieces of stone (photograph 28 B). One sample of clay plaster contained sand grains, silty clay/mud with alluvial sand and sandstone as illustrated in Table 11.

A sample of limestone from Luxor was slightly different, with dark hardened edges surrounding the sample, which was different from the inside, resulting from diagenesis by chemical changes. This was confirmed by polishing the sample as the edges remained, whilst the centre disintegrated. The structure contained microfossil of foraminifera (*Globigerina*) and rhombus were also present (Photograph 29). Samples of mud brick showed quartz sizes from very fine sand to silty grain, and plant fragments. The structure was very silty (table 11).

The identification of **clay plasters/mortars** from several samples were very similar containing crystal grains, silt/mud, clasts (cooperated in silt), the main body in the material being fully carbonated. The structure was full of microfossil of foraminifera and shell fragments. Young plant material, as well as lumps, rounded in shape, were also visible (photographs 30).

The examination of samples of **finishing plasters** from T. 192 and one sample from T. 255 under PLM has shown that all samples were reddish and their structures contained sand grains in different sizes, calcium carbonate, foraminifera, brownish materials, which may be iron rich rock (hæmatite), different sizes of quartz grains and lumps of rock fragment. One of the samples had a different structure, containing four layers, the first was reddish (photograph 31 A), with a structure the same above.

The second layer was a lumpy layer, which showed slight differences, containing straw, fine grain matrix contain calcium carbonate and silt/mud (Photograph 31 B). The lumpy layer was not uniform. The third layer was reddish and the fourth layer was lumpy layer, as before. This sample gave evidence that the application of the plaster, had been in two layers and the reason for finding such reddish/lumpy layers was that floating up of the thinner layer on the surface as a result of precipitation the mixture.

Once again microfossils of foraminifera (*Globigerina*) were identified in a sample from T. 255; angular pieces of rock rich in calcium carbonate, and iron materials, which may be goethite, were also present. Another sample of finishing plaster contained only gypsum with a very little calcite, indicating that good quality gypsum had been used. The second showed gypsum with calcite and quartz, as shown on Photographs 32 and 33 , which gave indication of the impurity of the gypsum which contained these minerals (See also the identification of finishing plaster, Appendix 16.11-12). The identification of samples is given in Table 11.

The identification of the samples from the **Alexandrian sites** has shown that the structures of the plasters/mortars and limestone contained oolitic and superficial oolitic limestone, shell fragments, calcium carbonate "fossil" spines, foraminifera, isolated shells, ummulites, micritic calcified alge, intraclasts "lithclast", echinoids, quartz grains, calcite spar and feldspar (photographs 34-36). The structures not only belong to the first ridge but also, to the other three ridges 2, 3 and 4, which contain similar material. In addition, the structures were of fully cemented materials on some samples, indicating the oldest ridge. The identification of the samples is given in Table 12.

Table 11, Polarised Light Microscopy (PLM) Identification, Saqqara and Dahshur

Sample name	**Identification of the Internal Structures**
	Limestone
S.52	Angular grains of different size ill-sorted calcareous mudstone, composite structure.
S.54	Very angular, uniform and very fine grain sizes with calcite.
	Mud Brick
S.50 S.51	Sand grain to silt size quartz, ill-sorted straw and limestone (carbonate).
	Clay Plaster/Mortar
S.143/1	Sand grains silty mud with beach sand and sandstone.

Luxor

	Limestone Identification
S.27/3	Limestone full of organisms foraminifera (*Globigerina*) and shell fragment of a tiny rhombus were visible.
	Mud Brick
S.131 S.129	Rounded quartz sizes (pebble) from sand to silt grain, very fine, calcite, feldspar and straw, internal cracks with pebble. See Appendix 16.5 identification of X-RPD.
	Clay Plasters/Mortars
S.121 S.111 S.127	Crystal grains, lumps: rounded in shape, clasts co-operated in silt, sand size grain, young plant materials, microfossil e.g., foraminifera and shell fragments. See identification of clay plaster. Appendices 16.9-11.
	Finishing Plaster
S.27/1 S.27/2	Sand grain and calcium carbonate (calcite).
S.29 S.31	Foraminifera, iron rich rock (hæmatite), different sizes of quartz, lumps of rock fragment and gypsum.
S.33	The reddish layers contained quartz and calcite. The second was of silt/clay, quartz, foraminifera full of calcium carbonate, plant and lumps of materials indicated that the process was incomplete of making the plaster.
S.120	Microfossil, angular pieces of rock rich in microfossil, iron material (goethite), foraminifera (*Globigerina*). See the identification of finishing plaster Appendices 16.12-14.

Alexandria gave an indication about the formation of the Pleistocene rocks. According to Said, the tombs are excavated through the Pleistocene layers. The Mediterranean Pleistocene sediments to the west of Alexandria consist of elongated ridges. There are more than nine ridges which range in age, on the basis of their elevation and foraminiferal content. They have been deposited in the form of offshore bars in the ancient inland extension of the Pleistocene embayment of the present Gulf of Arabs and were separated from the landmass by lagoons (1962: 19, 26).

The quaternary levels are from ridges 1, 2, 3 and 4 from the coast, which are considered the youngest to the oldest ridge and are parallel to the present day seacoast. Every ridge has its own composition and structure. The first, which is on the coast and is 5 to 10 m above sea level, consists of less cemented oolitic limestone, isopachous cement indicating a fresh water phreatic zone, most ooids are micritized with quartz grain, porosity and oolitic grain stone to pack stone. Ridge 2, which contains oolitic bioclasts, moderately cemented by isopachous cement, micro sparite or drusy sparite, is about 20 to 25 m above sea level. Ridge 3 is 35 to 44 m above sea level and its bioclasts have a thick micrite envelope, some parts are completely micritized forming peloids, which are well-cemented rocks. The fourth ridge is 55 to 65 m above sea level and consists of bioclasts algae (El-Shahat 1996: 265-275). The ridges 1, 2 and 3 are of Pleistocene age, as has been indicated. The Mecoastal ridges are mainly oolitic types and run parallel to the Mediterranean Sea coasts. They are separated from each other by longitudinal depressions (Said 1962: 26).

Geological studies indicated the similarity of the geological structures of the Mediterranean Sea area. It was interesting to notice the similarity between samples from Libya, Cyprus (Dr. Morgan's collection), and Anfushi. This led to the analysis of the rocks showing not only the same structures but also the same age especially the Libyan samples. It was noticed that the Anfushi and Libyan samples had the same black voids pointing to the youngest ridge (photographs 37-38). The sample from Cyprus was fully structured because it was geologically older (photographs 39).

Table 12, Polarised Light Microscopy (PLM) Identification, Alexandrian Sites

Sample name	Limestone identification
	Mustafa Pasha T. 1
S.4 S.14	Quartz grain, isopachous, calcium carbonate "fossil", micrite, shell fragments, Sandstone, cemented shape, calcite spar and superficial oolitic, well sorted.
	Anfushi T. 3
S.19	Foraminifera, calcareous alga and superficial, shell fragments, very porous structure, indicating the youngest ridge.
S.21	Calcareous alga, foraminifera, intraclasts, echinoid, cemented and lumps.
S.23	Oolitic limestone, superficial, shell fragments, and finer grain size.
	Libya (L) and Cyprus (C)
S.1 L	Foraminifera, shell fragments, calcareous alga, ferruginous and cemented material, very porous.
S.2*C	*Globigerina*, foraminifera, isolated shell, nummulites and quartz, full structure.
	Plaster/Mortar identification
	Mustafa Pasha T. 1
S.2	Foraminifera, quartz, oolitic and spiky, the same as S.14, well sorted.
	Anfushi T. 1
S.4/3	Echinoid, feldspar, calcareous alge, intraclasts and shell fragments.
S.4/5	Shell fragments, calcium carbonate, intraclasts, alge and lumps with clay/silica.

Polarised light microscopy of samples from **Piddington** has shown variation in the mineralogical structures of lime based plasters/mortars. Sections showed that the *intonaco* layer, was covered with a very thin calcite film (photograph 40 A). Crushed limestone/chalk with microfossils, such as foraminifera and coccolith, as well as other shell fragments were observed in the sections (photograph 40 B). Sheared quartzite and single crystals of limestone were identified in a few sections. Flint/chert was present in very fine grain sizes. Various shades of reddish-brownish brick were visible due to the use of different temperatures for baking brick. Some were deep brown, due to baking in the absence of air. Sandy clay was used as it was noticed that all bricks contained very fine grain and different grain sand sizes in the mixture. Oyster and various other shell fragments were present in all sections (photographs 40 C, E and F). Quartzite, ironstone, sandstone and quartz sand were identified in the mixture (photograph 40 D, G and H). Very fine cracks were observed in some samples and it was interesting to identify many fossil shells and oolitic ironstone in the mixture of the lime plaster/mortar which would have dissolved in acid.

The identification of some samples is given in Table 13. The presence of microfossils and shell fragments were visible, but with variation of the organism types found in the Alexandrian samples, due to the geological sources, as stated in Table 12.

Table 13, Polarised Light Microscopy (PLM) Identification, Piddington

Sample name	Plaster/mortar identification
S.1	Brick, limestone/chalk with microfossil and shell fragment, sandstone, other shell fragments and quartz.
S.2	Oolitic limestone, single piece of crystalline limestone, coccolith, sheared quartzite, brick, different shell fragment, limestone/chalk with foraminifera and other microfossil.
S.3	Painted layer and *intonaco* shown in section, ironstone, fossil oyster and the same as above.
S.8	Brick, sandy ironstone (opaque), fine quartz, flint/chert in very fine grains, limestone with microfossil, various shell fragments, rhombus, and fossil oyster in the limestone matrix.
S.9	Same as above with rounded fossil and coccolith and crystal lath inside the brick, *intonaco* layer which covered with calcite film observed (see photograph 40 A).
S.10	Same as S.2 and S.8, with oolitic ironstone, limestone with fossil and foraminifera and calcite fill the space of dissolved shells. Metamorphic rock was present as well.
S.11	Similar structure with ironstone rich opaque.

Coccoliths are found in calcareous rocks formed in various geological ages from the lower Jurassic to the recent. They have been found in rocks as old as the upper Cambrian, and are especially common in some Cretaceous and Tertiary formations, such as chalk and limestone. Each animal consists of a spherical cell, which is covered by the microscopic bodies of calcite called coccoliths. These have several different forms and the most common are perforate or imperforate disks, star-shaped bodies and paired disks with a single perforation. The diameters of the disks range between 0.002-0.01 mm. They grow near the ocean surface as they are plant forms and need the light for photosynthesis and growth. In some instances they were not recognizable, because the very small particles of calcium carbonate of which coccoliths are constructed are apt to re-crystallise, if the enclosing rocks are subjected to heavy loading or to shearing stresses, then the characteristic shape of the coccolith is lost. Also, during the conversion of marl into argillaceous limestone, the growth of cementing crystals may so encrust the coccoliths that they become unrecognisable. Black (1965) gave a good illustration of the range of the structures to be seen in coccoliths of different species.

Foraminifera, are very small floating forms found in shallow marine water. The sands of many tropical beaches are made up almost entirely of foraminiferal tests, which consist of different animals; some are composed of chitinous matter, agglutinated particles cemented together and some are calcareous materials (Moore *et al.* 1952: 39, 76, Black 1965 and Gettens *et al.* 1974).

Some studies have been carried out on the geology of Egypt, including Alexandria. Said gave complete details of all the areas in Egypt, their geological structures and composition (Said 1962). Another study concentrated on Alexandria and the north coast of the western desert in Egypt from Alexandria to El Salum, on the quaternary carbonate dune deposits facies that overlie the Miocene deposits away from the Mediterranean sea coast (El-Shahat 1996). In addition, the study of the Mediterranean coast between Rosetta and Bardia by Shukri *et al.* (1956), which show more or less the same as found in the Alexandrian samples.

The identifications of various foraminifera, shell fragments and other organisms in addition to iron oxides and quartz has been observed in a previous study of samples from Maadi town Temple in Egypt by Pinnska & Rabie (1999: 622).

Thin sections can show the composition, the rendering mixture used and identify whether it was lime or gypsum. The identification of samples from the Egyptian sites has shown variation in the structures both of the limestone and the plaster/mortar, in particular the Alexandrian samples.

3.4-Scanning Electron Microscope and Energy Dispersive Spectrometer Analysis

Scanning electron microscopy was carried out on different samples to examine the surface structures and identify the elements present on the surface. As a result of using different generations of SEM and EDS (see Chapter 4, 2.3.4), EDS analysis showed only the elements present without their counts, in particular in the Luxor samples. Scanning electron microscopy was also performed on polished cross-sections, to look at the surface structure (as shown on photographs 20 C and 23). That showed that different thicknesses of calcite film covered the painted layer. The sections were coated with carbon before SEM and EDS analysis. Several samples of limestone, plasters/mortars and salts from different areas and tombs were examined (see Appendix 7 analytical techniques).

Although the samples from Saqqara and Dahshur were very fragile, it was possible to obtain an idea of the surface structure, as well as the elements present in some samples of mud and clay plaster/mortar and gypsum plaster (photograph 42) (see also Photographs 68-71, Part II). The surface structure of small pieces of gypsum plaster had a very crumbly surface, as mention before (see photograph 8 B).

Scanning electron microscopy of different samples from Luxor showed that the natural limestone was full of organisms such as coccoliths, foraminifera and shell fragments, as shown in Photograph 41 B (see also Photographs 72- 75, Part II). The same organisms were found in the mixtures of the plasters, as shown in Photographs 43 B and C. Examination of the clay plaster/mortar showed very porous structures, full of holes, as shown in Photograph 43 A.

Microanalysis showed variation in the microstructure of the surface deterioration, in particular samples from Alexandria and Piddington. Different samples have shown the same surface structures, which gave a general idea about the deterioration and decay that affected the wall paintings (photographs 44 and 46) (see also photographs 76-77, part II) and showed the different shape of the salt crystallisation as well (photographs 47-49). SEM & EDS helped to determine the form of the surface problem, such as micro-organisms micro cracks and the shape of the crystal, as well as determining the micro-fossils found in the mixture (photographs 43 C, 44 B and 46).

SEM has shown that all of the samples from Piddington were covered with layers of calcite film as shown on Photograph 46 A-D. Also shown is the fungus on the surface illustrated in Photograph 46 B. The surface was covered with a layer of calcite film in different forms, with the fungus growing on top (photograph 44 C & D). This indicated a wet or damp environment suitable for the biological growth, as stated in Chapter 3.

Cracks were visible on all samples not only inside the stone (see Photograph 72, Part II), but also on the plaster as well, due to mechanical stress (see Photographs 74- 77, Part II).

A few samples of the salts from Luxor and Alexandria were examined to determine the shape of the crystals (shown in Photographs 46- 49) (see photographs 67- 78, part II). As indicated in Chapter 3, the different kinds of salt crystallisation and their shapes are needles, tubes or whisker etc. It is reasonable to assume that the salt crystallisation found was as a result of certain conditions which allowed salt to grow and dehydrate, in addition to encrustation in some instances as mentioned before.

EDS analysis showed that the most significant salt found was chloride, in some instances as sodium chloride (halite), which was noted to cover the surface of the limestone in some samples, as shown in Photograph 42 A. As mentioned before, salt contained in the stone itself may migrate to the wall plaster as well, giving the possibility of finding salt inside the mixture, as noted in the examination of the plasters/mortars, or decomposed on the surface as a result of salts re-crystallisation, as happened at the Alexandrian sites.

Energy dispersive spectroscopy (EDS) gives a measure of the leaching of the soluble ions. Wavelength dispersive analysis gives more information concerning the mechanism and the rate of solution. Analysis gave a full picture of the elements present on the surface. The results of EDS are given in Appendices 18.1-18.

Analyses of samples from Saqqara provided the elements present on the surface as stated in Appendices 18.1-3, however, most of the elements present were not in large quantity. The most important elements shown were; calcium (Ca), silicon (Si) and sulphur (S), being found on most of the samples.

From the results of the Alexandrian sites it can be concluded that the most important elements present on more or less all of the samples, were; calcium (Ca), sulphur (S) and chloride (Cl) as well as silicon (Si) and aluminium (Al) on some samples. On some samples from Anfushi tomb 2, the surface structure has shown that calcium (Ca) and potassium (K), were the most common elements and were present in huge counts. Other samples had calcium (Ca), sulphate (SO_4) and potassium (K) (see Appendices 18.6-13). It can be assumed that the presence of calcium and sulphate was due to the transformation of calcium carbonate (plaster/mortar or limestone) as a compound of the mixture, into a calcium sulphate, resulting in the process of the deterioration and decay that affected the rendering (see Chapter 3, 3.3.2).

EDS gives information about the elements contained in each sample, with concentration of some elements being much higher than others. Silicon, calcium, phosphorus and aluminium appeared to be the most important elements on the Piddington samples. It can be seen that iron and potassium were present as well on some samples. Calcium showed the highest level, then silicon and aluminium, as reported in Appendices 18.18-19, in addition to potassium and manganese, whilst iron was not commonly found. Potassium, iron and manganese in the rest of the samples had higher counts. Sulphur was also found in some samples. It can be assumed that the presence of these elements resulted from fragments containing them being buried in the soil. The porosity of the mortar helped to absorb the elements because the mortar mixture contained different types of aggregate, and allowed them to accumulate on the surface in the form of layers of calcite film.

An important item to be considered is that the surfaces were not plain. The layers, which were found in some instances covering the surfaces, meant that there might be elements on one area, but not on another, although the sample was one piece. Several analyses were carried out on each sample to obtain a general idea on the elements present. The results gave variation of their counts, as illustrated in the Appendices 18.1-19.

3.6-Summary

Physical methods provided useful information on the mineralogical compounds and the surface structures of samples, not only determining the different causes of deterioration and decay, but also showing the crystallization form of salts.

However, the results of X-RPD analysis must be considered to be approximate, as these samples were not pure and may have contained different elements, (such as S.46, in terms of the presence of cerium carbonate hydrate/lanthanite as stated in Appendix 16.8), which may have been formed by chemical reactions under certain conditions. It should also be noted that analysis of the powdered samples, by ICP, gave the total acid and water-soluble metal ion content. The water extract contains the water-soluble ions only. Therefore, using various methods in the identification of materials are necessary in order to obtain more information about the objects which would help to understand the structure of the materials used and this would be useful in the process of the conservation/restoration.

Chapter 7

Cleaning, Consolidation & Lifting-Case Studies: Results of Examination & Analysis

1-Introduction

Examination, analysis, cleaning, and consolidation was carried out on some fragments from Piddington. Analysis and examination gave a full understanding not only about the manufacturing techniques of the plaster, but also the deterioration and decay that affected the surface of the fragments. The cleaning methods used are an important operation for conservation. Knowing about the materials for cleaning, one can select and avoid factors which may cause damage. The preparation of cleaning materials and their application methods depend on the nature of the painting and the environmental conditions, and these must be considered in order to prevent any decaying action in the future. The testing of the cleaning materials must take place, especially in the case of sensitive pigments, e.g., those which are copper based, in addition to the use of the right application techniques.

2-A Review of Cleaning and Consolidation of wall painting

A variety of materials were used in the 19th century, such as soap, kitchen salt, vinegar, potassium hydroxide, hydrochloric acid etc. (Cronyn 1990: 127-129). The soft part of bread was also used. For more difficult situations, spirit of wine and benzine (petrol) were suggested. The traditional materials were still used, but, most were replaced by synthetic chemical in the last decades of that century, e.g., oxgall, which was frequently used as an additive in animal glue to reduce surface tension, is now replaced by synthetic emulsifiers. Although lime and lime mortars are traditional materials, they are most commonly used nowadays for filling gaps and replacing lost parts of plaster. Lime/*pozzolana* and lime/crushed brick are preferred for their better mechanical performance and their quality of setting in very humid environments. Some other traditional products, such as bees wax, paraffin wax, shellac and gums, are substituted with acrylic resin in diluted solution, which is used when the painted layer is powdering and needs to be fixed to the plaster (Tabasso 1993: 54-55).

Methods of conservation for cleaning and consolidation are varied. Different methods of **cleaning,** and materials for use on, wall painting are discussed by both Cronyn (1990) and Mora (1968). The cleaning material must be compatible with the paint layer; for example, wax or adhesives could cause alteration to the surface. Materials were darkened by animal glues or gum Arabic from previous restoration, organic deposits from bees, wasps etc. or vegetable growth produced from damp or tropical conditions. Each case requires a different solvent for cleaning and different application techniques. It is difficult to use just one method and solvent for cleaning because of the reactions, which could affect the pigments, colours or the plaster layers themselves. Sometimes it might be necessary to use two or three techniques for cleaning, depending on what needs to be done. The application of any particular cleaning method depends on the sort of deterioration and decay (See Science for Conservators 1992 and Mora 1968).

The most recent research for cleaning for particular application within the field of wall painting conservation is the use of different types of lasers; Nd-YAG, Krf excimer and Rhodamine 6G dye operating respectively in infrared, ultraviolet and visible wavelength ranges. These methods were applied to different replicas of *fresco*, linseed oil and egg *tempera*, to provide a wide range of typical wall painting stratigraphies for the trails. They were used to remove lime-wash layers and soot, which covered *tempera* painting. Shekede indicated that lasers appeared to be most effective when used for removal of coatings from lime white, or pigments mixed with lime white, and it was possible to remove soot as well.

The Nd-YAG laser removed wax from *tempera* or *fresco* layers with little physical damage. However, problems of pigments alteration have been observed due to the radiation. The effects of radiation on both inorganic and organic materials of paintings require greater understanding before lasers may safely be used for the cleaning and uncovering of wall painting (Shekede 1998).

Options for consolidation are varied and different materials can be used, not only because of the different chemical compositions but also with the conservator's choice, based on personal views and experience. I have found that there are no standard methods for conservation. Each case differs with the alteration and deterioration factors and the techniques of execution of the wall painting. It is also possible to identify the different organic binding media or materials being used in antiquity.

Consolidation may be used to fix the paint layer if it is flaking or tending to become powdery. Mora specified that must penetrate the colour thoroughly, be resistant to abrasion and must not modify the colours or change the general tonality (Mora 1968: 172). However, the consolidant has also to behave similarly in aging, thermal expansion and shrinking to the original materials. It should not change the refractive index or cause glossiness of the surface and not disturb the walls porsoity as well. In addition, it should not get an acid character by ageing (Peterson 1982: 55). Consolidant capable of coating the pores of the objects but not of creating a relatively impermeable surface film was requried, therefore repeated applications of the consolidant had to be possible to build up the necessary concentration in the objects (Tubb 1987: 388, 389). It may take several days, in damp conditions and may be necessary to use a hot air blower, using Vinamul 6815 (PVA-irreversible). This is a polyvinyl acetate resin, dispersed in water and can be used to strengthen damp objects. About 30% weight or volume concentrate should be diluted with clean water, which stirred well. Paraloid B72 may also be used; it is an acrylic resin and dissolves in acetone. It can also be applied with a spray gun on a dry or nearly dry wall painting (Watkinson 1972: 62-87). Each consolidant has some advantage and disadvantage when used e.g., Raccanello E55050 and Dow Corning Z6070 acrylic silane mix which employed for Neolithic statues dated to 6000 BC, based on lime plaster (Tubb 1987).

Injection methods can also be employed in areas of cleavage, as filler. It is necessary to drill holes from the surface in to the lacunae to inject the consolidant (see Mora 1972). Ferragni *et al.* (1984) give good guidline in tasting injectable mixtures e.g., lime casein, lime-synthetic resin emulsion, cement, etc. In addition the grouting mixtures which was tested by ICCROM; binder, filler, etc. providing data on the laboratory tests (See Ferragni *et al.* 1984: 110-116 and also Böttger & Knöfel 1993).

Limewater has also been used for the consolidation of wall painting. The methods of consolidation with lime water works by replacing lost calcium in brittle or friable plaster/mortar, or in flaking/powdering paint layers. The consolidant retains its original properties and behaviour, not as an addition of a compound of different character into the original materials. With some consideration, in case of the *tempera* techniques, one could use calcuim bicarbonate solution which has a lower pH. The methods of lime water and calcium bicarbonate as well as the application techniques are described by Peterson (1982: 55-59) The solubility of lime in lime wash is very low (1.7 mg lime in one litre). Davis said that the biggest misconception is in the amount of limewater necessary; complete saturation is needed and the wall must not be allowed to dry out between applications. The spraying stops when the surface causes absorption. Varying methods of applications for both consolidation and cleaning seem to be necessary in some instances, depending on the condition of the wall painting. Pre-consolidation is required in some cases before cleaning. Limewater and ethyl silicate was used as a mixture for pre-consolidation, injected behind the flaking areas of paint layer prior to cleaning. Limewater was used to act as a catalyst to make the silicate set behind the paint layer. A second consolidation with ammonium casein was applied on the rest of the paint (Davis 1997). The conservation methods applied during previous treatment are an important consideration. The methods of conservation do not relate only to the materials used for conservation, the kind of the objects and the methods of application, but are also related to the environmental climate with which these materials may be compatible, as an example of this is as follows. One of the most recent and interesting studies was carried out on the use of Soluble Nylon (white powder, which dissolves in water/alcohol and becomes transparent when the solvent evaporates). It was fashionable for treatment of the art and cultural materials.

Soluble Nylon as mentioned by Plather, L. (1997) was recommended as a conservation medium by A.E.A. Werner, for the gluing of paper, consolidation of fragile textiles, pottery, wall painting, ethnographic materials and metals, ivory and for impregnation porous stone surface before de-salting. It was described for fixing *tempera* painting on a fragment of an Egyptian tomb and the impregnation of the Egyptian limestone fragments with inscriptions, before these were laid in water for washing out soluble salts. It has shown good flexibility, did not shrink on drying, had a matt appearance and good adhesive properties. It was widely used for nearly 20 years in the 1950s to 1970s before it was considered necessary to subject it to laboratory examination. The first signs of concern regarding Soluble Nylon came in 1975 from Eddy De Witte at the IRPA (De Witte 1975) after the treatment of an Egyptian Mastaba with Soluble Nylon (Calaton) before washing out soluble salt from the stone. After two hours, the Nylon film became grey-brown. This led to tests for elasticity and solubility and it was found that Nylon-film became insoluble after some hours in acidic condition.

Plather, L. (1997) indicated some studies that investigated a range of situations in which Calaton/Soluble Nylon had been used in the 1950s to 1970s. The objects had attracted an uncommon amount of dust and dirt, and the objects were difficult to clean because the dirt had entered pores in the Calaton-film and it was no longer flexible, but brittle and insoluble. The study concluded with advice against the use of Soluble Nylon in conservation and recommended replacing it with other substances, e.g. Paraloid B72.

He discussed the history of the conservation of the roof painting, from Stavkirke church in Ål. The church was built in 1160-1170 and it was torn down in 1883. The decorated roof, which was executed in distemper, was transported to Oslo. In 1905 it was re-assembled in its present place in the University Antiquities Collection's mediaeval department in the newly constructed historical museum on TullinØkka. The roof was impregnated with Soluble Nylon in 1965. The Soluble Nylon from the roof has not yellowed recognisably, and troublesome quantities of dust do not seem to have stuck to the surface. Plather, U. (1997) stated that the painting samples are covered with a coating of Nylon, and had a silky, gloss finish. It does not appear that movements in the Nylon materials have deformed the painting structure and the film has a flexible consistency and is elastic. Since the roof became consolidated in 1965, there has been no contact with water or solvents, illumination has been subdued, and the microclimate stable due to the favourable condition of preservation in the museum. Plather assumed that in 1965 the focus was strongly on the permanence of conservation materials. Today it is enough that the most important requirement for the impregnation medium is that it does not change permanently the artwork and the material's character (Plather, L. 1997 and Plather, U. 1997).

The methodology of **cleaning and conservation** based on the use of ammonium carbonate and barium hydroxide was developed to address the problem of sulphation, in particular the effect of gypsum, which may have been affecting the stability of the rendering. As a result crystallisation, may occur beneath the surface as mentioned in Chapter 3 (see3.1.5).

Barium hydroxide was used in the past for the conservation of porous materials, particularly stone artefacts, but overall had never been considered as a consolidating agent for application to mural paintings. The method combined two operations: cleaning with ammonium carbonate and consolidation using barium hydroxide. Matteini & Moles (1984) gave two fundamental steps using these methods. The first step using ammonium carbonate as a preliminary transformation of gypsum to soluble sulphate. The second step is that barium hydroxide converts the remaining soluble sulphate into inert and no longer soluble barium sulphate. Generally, it consolidates *intonaco* and the painted layer through a slow crystallisation of the barium carbonate. This was considered as a complete conservation method. Matteini (1987) mentioned that when the treatment is undertaken carefully, the advantage of the method is fundamental to the control of sulphation by means of the conversion of gypsum to ammonium sulphate. It does not contaminate the painting since the products auto-decompose, and it facilitates the removal of proteinaceous coatings applied in previous interventions by softening them to a gel form.

This method has been applied on *fresco* and *tempera*, and in such cases, may lead to the use of lower concentrations of chemicals and shorter time-scales for operations with the colours containing copper, e.g. the greens and blues, can be altered by the agent (see Chapter 3, 3.3.3). It is recorded that the best season for the application is Autumn, when the walls are reabsorbing the humidity lost during the hot summer months, or from Autumn to spring (Matteini & Moles 1984, Botticelli *et al.* 1984 and Matteini 1987).

Understanding the chemical composition of the conservation materials is required in order to avoid any problems over time. Tabasso discussed important points of the evaluation of products for conservation treatments, whatever they are, industrial or traditional, if proposed for the conservation of historic and artistic artefacts. Different types of artefacts and constituent materials require different products for conservation treatment. In porous building materials or wall painting conservation, there is a similarity to chemical, mineralogical and deterioration decay. They must be carefully tested in advance unless there is sufficient experience for the specific constituent materials and use intended. The performance of a product is related to the specific use, the method of application and the nature and state of conservation of the original constituent materials of the artefacts to be treated. These should be considered in terms of testing materials for conservation whether in laboratory tests or on site, although each may have some advantages and disadvantages (Tabasso 1993).

I had the opportunity to investigate the methods used for conservation, during field research in Italy, visiting a few work sites (see Appendix 19). I found some similarities and differences between the north and south of the country not only with the technique used, but also with the methods and the materials used for conservation as described in Appendix 19. The strategy of the conservation was also related to other aspects, such as the techniques applied, the methods of application and the materials used depending on the conservators' own point of view and their experiences of application.

As stated above, there are various materials which could be used for cleaning and consolidation and care must be taken, because some of these methods are suitable for wall paintings as the deterioration and decay as well as the techniques of execution can differ. It was necessary to investigate and carry out analysis to understand the artist's techniques and their materials before doing any conservation.

3-Cleaning and Consolidation of Fragments of Wall Painting from Piddington: Methods Used for Cleaning of Fragments

It seems to be essential to determine the type of decay and deterioration, as this varies between the deterioration of pigments and/or deterioration of the supporting structures, etc., as stated in Chapter 2 (see 3.3). Investigative cleaning was carried out to assess the decay of the painting, which had affected not only the painting surface, but also the *intonaco* layer and the mortar. This made the *intonaco* layer and the mortar on some areas more fragile. Through primary examination it could be seen that all the fragments were covered with a very thick and hard layer of calcite film, as mentioned before (photographs 50 A & C, and 51 B, before cleaning). Although the painting on some layers was in good condition, executed using motifs and splashed with different colours on the surface, it was noted that some fragments were infected with fungus (photographs 45 B).

The reaction of the materials used for cleaning or consolidation of wall paintings can loosen the painted layer. The methods of cleaning can be divided into three types; brushing, solvents and mechanical cleaning using scalpels. The fragments of wall paintings cleaned were mostly from rooms 53, 56 and 57 (plan 6). Cleaning was carried out under a microscope (10X), which was very difficult and time consuming. Fragments were then washed, (a soft brush or fingers were used for washing, because a hard brush would have caused abrasion and deterioration) and left to dry. As a result of this investigation and analysis, mechanical and solvent cleaning took place.

The following approach was used in laboratory. Different methods may be necessary in the field when treating material that remaines in *situ*.

Mechanical methods (scalpels with different blades) were used for cleaning. On some fragments there was more than one layer with different forms. Sometimes it was very soft and covered with a hard layer, as shown on Photographs 53 and 54 A, of a deep yellow colour, in form of ancient dust as showon on Photograph 27. It took a long time to clean them and care had to be taken because it was easy to destroy the surface of the painted layer. Solvent chemical cleaning was carried out with ammonia solution, 10% in distilled or deionised water. This was left overnight before use.

A test was carried out on the effect on the colours of concentrated ammonia and also hydrochloric acid. In the case of some changes occurring to the colours, then one should not use chemical cleaning, but only distilled water and mechanical methods. Some pigments are very sensitive to such cleaning agents.

3.1-Investigating Cleaning

The fragments were covered with different layers of calcite, varied in colour between light and deep yellow as shown on Photograph 50 A & C, 51 A. Some marks remained after cleaning (50 B & C), which were not removed with ammonia. The colour of the painted layer has deteriorated being lighter on some areas, and it was difficult to know the original colour. The calcite films were white and some areas were covered with brownish layers of calcite. This was very difficult to remove, whereas the white layers were soft. A very thin layer of calcite film was left, because the painted layer would be removed by over cleaning. On this fragment (photograph 51 B) the scheme of the colour used can be seen, which was a panel/band in red.

It was noticed that there was little remaining of the painted layer on some fragments (photograph 52 A & B). The colour was a mixed green blue. The painted layer was covered with a very thin calcite film, which was easy to remove. The mixture of the plaster was found to be brick, flint, sand, etc. and shows clearly on Photograph 52 A as the main mixture of the mortar as illustrated in Appendix 12.3 & 12.6. The colours were; red, green, black, brown, pink, grey (made by mixing black and white) and blue (Egyptian blue) (see chemical analysis of the pigments, table 6). These colours were widely used in Roman times and have been identified in previous studies (Wallert & Elston1997, Járó 1997 and Rozenberg 1997).

These fragments (photographs 52 B & C and 54) showed two different types of alteration caused by fungus. As a result, some specks produced had had a bad affect on the surface as well as the *intonaco* layer, which turned brown instead of white. Another fragment was covered with a very thick layer of calcite on some areas and other fragments were covered with white calcite, which left some marks after cleaning (photograph 52 C on the left). The biological growth damaging the painted layer was also found on other fragments (photographs 53).

Deionised or distilled water was used as the first cleaning stage, but was not strong enough, so some tests were carried out on the colours, using ammonia solvent which was exposed for 12 hours before use. The result was that some black specks came off. A scalpel was used to complete the cleaning. It was noticed that the painted layer was covered with a very thin calcite film and some cracks on the surface had fungus in them, which was difficult to remove with ammonia. The calcite film was covered with dust, which was very soft on the surface.

On other fragments the calcite film was completely different, as shows on Photograph (53). It was a very thick, deep white calcite, covered with different areas of yellow. On the top right it was pure white with three different forms of calcite on the surface. A trial was carried out to remove the marks using 20% ammonia solution but the marks remained. The original cream colour turned yellowish on some areas and was very difficult to remove. It was noticed on that fragment that after cleaning with scalpels, the ammonia was not as effective. The white calcite film was removed very easily and also the yellowish layer, but the thick white had been very difficult to remove because it had more than one layer. Care had been taken to clean the surface without damaging the painted layer, which was applied as a paste and not by brushes. The *intonaco* layer was the same colour on the top (red ochre), but it seemed lighter than the painted surface. There was no white *intonaco* layer, as was usual in Roman technique. It was polished to get a very smooth painting, as described by Roger Ling (1991). This was noticed on several different fragments of wall painting from Piddington.

Another example of the fragments (photographs 55 A & B) had a very complicated deterioration and decay, not only on the surface, which included a different sort of calcite, but also the back of the mortar, as is shown on photograph 55 B. The surface of the painted layer was covered with a white calcite film and covered in some areas with a very deep brown layer of the ancient dust, which was very hard and thick on some areas.

The calcite film on that fragment (S.10) was yellow and glassy as well as easy to damage the surface. That kind of alteration might be from acid rain or by chemical reaction, which affected the back of the mortar and dissolved the binding lime.

Mechanical cleaning was the most suitable method, although it was time consuming. The marks left could be from calcium carbonate and its structure, as it contains some elements such as iron, magnesium and others as shown with ICP analysis (Appendix 17.13). For these reasons, I believe that under-cleaning (not cleaning down to original surface) is much better than over-cleaning (may cleaning down to below original surface causing damage) (see also Science for Conservation 1992), if the material is to be preserved for a long term. In my opinion, the cleaning should leave a very thin layer of calcite on the surface, as it is very dangerous to the painting if the film is removed completely. The aim is to treat not to damage. Ammonia alone cannot used to clean. It was suitable for cleaning some fragments, and could be used if the calcite film was produced recently.

3.2-The Lifting of Wall Painting

It was interesting to investigate the Roman technique of applying more than one layer of plaster, which I noticed on the fragments of the wall painting from Piddington. Lifting off wall paintings is considered as a method of treatment, but it is not recommended, except under special circumstance. The methodologies of transferring/lifting of wall painting must be based on the technique and state of preservation. The enthusiasm with respect to the *stacco* and *strappo* remained unchanged until the 1960s or later. However, at the beginning of the 1970s criticism against the too unconstrained practice of removing wall painting from the original support started to be expressed when considered by Mora and Philippot to be the "last resort". A mural painting is an integral part of the architecture it completes and any separation of the painting from its original support constitutes an irreversible alteration of both (Tabasso 1993: 54). (See also Schimd 1993).

Methods used depend on what is being lifted; the paint layer only (*strappo)*, the plaster with paint layer (*stacco)*, or both with a part of the supporting structure (*stacco a masello)* as happened at the tomb of Nefertari where the crystallization of salt pushed the plaster off. These methods are described by Mora; each is used for a particular case (1968: 176-188). The technique of mounting as well as lifting plaster in *situ* has beendescribed by Allag (1982).

The method used on a fragment (S.9 B, photograph 54 B), was *strappo* technique, to take the later painted layer off after cleaning. Consolidation of the surface was done before lifting, using Paraloid B72, 5% in acetone, to consolidate the painted layer, which was very fragile. Glue was used to paint the surface, after cleaning and consolidation. A first layer of muslin was applied over the glue, care being taken to press the muslin on the surface, to make sure that the glue covered the muslin. It was then painted with more glue and a gauze layer was applied and painted again with more glue. However, if the surface is left for a long time, it should be kept damp by covering it with plastic film. After drying, a saw was used to separate the layers. Care must be taken at this stage so as not to damage the earlier painted layer. This was cleaned from the traces of the mortar of the later re-plastering as is shown on photograph 56 A & B fragments upper right, with yellow paint, and a very thin black band at bottom, with brown on the top. This was discovered underneath fragment S.9 B, photograph 54 B, when it was removed.

The pick-marks to key the new plaster, which were cut before applying the second layer, the back of the second layer, which was removed, was cleaned carefully to expose the fragile *intonaco* layer. After finishing, a thick mixture of 10% Paraloid in acetone with added kaolin was applied as a first layer on backing of the mortar. This must be allowed to be dry before applying the second layer, a mixture of 10% Elvacite 2013, mixed with sand and painted on. After drying, a layer of fibreglass with polyester resin was applied. Finally, the plaster was reversed and the muslin taken off with warm water. The original surface was cleaned of the traces of glue.

3.3-Consolidation of fragments

A 5-10% solution of Paraloid B72 in acetone was used for consolidation of some fragments. The mixture was used to consolidate the ground not the surface of the painted layer, for which a more dilute solution should be used, e.g., 5% Paraloid B72 (ethyl methacrylate) 1, 1, 1 Trichlorethane, however, if the mortar pariculary fragile, one can dip the whole fragment in the solution (Allag 1982: 85). Paraloid B72 is widely used not only for archaeological objects, but also for consolidation of wall paintings, in standing building, e.g., in Italy.

A previous study has been carried out on the use of Paraloid B72 and it has been shown to provide the fast solvent release needed for the successful application of an acrylic adhesive. This described the best methods of Paraloid B72 preparation as well (Koob 1986). As said above Paraloid is a colourless synthetic resin and can be applied by spray or brush; it gives better penetration because the evaporation is slower as well. However, if used in too concentrated a form it leaves a skin on the surface, which may cause damage to the paint layer as it contracts on drying (Mora 1968: 172).

Consolidation depends on the condition of the object and what needs to be consolidated. Currently one should seek new materials for consolidation to avoid chemical hazards but should determine the kind of materials, not only for of the conservator's health, but also for the good of object (see Pascoe 1980). The main requirements in testing a consolidant, are to determine the advantages and/or disadvantages of the materials to be used.

Chapter 8

1-General Conclusion
The Differentiation of Egyptian and Græco-Roman Wall Plasters and Mortars

This thesis has clearly demonstrated that the wall plaster and paint samples studied from two distinctive periods of Egyptian history, Egyptian and Græco-Roman, can be distinguished by scientific analysis. Art historical, stylistic, technological and geological variations can also be distinguished. The most notable difference is the use of lime in the Græco-Roman period compared with the use of "mud/clay" is the various Egyptian phases. The study of the materials has shown that no one scientific technique also can supply sufficient information either regarding the structure and composition of the materials or the decay processes which have affected them. A prior knowledge of the local geology and environment is necessary not only for the interpretation of surviving materials, but also for the planning of future conservation and restoration.

The current research was carried out to differentiate the Egyptian and Græco-Roman wall plasters and mortars used in ancient and classical times, using a range of methods to obtain precise details of the manufacture of the plasters and mortars, by defining their structures and determining their mineralogical composition.

Examination and analysis is an important stage in the process of conservation and restoration. It is essential that adequate sample sizes and numbers are taken if accurate analyses are to be obtained. Analysis allows us to understand the materials that were used for making plasters/mortars, and even determine the mineralogy of the compounds, which provides information about the main source of the materials used. However, analysis tells us what is present now, not what the materials originally were.

Microscopic examination has proved extremely useful in identifying materials and determining the degree of deterioration and decay. The examination and analysis of the plasters and mortars has shown variation in the preparation methods used for the walls, and has found that components in the mixture of the plasters and mortars varied, according their use.

Research was carried out on the various layers, which were examined as broken and cross-sections, to determine their structures and basic characteristics. This examination gave approximate measurements of the thickness and numbers of the layers of plasters and mortars applied, as stated in Appendix 6, thus indicating the stratigraphies of the wall plasters applied in Egyptian and Græco-Roman times, as illustrated in figures 7 and 8. Summaries of the types of Egyptian and Græco-Roman plasters/mortars from different sites, periods and tombs are given in tables 14 A-D.

Figure 7

The Stratigraphy of the Egyptian Wall Plasters/Mortars
Diagrammatic; not to scale

(7.1)

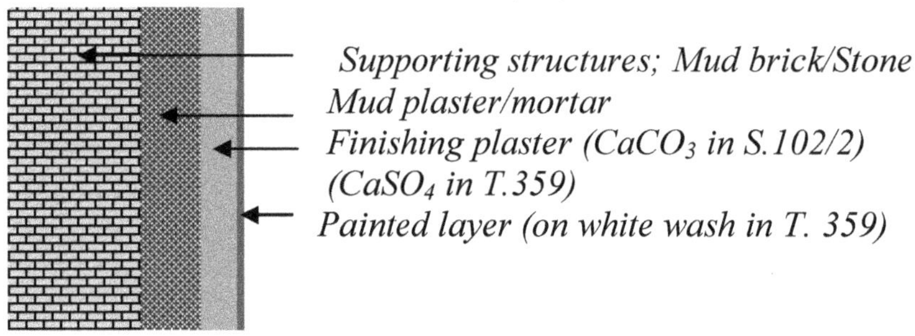

(7.2)

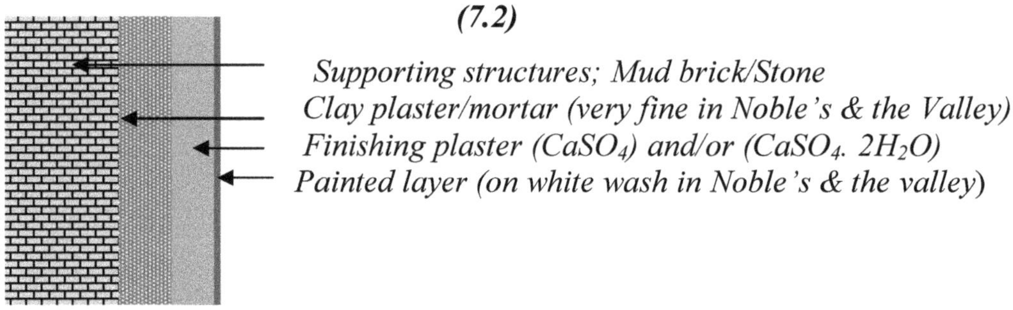

(7.3)

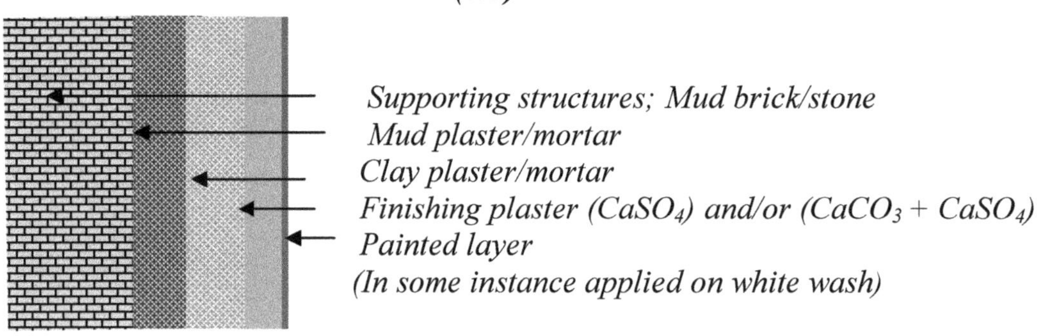

Figure 7.1, the structure found in first Dynasty (E E) (supporting structure; mud brick) Saqqara and Deir El Medina T. 359 and T. 414 (supporting structure; stone) Luxor.

Figure 7.2, structure in first dynasty (N E), Dahshur Mastaba 4th dynasty (supporting structure; mud brick), Noble's tomb T. 56 & 55 & the Valley of Kings (supporting structure; stone) Luxor (see Appendix 6.1, the measurements of cross-sections).

Figure 7.3, the structure found in Ka-Nefr 5th dynasty & Idy tomb 6th dynasty (supporting structure: mud brick and in some instances; stone) Drah Abu El Naga T. 255 & T. 13 and Khokha T. 192 (supporting structure; stone) Luxor.

Figure 8

The Stratigraphy of Græco-Roman Plasters/Mortars
Alexandrian Sites and Piddington
Diagrammatic; not to scale

(8.1)

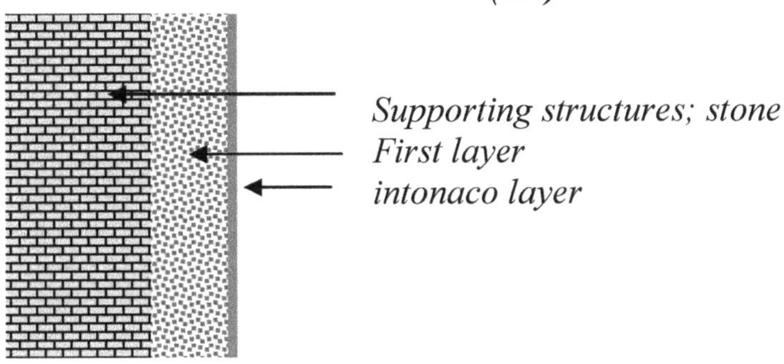

(8.2)

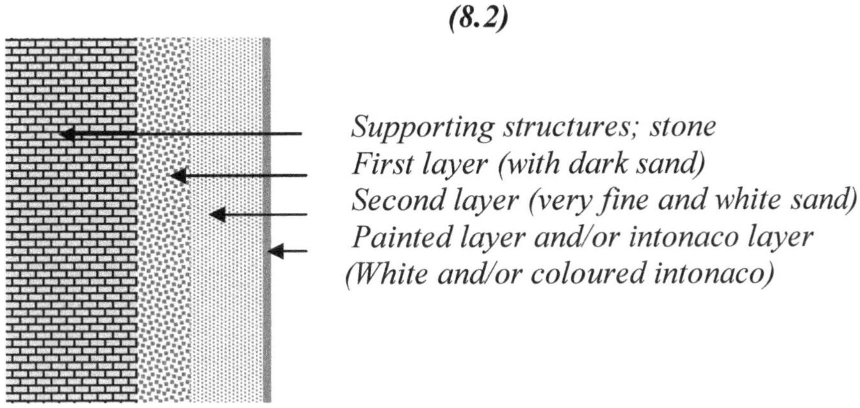

(8.3)

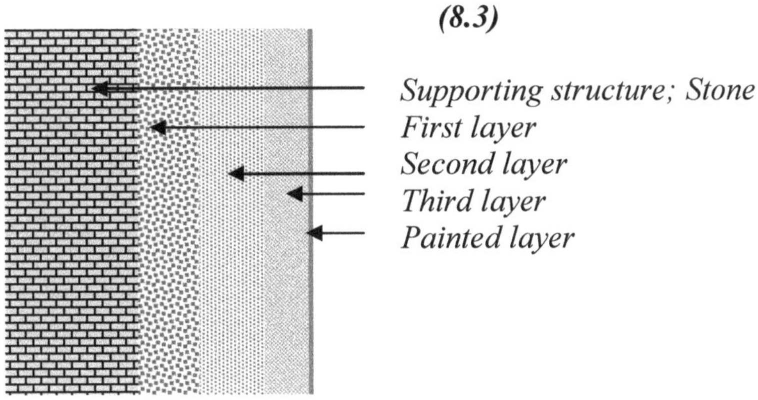

Figure 8.1, found in Alexandrian sites e.g., Mustafa Pasha T. 1.

Figure 8.2 & 8.3 identified in Alexandrian sites and Piddington wall plasters and mortars (see Appendix 6.2-3, the measurements of cross-sections).

Table 14 A, Summary of the Types of Plasters/Mortar from Different Sites, Periods and Tombs

Egyptian Plasters/Mortars

The Stratigraphy of the Plasters/Mortars	Tomb	Period	Site
Stratigraphy 1			
Supporting structure (mud brick) Mud plaster/mortar (see Appendix 16.6) Finishing layer ($CaCO_3$) (see Appendix 16.11) Painted layer	Herneith	1st Dynasty (E E)	Saqqara
Supporting structure (limestone) Mud plaster/mortar Finishing layer ($CaSO_4$, by chemical test) Painted layer (on unidentified whitewash)	Inherkhau T. 359	20th Dynasty	Luxor Deir El-Median
Supporting structure (mud brick & limestone) Mud plaster/mortar (see Appendix 16.7) Finishing layer ($CaCO_3$)	Ankh-Hour T. 414	26th Dynasty	Luxor Asasif
Stratigraphy 2			
Supporting structure (mud brick) Clay plaster/mortar (see Appendix 16.9) Finishing layer ($CaSO_4.2H_2O$) & ($CaCO_3$) (see Appendix 16.11)	-	1st Dynasty (N E)	Saqqara
Supporting structure (mud brick) Clay plaster/mortar Finishing layer ($CaSO_4.2H_2O$) (see Appendix 16.11)	Dahshur mastaba	4th Dynasty	Dahshur
Supporting structure (limestone and in some instances, mud brick) Clay and/ or mud plaster/mortar Finishing layer Painted layer (on unidentified whitewash)	Noble's tomb Userhēt, T. 56 & Ramose, T. 55	18th Dynasty	Luxor Sheikh Abd El-Qurna
Supporting structure (limestone) Clay plaster/mortar (see Appendix 16.10) Finishing layer Painted layer (on unidentified whitewash)	Queen Tausert, T. 14 Stethos II, T. 15 & Ramesses III, T. 11	19th Dynasty 20th Dynasty	Luxor The Valley of Kings

Table 14 B, Summary of the Types of Plasters/Mortar from Different Sites, Periods and Tombs (continued)

Egyptian Plasters/Mortars

The Stratigraphy of the Plasters/Mortars	Tomb	Period	Site
Stratigraphy 3			
Supporting structure (mud brick and limestone) Mud plaster/mortar Clay plaster/mortar (see Appendix 16.8) Finishing layer ($CaSO_4.2H_2O$) & ($CaSO_4$) (see Appendix 16.12) Painted layer	Ka-Nafr & Idy tomb	5th Dynasty Late 6th Dynasty	Saqqara
Supporting structure (limestone) Mud plaster/mortar Clay plaster/mortar Finishing layer ($CaCO_3$), ($CaSO_4$) & ($CaSO_4.2H_2O$) or ($CaCO_3$) & or ($CaSO_4$) (see Appendix 16.13) Painted layer	Roy , T. 255 & Shuroy, T. 13	18th Dynasty 19th Dynasty	Luxor Dra Abu El-Naga
Supporting structure (limestone) Mud plaster/mortar Clay plaster/mortar (see Appendix 16.9) Finishing layer ($CaSO_4$) and ($CaCO_3$) (see Appendix 16.13) Painted layer (on unidentified whitewash)	Kheruef, T. 192	19th Dynasty	Luxor Khokha

Table 14 C, Summary of the Types of Plasters/Mortar from Different Sites, Periods and Tombs (continued)

Græco-Roman Plasters/Mortars- Alexandria and Piddington

The Stratigraphy of the Plasters/Mortars	Tomb	Period	Site
Stratigraphy 1			
Supporting structure (limestone) First layer ($CaCO_3$ + sand) *Intonaco* layer ($CaCO_3$ + very fine grain sizes)	T. 1	2-3rd Century BC	Alexandria Mustafa Pasha
Stratigraphy 2			
Supporting structure (limestone) First layer ($CaCO_3$ + with dark sand) ($CaCO_3$ + aggregates in Piddington) Second layer ($CaCO_3$+ very fine white sand) ($CaCO_3$ + aggregates in Piddington) *Painted layer and/intonaco* layer ($CaCO_3$ + very fine grain sizes) (white and/or coloured *intonaco*)	T. 1 & 2, T. 1 & T. 2 & Roman villa	2-3rd Century BC 3rd Century BC & re-used in Roman Times 4th Century AD	Alexandria Mustafa Pasha, Anfushi & Piddington

Table 14 D, Summary of the Types of Plasters/Mortar from Different Sites, Periods and Tombs (continued)

Græco-Roman Plasters/Mortars- Alexandria and Piddington

The Stratigraphy of the Plasters/Mortars	Tomb	Period	Site
Stratigraphy 3			
Supporting structure (limestone) First layer ($CaCO_3$ with dark sand) ($CaCO_3$ + aggregates in Piddington) Second layer ($CaCO_3$ + very fine white sand) ($CaCO_3$ + aggregates in Piddington) Third layer ($CaCO_3$ + very fine white sand) ($CaCO_3$ + aggregates in Piddington) Painted layer	T. 1 Roman villa	2-3rd Century BC 4th Century AD	Alexandria Mustafa Pasha Piddington

Egyptian plasters and mortars were made from mud brick and/or stone covered with a layer of mud plaster/mortar and then another layer, consisting of clay plaster/mortar. In some instances the clay plaster, used to cover the supporting structures, received a layer of gypsum and/or calcium carbonate or a mixture of both as a finishing plaster. In the case of good quality stone, a layer of gypsum plaster was applied. The plasters and mortars were of different thicknesses, but the finishing plasters were thinner, as observed in the Luxor samples (Appendix 6.1). The methods used varied between the sites and areas of the samples, as illustrated above.

The mud plaster (coarse plaster) was of an alluvial mud base, mixed with other additives, the clay plaster was based on a mixture of powdered limestone and clay with other additives, with more or less the same additives in all samples, as stated in Appendices 9.3, 9.6, 10.4-5 and 10.9-10. **Clay plaster** (better plaster) had a very high carbonate content due to the powdered limestone used, which was found to be full of different micro-fossils, such as foraminifera, coccoliths and other shell fragments. This was confirmed by SEM (photographs 25, 41 B, 43 B & C, 45 and Appendix 19.3, photograph 16 B) and also revealed by PLM in the mixtures from the Alexandrian sites (photographs 37 and 38) as well as in the mixtures of the lime-based Piddington plasters/mortars (photographs 40 B) as stated in Appendix 18. These microfossils gave an indication of the composition and structure of the limestone formations. They are the natural forms of calcium carbonate, as calcite, which occur in sedimentary rocks like chalk. Calcite is the main constituent of some mollusc shells and fossils of the lower organisms. It also occurs as the skeletal materials of many forms of marine life, such as foraminifera and coccolithophoridate (see Gettens *et al.* 1974).

The clay mix of powdered limestone and clay used in the plaster/mortar samples, acts as a binder with other additives, it has a high plasticity and the cohesive properties of clay/mud enable it to adhere without any addition of building materials. This was noticed in the washing of a sample of mortar filler based on clay and crushed limestone, used to cover irregular surfaces (see Appendix 16.4, S.16). The high percentage of calcium carbonate found in clay plasters samples, compared with mud plasters based on Nile alluvium, of which the differentiation is attributed to carbon in the soil . This was different to the Alexandrian and Piddington plasters/mortars as the presence there of calcium carbonate was in the form of chalk/lime based materials.

Analyses have shown not only variation in the composition of the plasters/mortars used in antiquity, but also the percentages of the sediment types of gravel, sand and silt. In addition, the percentages of acid soluble materials differed due to the use of powdered limestone in the samples, in particular, clay plasters/mortars. The approximate proportion of binder to aggregate and the chemical composition of the binder were determined by quantitative chemical analysis, as illustrated in Appendices 9 -10.

The purpose of organic materials such as straw, used in the Egyptian plasters and mortars, was to bind the aggregates, due to their natural capacity for cohesion, which gave more strength to the mixture when mixed with clay and/or mud alluvium. The straw used was brownish, differing from the original white straw mixed with other additives, which was found in some instances. This suggested that they may have used charred straw as well. The reason for that can be assumed by the need to use dry material to ensure the correct mixture. The use of charred plants was reliably confirmed by their presence in the mixture of the plasters/mortars.

Bricks or their ceramic material (fired clay/burnt mud brick) were found in the mixtures of both mud bricks as well as mud/clay plasters. Brick has also been found in the mixtures of gypsum mortars from the 4th dynasty pyramid of Cheops in Giza, (2900-2750 BC) and a Roman theatre in Alexandria, dated to 2nd–3rd century AD. They indicated that the mortars used were mainly gypsum containing ceramic (brick) and fillers, the binder being crystallised gypsum it is possible that the burnt brick material may be kiln residues from burning of gypsum but the original sources is unknown. Gypsum mortar had been used in ancient times and the technology of gypsum binders spread from Egypt to Babylon, and via Asia Minor and Greece to Rome, where similarity may be found in to the gypsum mortar used in ancient and classical times. It was assumed that Egyptian gypsum was burnt at a lower temperature, due to the application of quick-burning vegetable fuel rather than wood, the calorific value of which is much higher. In addition, the preserved carbonized organic skeletons in the grains of ceramics in the Egyptian mortars indicated the relatively low temperature of clay-heat (Penkala and Bralewska 1981).

The analysis of **the finishing plaster** did not show any lime. From X-RPD analyses I have found that gypsum was in the form of hydrate and anhydrite, as stated in Table 10. In some instances, the plaster was a layer of calcium carbonate only, as in sample S.102/1 (see Appendix 16.11). This sample in particular, dated to the first dynasty, gives a clear idea of the development of the manufacture technology for making plaster compared with other samples as illustrated in figure 7.

The finishing plaster was often found to be a mixture of gypsum and calcium carbonate, but there is a suggestion that this may be natural gypsum in the form of an impurity. Micro-chemical tests of some samples gave a little effervescence of calcium carbonate and the residue showed some other minerals in very fine particle sizes, giving evidence that gypsum may have been of poor quality, as confirmed by PLM (compare photographs 32 and 33).

A recent study carried out on the analyses of mortar from the Old Kingdom in Giza (Sphinx, Cheops and Chefren pyramids) indicated that the mortars were made of gypsum, sand and limestone. Others were made from pure limestone, while mortars from the Chefren pyramid gave evidence for the presence of gypsum only. Although it is assumed that kiln temperatures around 950 °C could be reached in Egypt during the period of the Old Kingdom, e.g., for copper alloy casting and pottery production, there is a doubt whether limestone was calcined and that there was re-carbonization or a formation of cement by secondary reaction, as the re-carbonization leads to fixation of CO_2. The problem of the differentiation of natural limestone and processed limestone is rather difficult, as there are no clear-cut differences between the morphology of natural limestone particles and calcined/re-carbonated particles (Reller *et al.* 1992). The presence of intact coccoliths in many samples is an indication that they were not burnt. The burning of limestone would normally damaging them.

I think that the use of gypsum, rather than lime plaster, was not because the Egyptians could not reach the kiln temperature required to make lime (700-900 °C), but was for two other reasons. The first was that they understood that the very dry weather in Egypt was suitable for the use of gypsum as a finishing plaster and not lime, which is demonstrated by the fact that the paintings have survived until now. In addition, the use of distemper rather than *fresco* techniques, led to the development of the technology for gypsum. The second reason was related to their customs, which demanded the decoration of tombs and temples. It would be expensive to produce and prepare such a mass of lime based plaster/mortar. The majority of the bibliographic sources indicate that gypsum plaster was used by the Egyptians until the Ptolemaic period (3rd century) when lime started to be produced in Egypt, which this research has also shown. However, the presence of magnesium carbonate hydroxide hydrate/hydromagnesite ($Mg_4(OH)_2(CO_3)_3 3H_2O$) in sample S.46, gave evidence that the Egyptians could know how to produce lime, although it may have been unintentional. It was known at an earlier date than mentioned by Davey (1961: 94) and Lucas (1989: 77-79), but the use of gypsum was preferable for the reasons mentioned above. Therefore, it could be possible to find an example showing the use of lime plaster/mortar elsewhere. The use of lime at a very early date, c 9000BP, is attested by the finds at Ain Ghazal (Tubb 1987).

Græco-Roman plasters and mortars, as illustrated in Figure 8, were based on lime/chalk and/or marble-dust as found in the **Alexandrian sites** and described by Vitruvius. The underlying layers found in the Alexandrian samples were normally sand and lime, as a two-or three to one mixture of different thicknesses as illustrated in Appendix 6.1-3 (see also photographs 20 E and 21 B & E). The finishing layer was fine white lime for the application of the wall painting called the "*intonaco*" (see photographs 20 A& B and 21 E). This technique was also found in the Piddington samples, although dated to the 4th century AD (see photographs 22 B and D). However, the plasters and mortars used in **Piddington** Roman villa were built up from layers based on slaked lime (calcium hydroxide). The plaster and mortar from Piddington Roman villa were of different thickness (see Appendix 6.4) and showed purely the mixtures of the mortar used in Roman times, in which crushed brick was used extensively in the mixture with other additives. The use of lime/chalk was to act as a binder.

The Græco-Roman plasters and mortars normally consisted of one, two or three layers according to the location as showed in the present study of Piddington wall plasters. The stratigraphy of the plaster could differ according to whether tomb/domestic or house. A study of Roman wall painting from Cyprus showed that the stratigraphy of the plasters used, was different in houses (private/public) and tombs. In houses, an elaborate *intonaco* consisting of 3-4 thin layers of alternating pure calcareous aggregates and a mixture of calcareous and rock fragments was applied. Whilst, tombs were executed in much less complicated ways. One final layer (*intonaco*), consisting primarily of calcareous aggregate, was applied over the rougher layer made of calcareous aggregates and rock fragments (Kakoulli 1997: 140). I never found six layers as described by Vitruvius and stated in Chapter 2 (see 3.2.4.1). However, plaster samples from Settefinstre (Tuscany) revealed that they were composed of five layers and river sand has been identified in the mixture as well (De Vos *et al.* 1982: 21, 22). The aggregates used in traditional lime mortars were natural sand and gravel and crushed materials such as shell, various types of rock, brick and old crushed lime mortar, as shown in the Piddington samples and indicated also by Morgan in the study of Romano-British plasters and mortars (1992), and also Béarat *et al.* (1997).

The examination of the plasters/mortars has shown that the techniques used for painting were of ***tempera/*distemper** for the Egyptian plasters/mortars, after the painting of the surface with whitewash; however, it is not clear if that layer was lime or gypsum. It was very difficult to distinguish the original material applied because the whitewash was in a very thin layer and adhered to the pigment/plaster, as observed in some instances. Lucas stated that fine quality of gypsum as a mixture of whiting and glue was used to fill the pores and smooth the surface (see Chapter 2, 4.3.2).

In classical times both ***buon** fresco* and ***fresco secco*** was applied in the Alexandrian sites. In Roman times, at Piddington, the technique used was *buon fresco.* A polishing/burnishing technique was observed in the Alexandrian sites, as on an example of Græco-Roman 2nd–3rd BC material, and at Piddington in the 4th century AD.

The results of analysis of the pigments indicated that the colours used were traditional (see Chapter 6, table 6, the identification of the pigments from Piddington, and X-RD identification, table 7) and were widely used in both ancient and classical times, as discussed in Chapter 2 (4.1.1) and have been identified (Béarat *et al.* 1997)**.** However, no presence of cinnabar was detected in either the Alexandrian or Piddington samples studied, although it (HgS) was reported from other samples from Piddington (Morgan 1992). Egyptian blue, a synthetic pigment, was identified in the Alexandrian sites as well as at Piddington, and could vary according to the manufacture of the pigment.

Differentiating between *buon fresco* and *fresco secco* was a hard task. Hydrochloric acid dissolves lime with effervescence, which indicates *buon fresco*; on the other hand, as Doerner mentioned, *secco* colours may contain lime or marble dust which also effervesce. *Fresco* can be distinguished because it does not dissolve in water whilst *tempera* does, in the case of glue distemper, but not emulsion *tempera*.

I did not see any traces of joins in ancient *fresco* in the Alexandrian sites, which is necessary to determine the technique. It has been said that the works were executed by a group of painters instead of one (the same as Egyptian paintings), an example of unfinished decoration is present in the house of Iliadic Shrine at Pompeii described by Ling (1991) show that the work was in progress on the last six separated rooms and had reached varying stages of the completion (1991: 215-217). In addition, the thicker plaster layers were the reason for the wall retaining its moisture for a week or more. One theory is that *fresco* had its beginnings in Byzantine mosaics, according to the execution technique described as day-work or *giornata di lavoro* (Doerner 1969: 304, 336).

The mineralogical and chemical analysis of the plasters and mortars yields information about the type of binders, the type and grading of the aggregates and their quantitative composition. The results of these investigations give information for the implementation of plaster and mortar preparation, giving approximate calculations for the binder and aggregate ratios used in Egyptian and classical times, as summarized in Table 15 and illustrated in Appendices 9-12, and to identify the mixtures, which can be used for the repair of wall paintings by modern conservators.

Table 15, Summary of the Ranges of Percentages of the Sediment Types Found in Wall Plasters and Mortars

	Mud Plaster/Mortars	
	Old Kingdom	**New Kingdom**
Acid Soluble	2-14	13-23
Gravel	3	2-6
Sand	12-53	8-22
Silt	43-87	58-92

	Clay Plaster/Mortars	
	Old Kingdom	**New Kingdom**
Acid Soluble	9-52	16-84
Gravel	1-25	5-38
Sand	18-74	8-50
Silt	25-81	50-88

	Lime Plaster/Mortars
	Alexandrian sites
Acid Soluble	36-73
Gravel	0.5-2
Sand	17-82
Silt	18-40

	Piddington		
	First Layers	**Second Layers**	**Third layer**
Acid Soluble	37-57	39-57	49
Gravel	6-40	4-38	23
Sand	35-68	38-59	47
Silt	16-38	20-38	30

The results of the analysis of re-plastering samples show that the plasters were not compatible with the original mixtures. Analysis allows us not only to identify the mixture used, but also shows the percentages of sediment types found in accurate proportions, as stated in Appendix 10.13-15. The aggregate grain sizes used were too large and made the plaster porous, in addition to the cracks which were observed in *situ*. This may have been as a result of the use of that particular sand and/or that too much water was used in the mixture. Cracks occur if sand contains clay and may cause efflorescence on the surface as well. New plaster has high surface tension because of the use of strong additives, like gravel, cement-based materials, etc. Another thing observed in *situ* was that a thin layer of lime/cement particles on the surface of the new plaster was generally caused by the mortar being over-worked or by using a mortar that was too wet. This led to a loss of surface porosity in the materials.

It is worth mentioning that the analytical methods used in the examination and analysis of the plasters/mortars and the wall painting fragments, helped to identify **the main causes of deterioration and decay** which affected both the mortar and the paint layers. The formation of the crystal deposits identified by X-RD, as gypsum crystals, inside the mixture of clay plaster/mortar was a result of decomposition caused under certain conditions. The loss of organic binders such as straw, eaten by insects, or other organisms such as bacteria, in addition to the dry atmosphere was responsible for the plaster being so fragile. Bio-deterioration was noticed inside and on the surface of the plasters as well, and was confirmed by the examination of the surface using SEM (photographs 44 C & D & Appendix 18.6). The decay of wall painting fragments from **Piddington** was due to different alteration, resulting from the dissolution of materials and the promotion of fungal growth, attributed to wet conditions (photographs 45).

The analysis of the water-soluble salts indicated that the most significant salts were chlorides, sulphates and nitrates, and interestingly, in some instances, nitrites as illustrated in tables 3 A-D. Salt was the main problem found in the Alexandrian sites in particular. SEM identification determined the different shapes of the crystals as shown in Photographs 47-49 (see also Appendix 18.5). As mentioned before, wall paintings usually contain soluble salts as impurities in the materials used or formed by chemical reaction with the surroundings under certain conditions.

It is of particular interest that Smeaton and Burns said that the presence of sodium chloride (NaCl) may well help catalyze the dehydration of gypsum to anhydrite (Smeaton and Burns said 1988: 303). The presence of sodium chloride found in the stones, containing gypsum as a part of its formation, may have been responsible for the presence of anhydrite, particularly in the clay plaster. These stones were used in the form of powder for making mortar/plasters. However, the movement of salt from the supporting structure through the wall to the finishing plaster may have been responsible for finding some samples of gypsum plaster in the form of anhydrite, in addition to other factors as stated before.

The total amount of water and soluble metal ions present in samples was determined by ICP. The results are reported in Appendix 17, and confirmed the micro-chemical results which indicated the actual salts present, in particular sodium chloride in Egypt. As previously mentioned in Chapter 3, it can be summarized that the sources of salts are varied, either from different origins or from the chemical reaction of the materials. Salt crystallization occurs under certain conditions and, in the case of water containing a high enough concentration of carbon dioxide (CO_2), lime/limestone could be dissolved as soluble bicarbonate. This may dissolve in water in the wall in humid conditions. Periodical changes of the temperature allow the walls to dry and then the bicarbonate solution comes to the surface by evaporation, resulting in the equilibrium between various substances and the formation of insoluble calcium carbonates, which deposit on the surface in the form of salt crystallization. This happened in some Alexandrian sites, in particular the Anfushi tombs. In addition, other factors, particularly ground water, resulted in the production of different salts, as well as different characters in the form of crust or powder.

Mineralogical analysis of the plasters/mortars by thin section analysis with the combination of **the results of X-RPD** give a view of the mineralogical compounds in the mixtures and determine the characteristic of the clay/mud used in Egyptian plasters, as discussed in Chapter 6 (see 3.1.3) and stated in Appendices 16 and Tables 11-13. **The identifications by PLM** of different kinds of micro-organisms, micro-fossils and other shell fragments as found in the mixture of the Egyptian and Græco-Roman wall plasters and mortars, where found similar in some instances, are shown in Photographs 28-40.

The physical methods used, such as non-destructive techniques, are very useful in confirming the results obtained using chemical and micro-chemical tests, and in addition, these tests also allow the reuse of the samples for future work as well. However, both chemical and physical analysis is required to understand and identify the materials used, in order to obtain a full picture of the manufacturing technology used and help to determine the causes and source of deterioration.

2-Recommendations

There are many basic ethical points about conservation and restoration which should be considered. Conservation and restoration is an interference with the structure and history of the materials and every treatment must be fully considered and recorded. Ideally, in any conservation/treatment, added material should be removeable at a later date, allowing mistakes to be rectified. Any repair most be obvious, see below. Great care must be taken with compatability when using new material to conserve the old: such as problems as; differential shrinkage, strengths, bonding power, resistance to decay and porosity must all be considered before their use.

The advantages of conservation are that although conservation may affect the appearance of the wall or plaster, it may strengthen and help the preservation of the original. Conservation and restoration should help to preserve the structure for future generations. Disadvantage of conservation and restoration occur with problems in matching supplies and that new material will not be weathered or decayed like the original and unless great care is taken new repairs may mask original work. There are recommendations for the distinction of old and new work e.g., at one metre it should not be obvious to the casual observer, but at less than one metre clearly obvious to the expert. This project has shown that lime was not used in Egyptian plasters but that they were based on Nile alluvium and clay in the form of powdered limestone and clay. Gypsum plaster was used as a layer of finishing plaster. The preparation of new plaster for restoration work, must be more or less the same as the original materials used in antiquities in order to achieve correct restoration/re-construction of the materials, therefore lime should not be used for Egyptian plasters, except for those of Græco-Roman origin. I think the preparation of new plasters must be more or less the same as the original. Unless it is necessary to use only standard methods, it seems to be very useful, to identify and determine not only the sediment types used, but also the approximate proportions of each mesh size, which could result in it being possible to match the original mixture. However, there are some important considerations.

Firstly, investigation of the original materials used for the preparation of the new plaster is essential. If the aggregate contains a considerable amount of fine argillaceous material, this causes cracking. Very porous mortars are not required, because if the mortar is very porous, excessively rapid evaporation will take place and salt will accumulate on the surface. Sand is important in the mixture of the plaster for its quality, quantity and grain size. If it contains clay, it causes cracks, so it must be washed, dried and not stored on open ground. **Secondly** the methods of application, including tools, the process of preparation of the walls, the time scale of application, etc., must be sufficient for the preparation of new plasters/mortars. **Thirdly**, Documentation should be addressed in order to differentiate the old/new re-plaster and the original ones because this is very important consideration in the process of restoration/re-construction.

●-**The function of sand** is to strengthen the plaster by eliminating shrinkage during the hardening process. When the correct amount of sand is used, the coarse grains touch each other and the finer particles of lime surrounding them fill in voids between them. A porous condition in the plaster may be a result, but the mass content cannot shrink. In contrast, rounded grains roll easily over each other and too many rounded grains will give rise to a poorly bonded mortar. So, a mixture of sharp and rounded grains will form an equally well-knit permanent mass. However, their freedom from soluble salt impurities is a consideration of great importance. Sea sand is not good because of the salt unless thoroughly washed and dried. Vitruvius gave different recipes for the use of different sands, as stated in Chapter 2 (see 4.5.1). If the sand is wet, lime putty will not adhere to it. Marble dust is recommended in place of sand, because of its angular particles.

●-**The use of extra water** with mortar made from traditional lime putty must be considered, because the water contained within the putty will be released on "knocking up" and that should be sufficient to produce a good workable mix. If there is any need to add water, it should not be from sources which are likely to introduce salts.

•-In the long term **hard/brittle mortar** will be susceptible to cracking, as a result of seasonal thermal or structural movement. In addition, it will inhibit evaporation and can trap moisture in the building materials. This may lead to extensive micro-cracking, which will admit water under wind pressure and capillary action. Soft mortar developed cracks if used too wet or over worked in *situ* (see Doerner 1969: 296-271, 340-342 and Historic Scotland 1995: 8-13 & 40-51).

•-**The use of cement** is not recommended in the preparation of a new plaster, because the soluble salts present in the cement causes destructive efflorescence or crystallization upon evaporation, in addition to its low porosity as previously stated in Chapter 3 (see 4.1). Modern cements also are much harder than the original materials. However, Torraca mentioned that low alkali cement should be used to reduce the formation of soluble salts. Another way to match the old mortar is to dilute cement with calcium carbonate (Torraca 1988: 80, 81). See also Holmstrőm (1982).

•-**The use of lime** can be acceptable if the original mixture is lime based, e.g. in the Alexandrian sites. The preparation methods are an important aspect in order to achieve high quality new plaster, and this depends on the quality of the materials, such as aggregates, lime, etc. In addition, the environmental condition for application should be considered wet/dry etc.

•-**The quality of the lime** depends on the original composition of limestone. White limestone is preferable because of the high calcium carbonate content, about 95%, which produces high quality lime. One should avoid grey or yellowish and brownish limestone as they contain iron oxides producing coloured lime. The process of slaking the lime, including temperature and the amount of the water used, are important. Too much water will produce a thin mixture, instead of drier plastic putty, and results in the temperature being too low for effective slaking. In contrast, too little water causes burning and will not produce the firm bonding of sufficient lime. The fire used also has an important effect on the correct slaking. As mentioned before, a wood fire is recommended rather than coal, because lime will absorb sulphuric acid from coal, which will cause problems, with the setting time and efflorescence similary fuel oil in modern commercial lime production will add sulphur. Ageing of lime putty is necessary to improve the plasticity, so it may be prepared from six months to a year before use. As mentioned before, in some instances lime contained un-slaked particles of quick lime which, when applied, complete their slaking on the wall. This causes defects on the wall resulting in "popping or blowing". Lime slaking must be carried out carefully and thoroughly, to avoid leaving small particles of un-slaked lime in the mixture and the lime should be sieved to prevent any sediment from being transferred into the fresh supply of lime putty.

•-**The use of appropriate hydraulic lime** (influenced by the existence of certain impurities in the minerals, such as silicates or aluminates present in the original limestone) gives the plaster some ability to set in wet conditions in such environment. There is good reason to use this lime, instead of cement in the case of quick setting or additional durability, which is required in some instances. This is more likely to result in a more sympathetic and potentially less damaging mortar than one containing Portland cement (Historic Scotland 1995: 10, 19).

•-The preparation of a new mortar is an important aspect in re-construction. It is necessary to fully understand all the materials used. Historic Scotland (1995), Peroni *et al.* (1982) and Holmstrőm (1982), give general principles for making mortars, their preparation and provided simple tests for examination, which can be carried out to evaluate the suitability of sand and other materials which may be used for preparation of new mortars. This helps to achieve a correct mixture and must match the original used in antiquity, with some consideration according to the original materials used. These results help to avoid problems that can occur and develop in the future.

3-Future Research

•-Analysis Plaster and Mortar

Although I carried limited testes on lime mortars I carried out several trials on the preparation of a new mortar at the Woodchester Mansion Trust "*Lime mortars*" and Orton Trust "*Lime mortar in the maintenance and conservation of stone buildings*". The use of lime putty, which was made a year earlier, gave good results without the addition of any water. I also experimented in preparing different mortars with different types of sand grains, and the addition of hair in the mixture, as used in traditional mixtures (see Appendix 20). Further investigations need to be carried out in the future on materials used in Egypt.

Further work would be useful on testing different mixtures for the peparing of mortar and plaster. Another important area would be to compare materials and technologies from other sites, relating them to conservation problems such as salt decay. I have had in this project but I feel that it would be a very interesting area to pursue.

•-Identification of Organic Materials

My research was concentrated on the analysis of plaster/mortar, using different methods to identify the structure, composition and additives and determine the deterioration factors that affected the plaster/mortar, in addition to identifying of where some pigments were found on the plaster. The main difficulty that I found with the analysis of binding media was the lack of published research and the need to produce experimental information.

As I had not originally planned to use chemical techniques, staining analysis was not involved in this research, so that this aspect of the identification of binding media could be the subject of future work. This was because my samples were not large enough, and the paint layer was too dry and powdery, to carry out such research. However, throughout the research it seemed increasingly necessary to carry out some work on the identification of the binding media, as some studies indicated the presence of organic media, particularly in Alexandrian as well as Egyptian paintings, where several media were found and a few studies have indicated the use of various media.

Numerous studies have been carried out on the identification of binding media using staining tests on cross-sections, using different chemical and reagents, such as Amido black for proteins or Sudan black for oils, as well as paper chromatography for amino-acids, to determine the presence of organic materials or binding media such as: egg, glue, gum Arabic, etc., used alone or mixed with the pigments (Jones P. 1962, Elzinga-Tar Haar 1971, Johnson & Pakard 1971, Masschelein-Kleiner 1974, Mertin 1977 and Kockaert *et al.* 1988). The method was described by Plesters (1955-56) and van Asperen De Boer (1984) who gave a review of these methods. This involved the identification of the binding media, considered as the one aspect of the identification of the painted layer which should precede investigation with instrumental methods such as gas chromatography (van Asperen De Boer 1984). The problem of the analysis of binding media seemed to be more difficult to solve than that of pigments analysis for two main reasons: firstly the proportion of medium in a given sample is much smaller than the proportion of pigment, and secondly, physical and chemical changes may have taken place with time, so unambiguous identification would not be possible (Hey 1959: 183).

It was a great opportunity however, to use infrared spectroscopy (FTIR) in the identification of a few samples as standards of some organic substances, such as animal glue, whole egg, white and yolk of egg, gelatin, gum Arabic and glue in order to compare, and identify the original composition of the organic materials and determine the binding media used. This would be useful in future work.

Infrared Spectroscopy (FTIR) was used to identify organic substances. This provides different types of information from the different parts of the IR spectrum obtained for gaseous, liquid or solid samples (Ferretti 1993 and Gunstone *et al.* 1970). Although this technique is useful for classifying a range of materials associated with paint and plasters, including synthetic polymers, but produces non-specific identification of the materials, e.g. it can identify oil, but not the type of oil and there are problems in defining some classes of materials occurring in mixtures (Burnstock 1998: 7).

Gas Chromatography/Mass Spectroscopy GC/MS was introduced in 1952 and become established as an excellent method for the identification of organic materials by separating, detecting and quantifying the compounds. The advantage of this instrument is that analysis can be undertaken on a very small sample and gives the compound structure. Rose mentioned that the combination of GC/MS exploits the resolving power of gas chromatography and then the ability of mass spectrometry to identify the separate substances, giving the characterization of the mixture (see Rose 1990). (For more information see Mellon 1991).

It has been mentioned that there are different instrumental techniques that have been proposed in the literature for the identification of organic media, but the most reliable one is based on the hydrolysis of the protein present in the sample for the determination on the relevant amino acids, using high performance liquid chromatography or gas chromatography coupled with spectrophotometer or mass spectrometry. The quantitative determination of amino acids and the selection ratios between their concentration can identify the organic binder in the wall painting samples (Colombini *et al.* 1998).

•-Conservation Methods for the Treatment of Salt Crystallization

The study was focused on examination and analysis of the materials of plasters/mortars. However, it was essential to identify the salts, which were found covering the walls in the Anfushi tombs and appeared to be the main factors causing deterioration and decay. The deterioration in a wall painting is always related to some defect in the structure of the wall such as humidity, salt crystallization, etc., and its related effects, as previously stated in Chapter 3. However, to ensure the right methodology for conservation/restoration the main problems must be understood and the decay factors that affect the wall painting must be identified, both in the rendering and the painted layer.

From the investigation and analyses of different salts from the Alexandrian sites, in particular the Anfushi tombs, it was clear that the main problems of salt crystallization found in Anfushi related to many sources, such as underground water, chemical materials used for conservation, etc., as mentioned before. When salt is accumulated on the surface, removing it is a simple matter, but it will soon return under certain condition, as happened in Anfushi.

My interest is to find solutions to such salt problems and that would need the teamwork of different specialists. The need is to know how to control such problems and select suitable methods for treatment. These can vary with the type of building, its location and the significance of the accepted conservation philosophy of the country. In addition, the techniques of wall painting used should be considered, to avoid any future problems.

Some important studies have been carried out on the treatment of such phenomena (Arnold, 1984 and 1998, Hammer 1995 and also Van Hees & Koek 1995). In addition, the studies using the barium methods developed by Florentine (Matteini & Moles 1984 and Matteini 1987) gave good results in removing salts, in the case of sulphation. However, applying such methods is not easy, unless the main salts present are identified and the restoration techniques are used with great consideration to the previous techniques of mural painting. Hammer mentioned the treatment of mural painting and the surfaces of historic architecture deteriorated by humidity and salt, and described two interdependent aspects of radical and/or cosmetic intervention, which make invisible the decay phenomena without treating the causes of decay (Hammer 1995).

One method that could be used is that of the heating system; however, this depends on understanding how to employ it with mural painting. Beck and Koller have reviewed and described briefly a number of historic heating devices and presented many examples of treatment systems including many measurements of temperature and humidity within buildings and concluded that constant floor heating systems minimize damage (Beck and Koller 1980).

The stabilization of the relative humidity and temperature requires several different measurements, including the moisture contained in the wall as well as a range of environmental parameters. To achieve such work, it is necessary to investigate the problems on site and find out suitable methods of preventing such phenomena, which requires time. A useful study has been carried out by Valentini *et at*. (1994) of treating salt crystallization and their strategy of the treatment which could be carried out on site, in addition to some studies of damp buildings by Giovanni and Massari (1993) and the method of measurement of the humidity in the walls (Massari 1998).

My aim for future research, is to determine the methods which could be used for treating the problems of salt crystallization, and depending on my results, to determine the actual salts found in the Anfushi sites, which would help the search for convenient methods of application.

Part II

This Contains Appendices Part I
"The Study of Plasters and Mortars from the Four Sites: Saqqara, Luxor, Alexandria and Piddington"

Catalogues of Samples,
Tables,
Graphs
& Photographs

Appendix 1

1.1-A Summary of Previous M.Sc. Research

Mural Design in Ancient Egyptian Architecture "3100-1085 BC"

An insight is given in this appendix of the beginning of wall painting in Egypt and the style of the decoration which developed in ancient Egyptian times, from the principles and rules that were formed in the dynastic period. This appendix is from previous research carried out by the author on "***Mural Design in Ancient Egyptian Architecture "3100-1085 BC"***. M.Sc. 1992. I would like to give a summary of that research as the study of Egyptian wall painting was a part of the work that considered the methods of execution and techniques which were used for the design. Since design is important in the fields of arts, I directed my study towards this subject, especially mural design, which has a specific role, owing to its relation to the architectural model prevailing during the time of the design, and its characteristics, making it a true reflection of the different influences of human thought and artistic creative power of the period it expresses. The environment, architecture artistic style, intellectual or mythical conception and religion or religious beliefs, all had a great influence on the creativity of the ancient Egyptian artist.

The aim of the research was to discover the aesthetic values of the mural design in relation to the architecture, and to show to what extent the arts of the ancient periods were united and in harmony with each other. The matter which we are in need of is to restore that age in a way that goes along with spirit of our modern age, especially in our intellectual, cultural life and in our architectural pattern, which has lost a lot of these aesthetic values. That led me to concentrate on the mural design on the architectural surfaces and the artist's style in his design on these different models, and kinds of architecture. I also wanted generally to clarify the quality of the connection and close unity between the painting and the building itself, its architectural model, as well as period. Thus, I emphasized the importance of the unity between the mural painting and the architectural elements of the building, which has a role in the mural design, as we cannot separate between them and the walls they occupy. I did not deal with the mural painting of each period separately, but I studied the paintings alongside the other artistic activities of the period with their architectural components. These had a bearing on the mural design of the ancient Egyptian artist, showing the styles of that period and comparing it with the styles of other periods. I also dealt with the mural design of the surfaces, grounds and all the different architectural parts in the buildings, which ancient Egyptian artists considered important in their design, in addition to the artistic styles of every period, their development in comparison with previous periods, the extent of change as a result of the social change in thought beliefs, models of architecture, the level of civilization, the rise and fall of the dynasties and the flourishing and decline of their Kingdoms.

The differences in architecture enabled the distinctiveness of the designing style of the ancient Egyptian artists and an assessment of the extent to which they observed the principal and specification in their mural design with its different elements. That was seen clearly from my analysis of the chosen models of every period, which showed to what extent the design was connected to its architecture.

1.1.1-The Outline of the Research

The study was divided into four sections, preceded by an historical introduction. This was important in the ancient Egyptian heritage since the pre-historic and pre-dynastic period, and from the beginning of these to the pharaonic ages. The first section contained three chapters. The first, dealt with the essence of the design and the views formed about them, the components and elements of the design, its independence as a work of art, the factors which affect the design and their importance and what is involved in the artistic work as a design, as well as its connection to architecture. These were a basis for my analytical study of different examples.

In the second chapter I was concerned with the natural elements and their effect on the environment, the human being and on his art. As the study was related to ancient Egyptian art, the ancient Egyptian environment and its natural aspects must be considered as having a great influence on the Egyptian thought, its mythical concepts as well as religious beliefs, in addition, the materials gained from nature, which had importance in the execution of the designs of mural painting. The religious elements were important in directing my research to deal with the concepts the ancient Egyptians had about natural phenomena and their belief in resurrection, eternity, death and the after life.

The third chapter had a different aspect due to its material, and consisted of two parts. The first was about architecture and its place among the arts and Hegel's opinion about it. I considered Hegel's view very important in the field of my study, especially since he is one of the few philosophers who linked his philosophy with arts and the connection between art, religion and philosophy. He explained his opinions about the ancient Egyptian architecture saying that it represents architectural symbolism. Hegel's architectural aesthetic basis was mentioned in my study due to the importance of these values, which had a great effect on the creativity of the ancient Egyptian artists on the wall of his architectural building with its available aesthetic basis (Abd El Salam 1992: 45-59).

The second section in the study dealt with mural design in the ancient Egyptian architecture during the pre-dynastic period until the Old Kingdom period. The third section of the study dealt with mural design in the architecture of the Middle and New Kingdoms. I divided the second section into four chapters. The first included two parts, the first of which dealt with the painting (on pottery) and the first wall painting in Egypt, which introduced my study in that section. The first paintings are important because they preceded the mural paintings and affected them during the beginning of the pre-dynastic period. The beginning of the dynastic period had a special importance because, it represented the period in which the artistic rules and traditions were formed, and became established in the ancient Egyptian artist's style and to which they adhered all through the following periods. Although every period had its style according to its conditions, the Egyptian artists continued to follow the rules and traditions until the end of the pharaonic period, with the exception of those periods which were distinguished with a specific style, e.g., the Amarna period.

The second part of the first chapter in the second section was concerned with the study of these rules, traditions and styles, how they began, their type of hieroglyphics and how the artists followed them, and how some inscriptions from these periods helped to determine and develop these rules, especially the hieroglyphic inscriptions. The hieroglyphic writing reached its maturity in that period, so it was important to shed some light on it because of its essential part in the mural design on the architecture surface, to see how the artist arranged it on his painting, the kinds of hieroglyphics used and the style of the decoration, how it was used in almost every painting and its part in explaining these paintings, in addition to its forming an aesthetic element in the design.

The second chapter of the second section, dealing with mural design in the Old Kingdom architecture, was in two parts. The first, was a study of the mural design of every dynasty, showing that each had its architectural model, artistic and designing style, as well as the painting. It also detailed the factors of change, which affected the creative and formative attitude in that specific period and appeared clearly in its designing style, as was used in the architecture of its temples and tombs, as well as houses and palaces. The second part was concerned with analytical study of some examples chosen from the architecture of the period. This contained two examples; the first was to study the false door, which is important in the study due to the fact that its architectural development was connected largely to the ancient Egyptian artist's style. It had an important part to play, whether it is found in the tombs representing the religious belief and philosophy of the ancient Egyptian, or found in the mural design of the tombs, especially since some of the paintings on the walls in the tombs are directed in their design towards that false door. The second example represented the tomb of Ti of the 5th dynasty, which had its impact on the mural designing style of its tombs; the tomb represents a model of that dynasty.

I chose for the analytical study only walls representing examples of the mural designing style, which the ancient Egyptian artists followed. It was necessary in studying this example to understand the style and to compare the mural design styles in their principles, what was accomplished by the Egyptian artist by his mural designing style on the walls of the tombs in connection to time, place, theme, artistic style, and the artist's arrangement of his items, what he adhered to, as a result of following the principles and rules, how he divided the walls, as well as his concern with choosing the media of execution, which suited the nature of the place and the theme of the design. That was followed as an approach in all my studies of examples in different periods.

The first chapter in the third section was in two parts; the first part of which dealt with the study of mural painting in the Middle Kingdom architecture, and contained two chapters. The first was an introduction to the first intermediate period and its artistic aspects, as it proceeded the Middle Kingdom period. Then there was a study of the mural design in the architecture of that period, in addition to the changes, which took place and the factors that affected the artistic design style of the architecture of this period. The second chapter in the first part was about the analytical study of the examples of this period. This was the Kiosk of the Sonosert I, because it represented a kind of architecture different not only in its model, but also in architectural function. It was neither a divine nor a funeral temple; it had a specific ceremonial architectural nature related especially to the Heb Sed feast, so it had its own design style. That exactly suited my aim in getting architectural examples which were different in their model or structure to identify the designing styles resulting from this difference. The other example was the tomb of Tehuti-Hetep from El-Barshe of the Middle Kingdom.

The second part, of the third chapter, dealt with mural design in the New Kingdom, which was important to the study. It contained four parts, each of them treating an example of the architecture. The first part was an introduction to the second intermediate period and the artistic aspects of the New Kingdom period, followed by a study of the mural design in its temples, the themes used outside as well as inside and the style adopted in each. The second part, studied the mural design in the New Kingdom tombs, which included the walls and ceilings of the tombs, which were distinct stages within the New Kingdom, in addition to the effect of the intellectual and mythical conception on the themes of the mural paintings in the tombs and the design style.

The third part dealt with the mural painting of the palaces and houses, shown by what remained of their monuments to indicate the design style of both. The fourth part of the second chapter ends with my analytical study of two examples, which are important to the architectural and mural design and the artistic style to which each of the two types belong. This emphasized the extent of the connection between the mural designs. The first example represented the tomb of Re-Kam-Rh, the vizier of King Tohoutmous III; the other example represented the tomb of King Seti I. These examples were from different periods and were different models as well as for different persons which shows the impact of these differences on the style of the mural design of each of them. The curator of monuments was so kind as to allow me to visit the tomb of the King's Seti I, especially as it was closed and was not open to public.

The fourth section, in conclusion of the research, dealt with the preparatory work for mural painting and the styles of execution in ancient Egyptian mural painting and design. That section contained all of the steps, beginning with the completion of work on the tomb and ending with the colouring of the design. It contained two chapters, the first of which dealt with mural styles and the impact of the factors of the civilization, the relationship between painting and coloured engraving as well as with the architecture. There was also a description of the steps followed in the preparatory stages until the mural design was placed on the walls. The second chapter concluded these steps by mentioning the style of execution, the artist's media and colouring materials, how he executed his mural painting and coloured engraving and how he used coloured pastes as showed on Photographs 64-66.

The result of the research was the outcome of my study, emphasizing the connection between the mural design and its place, time, theme, artistic style, the materials used in execution of the painting, the observation of the external factors in the artist's choice of the style of execution, the simplicity in the mural designs being seen clearly, as well as the appearance of the impact of changes and other influences on the style of every period on the mural paintings.

Although the ancient Egyptian artist did not take into consideration all these principles, which became clear by the study, he was not one of the followers of the theory of "art for art sake", but art to him was a result of his religious belief, which had a tremendous impact on his mythical thinking and his artistic creation and which found a suitable medium on the walls, ceilings and surfaces of the buildings, enabling him to embody his visions, by which he could climb to the ideal world.

1.2-Introduction into the Beginning of Wall Painting in Egypt from Previous M.Sc. Work

Wall painting in Egypt has a long history since its developments have been on walls throughout the Egyptian periods. My intention is to give an outline of the most important paintings that have been found at the beginning of wall paintings in Egypt, formation of the Egyptian principles of art that founded the style of the Egyptian decoration and how creating these designs were created or painted on walls, as well as some other points. I have found it is necessary to abstract from my previous study in order to have a general idea about the Egyptian paintings.

Wall painting in Egypt was started in the pre-dynastic period (about 3300 BC) where it was transferred from pottery surfaces to walls. Pre-historic art was known in Africa and on the Nile valley, which was at a later stage dated to the Middle of the Stone Age, the same time as the beginning of art in southern France. There was a connection between the east of Alberia and the East Coast of Spain, North Africa, the great desert and Egypt as well as South Africa. There were two groups in Egypt. One group inhabited the east and west of the mounds around the Nile valley; they were hunter-gathering peoples and their drawings were on the rock hills as well as the sides of the valley. The other group inhabited the villages and cultivated the land on settlements in the Nile valley and their drawings were found on pottery. They combined simple industries and agriculture; their drawings were varied not only because of the motifs used but also the painting techniques on the pottery and their decoration. This has been divided into five groups, according to where they lived; First developed to Deir Tasa, second to Badari, third to Nagada I and fourth and fifth to Nagada II and III in Qena, which was the most important of the pre-dynastic cultures of Upper Egypt. Each had a characteristic phase developing through its various stages. The drawings were transferred from the pottery surfaces to the walls by the second half of the 4th millenium BC (Abd El Salam 1992: 84-86).

The reasons for their drawings and painting were magical and religious, as most archaic cultures affected their drawing and painting. Another reason was that the Egyptian men wanted to find a firm surface capable of registering their motifs or drawings rather than carrying these on pottery, which could break up. In addition, his beliefs about survival after death, led him to think of the after world. His thoughts led him to decorate the tombs with the same simple motifs as were used on pottery (Abd El Salam 1992: 88).

One of the most significant sites in the history of Egyptian wall painting is the tomb at Hieraknoplis (Upper Egypt royal tombs about 3300 BC). This tomb is considered to be the first with a wall painting, executed on a surface of mud plaster which was faced onto the sun-dried brick used for building. The tomb consisted of two rooms and was dated to the pre-dynastic period. The colours used were; black, white, green "malachite", blue and yellow (photograph 57). The motifs used for painting had also been used on pottery. It has been noticed that there is similarity between these figures and the ones, which were used on pottery, especially the pottery of Naqada II, Deir Tasa and El Amary (figure 9).

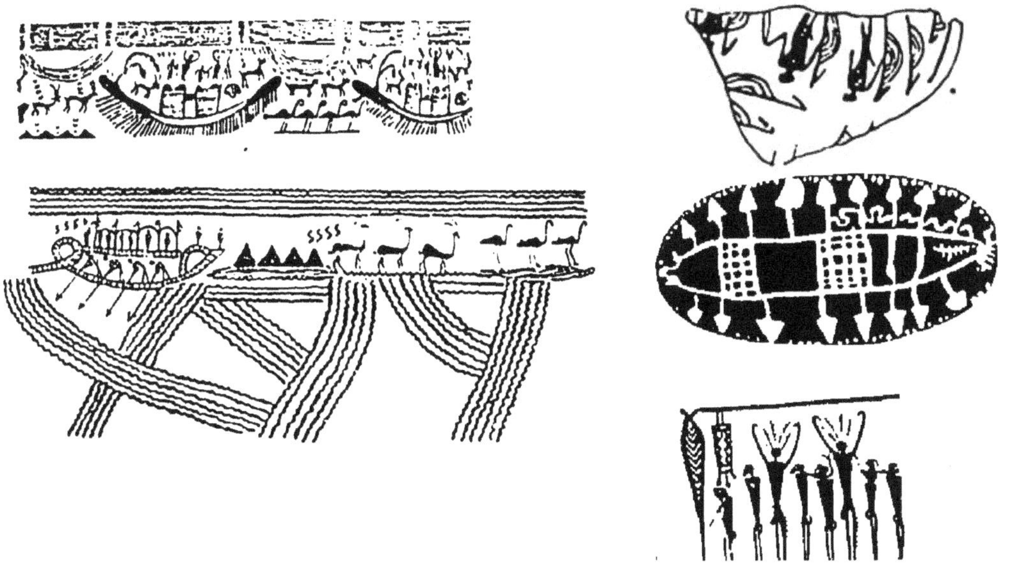

Figure 9, shows the motifs used on the pottery of Nagada I and II. (Abd El Salam 1992).

The motifs (designs) of the paintings were applied without any systematic decoration. However, it is considered that the first wall which was decorated in that period was applied on a rendering made of mud plaster. These techniques or methods of preparation were used throughout the Egyptian periods, with some consideration that when stone was employed for building; the methodology for the preparation of walls depended on the condition of those stones (Abd El Salam 1992: 91-95).

The Egyptian principles were formed by the time of the dynastic period, when the men settled down and started to realize the natural order of things, and noticed the natural elements as well as cultivating their land. These factors helped to form his principles of art. There were some important examples, which show how the Egyptian artists started to arrange the elements as well as the figures in horizontal strips and concentrate on the main figure, to represent it larger in size in comparison with the other figures. The art historians have divided these into four stages. The first stage is shown on (figure 10). The Hunters palette from the pre-dynastic period (about 3100 BC) shows scenes of hunting lions and other wild animals, and you can see that the elements and the figures were separated.

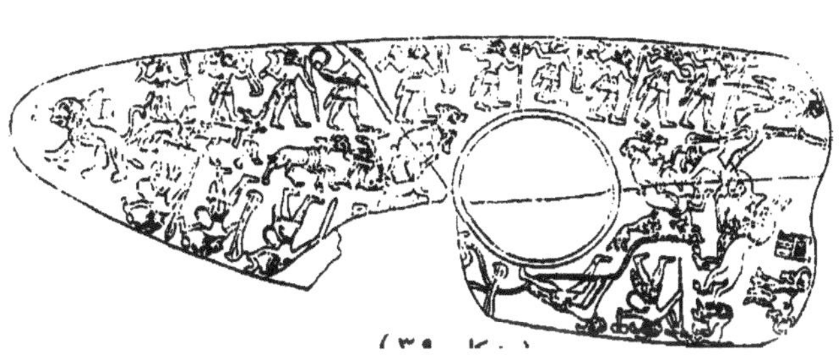

Figure 10, the hunters palette shows scenes of hunting lions and other wild animals, in which the figures were separated; pre-dynastic about 3100 BC. (Abd El Salam 1992).

The second stage is shown on (figure 11) the decorated handle of an ivory knife from Gable-El-Arak, with carving of animals, from the pre-dynastic period (about 3200 BC). This shows a sort of relationship between the elements, arranging the figures rather than spreading them, which was developed by the late pre-dynastic. This is also shown on the mace-head of King Scorpion (figure 12) which is considered to be the third stage of the development of the Egyptian principles of art. The fourth stage is shown on a palette of King Narmer, (figure 13 about 3100 BC), who was the first king of the first dynasty which saw the unification of the upper and lower Kingdoms of Egypt. By this stage the development of the Egyptian style, in which the principles of Egyptian art had been founded, is clear. These were both in low relief and are considered the most important and significant in early Egyptian art.

figure 11

figure 12

Figure 11, shows the decorated handle of an ivory knife of Gable El-Arak, and here you can notice a relationship between the elements and the arrangement the figures; pre-dynastic about 3200 BC. (Abd El Salam 1992).

Figure 12, the mace-head of King Scorpion. (Abd El Salam 1992)

Figure 13, ceremonial palette of King Narmer, on the left showing the king wearing the red crown of Lower Egypt and on right wearing the white crown of Upper Egypt, the first dynasty about 3100 BC. (Abd El Salam 1992).

Once these principles of art had been founded, they were then followed throughout the Egyptian periods with the following features: -

-The division of the wall in horizontal strips (registers) with some vertical or secondary strips in between.

-The difference in size between the figures is according to their position in life and their employment.
-The relation between the drawing and Heliography.
-The ignoring of perspective vision and the use of two-dimensional representations.
-The law of frontal presentation; head and foot in profile and the body represented face on.

These principles or rules were applied in all Egyptian art, both painting or coloured relief, whether it was found in tombs or temples, and continued throughout the Egyptian Pharaonic periods (Abd El Salam 1992: 108-128).

The grids were used according to the principal periods of Pharaonic Egypt. The grid lines in the Old Kingdom are in horizontal or vertical lines and on different levels (figure 14), but later on they were in equal horizontal and vertical lines forming a grid. These helped the artists in the transformation of the sketches (Ostraka) which followed these grids, according to the elements to be drawn, to see how things fitted as a final outline.

14 A

14 B

14 C

Figure 14 A-C, shows the grid lines used in the Old and New Kingdom. A shows on, the grid lines were horizontal or vertical and on different levels in the Old Kingdom. B shows the grid lines consist of 16 squares accommodating the sitting human figures, whilst C shows a different position of the human figures and the grid lines used were about 19 squares. (Abd El Salam1992).

Two grid lines were used; the first was about 18 squares in height accommodating the standing human figure and 14 squares in the case of the sitting human figure. Later by the 26th dynasty this changed to be 21 squares for the standing figure. That also gave them a chance for the correction of the drawing and scenes, which were then made up in a red colour before painting (photograph 60).

A black colour was used for the drawing but in the case of carving, the scenes shown in many of the tombs in the valley of king's were executed in relief paintings (coloured relief). In the case of drawings of small figures, such as birds, animals, heliography etc., the grid lines were in different sizes and used for the figures as well. I have noticed that different size grid lines have been found in some tombs during my studies, especially in Mair in Assiut, the Middle of Egypt (photograph 59).

The ancient measure of length used was the short cubit, which is equal to the length of the forearm, about 18 inches or 47.7 centimetres. Each cubit was divided into six palms (six times of the width of a palm and each palm was divided into four digits), or 24 digits, about 45 centimetres, (the digit being the thickness of a finger and the palm being the width across the back of the hand). Each digit was equal to about ¾ inch (1.9 cm). Each line of the square was a measurement of the length of a fist.

The squares numbers have been changed two or three times. The 18 squares accommodated a standing human figure, starting from the foot line of the human figure to the eyebrow, whilst the head above the eyebrow is left according to the head covering, e.g. a crown or a wig.

The short cubit was divided according to the rule of the measurement to four and half unit squares (about 6 palms) which accommodated the standing human figure. These rules were applied in the case of religious scenes, whilst they were a little less formal in the representation of the daily life scenes. It has been found that different numbers of the squares were used 15, 16 and 19 squares (photograph 58), some were 21½ and 23 squares as well, which indicated the use of different grid lines in the Egyptian times (Abd El Salam 1992: 478-481).

The building material used in the early dynastic period was sun-dried brick, which was covered with mud plaster, then covered with another layer of clay plaster, then a very thin layer of white wash, and then painted. Aldred mentioned that the wall paintings were only painted in red, black, yellow and blue on a white ground as an imitation of mat-work By the third dynasty, stone was introduced for building in the king Zoser region (Abd El Salam 1992: 113-127).

In the step pyramid of King Zoser, in his south Mastaba, there are a series of low relief panel carvings with a combination of faience (photograph 61). Another example of low relief is shown in the tomb of Hasy-Re in Saqqara, on a wood panel there are very fine and accurate details as shown on figure (photograph 62). No paint has been found in Zoser Mastaba, whilst in the tomb Hasy-Re, which was built up with sun-dried brick and plastered with mud, a white layer of plaster as a ground was painted in different colours. The painting of this tomb was a kind of imitation of mat-work (Abd El Salam 1992: 157-158).

In the fourth dynasty wall painting was rare, with some exceptions, one of which is considered the most famous as well as important in the history of Egyptian art. This is the Geese of Madium (photograph 63). The painting was a part of the wall decoration in the tomb of Nefer-maat and Atet. Some fragments of the Atet tomb are in the British museum. Nefer-Maat was a son of Snofro, the first king in the fourth dynasty, and in his region the drawings and scenes were in coloured or painted relief. The tomb was built of sun-dried brick where the painting was carried out. Stone was also employed in the tomb where it was used for the inlay process (photograph 64).

However the inlay technique was not a new invention in the fourth dynasty, but been used in the pre-dynastic period when Egyptians engraved motifs on the pottery surface. Straight or inclined bands or lines as well as small dots were the most decorative elements for the pottery and were in-filled with white gypsum paste. The technique used on the wall depended on the cutting out of the stone to a depth of several centimetres and then inlaying the coloured pastes. That method was not very successfully because the surface had fallen off after drying and shrinking. The artists then modified the method of fixing these pastes on the wall. They tended to leave some crossbeams in the larger areas of the stone and then divided those into small squares (photograph 65).

The remains of the tomb of Nefer-Maat are in the Cairo museum where it has been suggested that the reason for shrinkage could be that the mixture of the pastes used with a small percentage of adhesive, used was not enough to adhere it to the wall. In addition, the differences between the temperature in the day and night as well as the humidity in Egypt, helped to disintegrate the material used, which resulted in its loss from the wall. This led the artists to try other methods to decrease the thickness of the pastes used as paint on the stone surface or on the plaster. This was the same method, which was used before with success.

The inlay process continued in the tombs of the fifth dynasty where it was used to fill up the heliography names, and has been found in the 18th dynasty where they were used for the decoration of houses as well as palaces (Ramesses111), as shown on Photograph 66 (Abd El Salam 1992: 171, 509-551).

In the fourth dynasty King Khufu, stopped the decoration of the tombs and there was no longer any wall painting. The walls were painted in white without any scenes except the Stela, which were made of limestone and executed in coloured or painted relief. Also in his reign they used to cover the walls with fine limestone, coloured in imitation of marble. In the middle of the reign of Khafre, who succeeded the use of Stela, false doors and the decoration of the walls, was allowed which were executed, in coloured relief. In the reign of Mankaure who followed Khafre, the false door and sculpture as well as the representation of the scenes on the walls were introduced in the tombs (Abd El Salam 1992: 167-172). Bas-relief was a technique used in all the periods which continued until the end of the Egyptian pharaonic era. In the later period of the New Kingdom, another kind of the relief was introduced; Relief-encreux in different thickness (Abd El Salam 1992: 502-514).

In the Middle Kingdom, Egyptians were interested in painting on wood, as well as on the walls which had received suitable preparation, depending on the condition of the stones, e.g. the tombs of Bani Hassan, tombs of Meir in Assiut etc. In that period the painting developed not only because of the use of new colours, such as pink as well as different reds, but also the style of the decoration (e.g., length of the figures). The Egyptians had followed all the principles of the art with one exception, which appeared in the tombs of Mair in Assiut. This was that the scenes were not in horizontal lines as before (Tomb of Aokh-Hateb III). In addition, new subjects were represented on the walls, as well as the scenes of daily life (Bani Hassan) (Abd El Salam 1992: 252-259).

In the New Kingdom both painting and coloured or painted relief were applied according to the owners position in the society. In the valley of Kings, the tombs of Kings and Queens were decorated in "painted relief" and Noble's tombs were executed in painting, except for a very few tombs, e.g. the tomb of Ramosa (No 55) which contained both techniques. These tombs were cut into the rock so the preparation of the tomb was different from the preparation of sun-dried brick, with some exceptions associated with the condition of the stone and the techniques which were to be executed, whether in relief carving (painted relief) or painting. Limestone was used widely from the third dynasty, about 2600 BC. till 1550 BC. when it was replaced with sandstone. The best limestone provided the best relief, where it varied in the texture as well as quality (Abd El Salam 1992: 360).

The author included on the study of "*Mural Design in the Egyptian Architecture*" a section dealing with the mural preparation and execution styles in ancient Egyptian mural painting and designs. It also detailed all of the stages, beginning with the end of work on the tomb and ending with the colouring of the design or drawing. In addition, the style of mural painting and civilization factors which affected the style the arts. Also, the relationship between the painting, coloured engraving and the architecture. There was also a description of the steps followed in the preparatory stages till the mural design is placed on the wall.

Appendix 2

The Schemes of Decoration in the Greek World

The zonal style of decoration scheme used at Pydna, Niauste, Langaz, Vassiuim, Delos and Alexandria has been illustrated by Adriani in six pictures (photograph 11), which show there are some similarities. This style was also found in Macedonian tombs and at Marissa in Palestine, as well as in South Russia. Adriani compared the evidence of the application of the zonal style of decoration in the Macedonian tombs with that applied in Alexandria and assumed that the style applied in the Macedonian tombs was a primitive form of the zonal style. The wall is consistently divided into several horizontal parts or zones;

A-Plinth or pedestal base.
B-Line of orthostats.
C-Intermediate band.
D-Plain zone delimited in height by a small cornice. Smooth/plain/united zone, delimited or not by a band (in relief or in paint) below the ceiling.

Some of these zones were occupied by figure representations, others by purely decorative motifs and still others were left in monochrome. Sometimes the figure representations are arranged in the archaic manner in several registers or occupy almost the entire wall and are framed above and below by monochrome bands.

Adriani mentions that from the examples represented in Alexandria, Vassiuim and Delos, it can be concluded that the characteristics of the First Pompeii style depended on the division of the wall, the imitation of stone blocks or incrustation style with engraved lines of relief or polychrome, and the cornice/frieze at a certain height on the wall (Adriani 1936: 113, 119, 123-126).

However, the first style of decoration does not differ in principle from the Greek style, but is finer and richer in colours and more elaborated in detail. In the east, two styles were developed. The first one during the late structural style, the floral style emerged as a framework of architectural structure with the division by the *Dado* and the upper part of the wall as a background for decoration. The second was incrustation style, which originated in ancient Mesopotamia and Iran and not, as Vitruvius and Pliny mentioned, in Mausolus in the Greek world; the same fundamental system of dividing the wall into three parts was retained. The whole or some parts of the wall are covered with slabs of different coloured marbles forming geometric ornaments and separate figures or whole scenes in the upper part of the wall. The lower parts in such cases were covered with imitation marble as was observed in the later decoration of Pompeii (Rostovtzeff 1919: 150-152).

A B C

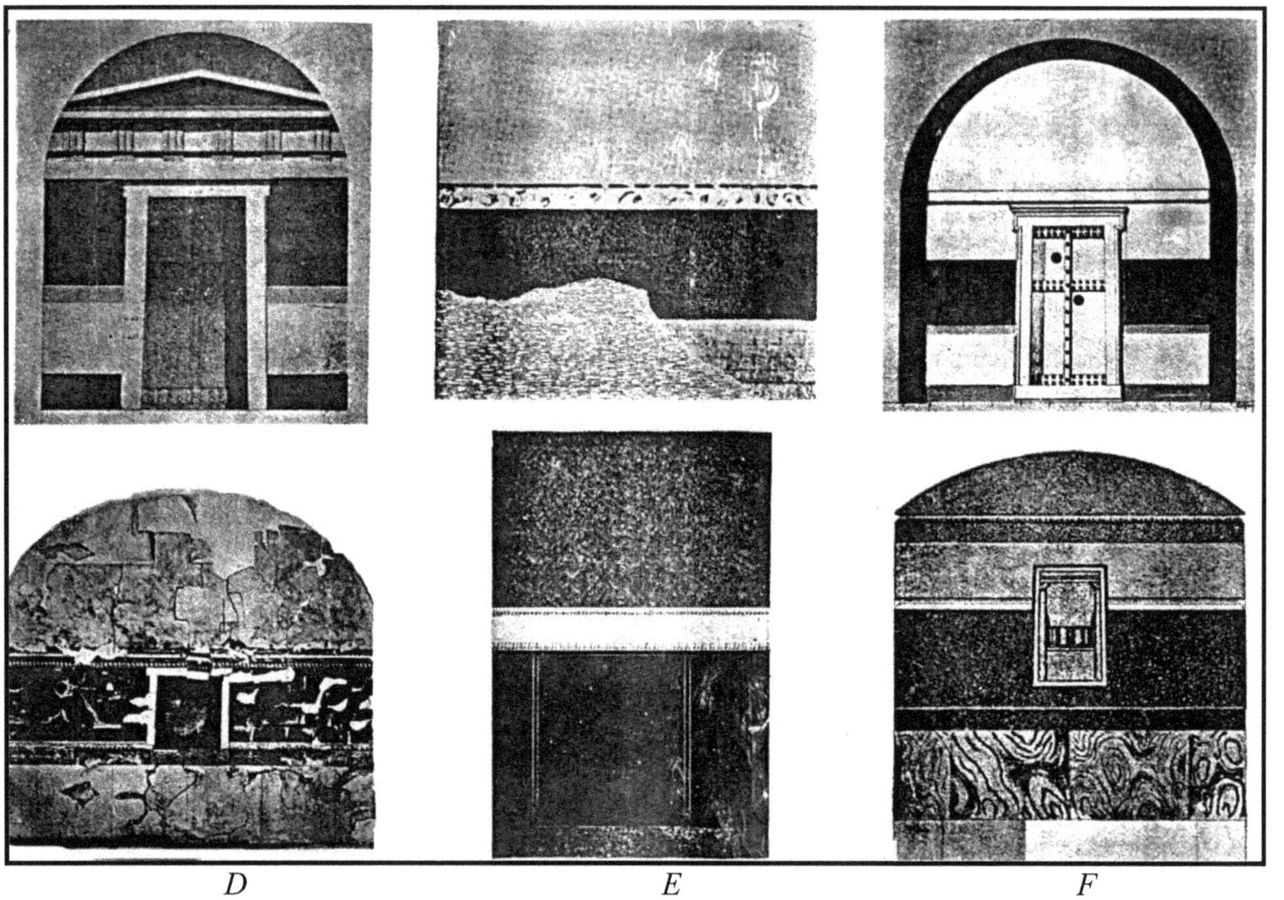

D E F

Figure 15 A-F, show the zonal style as described by Adriani which was used in different parts in the Greek world; A, Pydna,
B, Niauste, C, Langaz, D, Vassiuim, E, Doles and F, Alexandria (Sidi Gaber) (from Adriani 1936)

Rostovtzeff and Adriani have argued that the so-called Pompeian style was not the same everywhere and that three varieties can be distinguished; the Alexandrian style of Asia Minor, the South Russian decoration and the Pompeian style in Italy. He points to the differences between them. In Asia Minor the style was rich with elaborate details and many colours. In contrast, the Italian style was more strictly architectural and had developed both polychrome and architectural elements in which the use of horizontal divisions was remarkable, as in the Pompeiian wall paintings. From the *dado* to cornice, the walls were divided by columns, pilasters and half columns.

The frames of doors, windows and niches were richly decorated and painted walls had two lights reproduced in the mural decoration. He maintained elements occurred in Italy as the first Pompeian style and developed into the other three styles (Rostovtzeff 1919: 150).

The most important point made by Adriani, was that in Pompeii, the examples decorated in the first style were dated to the period between the second half of the 2nd Century BC and the beginning of the 1st century BC. Adriani states that the zonal style was a phase of the First Pompeiian style in which it was associated with the zonal abstracted style and was widely spread and known before the first Pompeiian style (Adriani 1936: 128-130). Rostovtzeff said that the tendency to vertical division was not peculiar to Italy and Rome because the Greeks cultivated it during the whole Hellenistic age. However, the richness in real architecture and paint imitation was a new development. In the second and first century BC, there was important development in Italy, where Greek artists were employed to adorn the town, due to the economic decay in Greece. As a result the development of the arts were in Rome, Italy and Gaul not in Greece, Asia Minor and Egypt.

When the east was under the Roman Empire, the architectural style of both the third and fourth Pompeiian styles were developed from the second, probably in Italy and the third in Alexandria (Rostovtzeff 1919: 160). However, it issaid that the thrid style originated in Italy under the reign of Augustus and ended circa AD 40-45 (Barbet 1982: 75).

Noshy also mentioned that the wall incrustation of pre-Hellenistic buildings in Asia Minor was the home of the first style. He points out that the inlaying of different coloured stone was known, and was in use long before the Hellenistic age by the Assyrians, Caldeans as well as Egyptians (Noshy 1937: 25). Adriani concluded that from the beginning of the zonal traditional style until first Pompeiian style, the evolution had continued to correspond with the application principle that was imitating the structure of the original wall. This consisted of the base, the intermediate part and the cornice, due to the materials used for building and led to the covering of the wall or stone with a *stucco* layer. The imitation of marble or stone already had appeared in the last phase of the zonal style. The first style only generalized it and extended it to the upper part of the wall, proving the erroneous theory which had the imitation beginning in the *stucco* dressing, in imitation of limestone walls or in white marble. Adriani gave different examples of the imitation of marble found in the Alexandrian tombs of Mustafa Pasha (Adriani 1936: 127, 131).

It is worth mentioning that Adriani says M. Rostovtzeff has contributed subsequently to a better understanding and definition of the system in regard to the tombs of southern Russia, and M. Pagentecher in respect of the tomb of Sidi Gaber in Alexandria. One still cannot say that the pre-Pompeian system of decoration has received full right of citation in the studies of the decorative painting of the classical world, for several experts have continued to ignore or fail to recognize its existence. The six new examples of Mustafa Pasha and the others that have been recognized at Shatby and Hadra in Alexandria augment, in Adriani's view, a remarkable contribution to the definition of the zonal system and to the question of its priority over the first Pompeiian style (Adriani 1936: note 120).

Appendix 3

Inductive Coupled Plasma Optical Emission Spectrometry (ICP-OES)

There are two main types of ICP; simultaneous and sequential. Simultaneous ICP allows the determination of a large number of elements during one integration period by using polychromatic a spectrometer equipped with a series of fixed channels. The ICP device at Leicester is a simultaneous spectrometer; the height and intensity of the wavelengths increase with an increasing concentration of the element. The procedure used hydrofluoric acid to dissolve powdered samples. There are about seven key stages in the analysis of a sample by ICP-OES, including: -

-Dissolution of samples.
-Nebulization.
-Introduction of solution into the ICP torch.
-Emission of light of different wavelengths by different elements.
-Separation of light rays using a polychromator.
-Measurement of the intensity of light rays using photomulipliers.
-Collection, processing and output of data.

ICP was used for both powdered and solution samples, but in the case of the powdered samples they were dissolved first; the solution samples did not require any dissolution factor, as I dissolved 2 g of sample in 20 ml of distilled water. The only difference between the two types being the need to multiply the numbers of the analysis by 10, in order to get ppm value.

Procedures for dissolution

1-100 mg of samples were weighed out and put into clean PTFE test tubes (blanks, standards and duplicates included), then 2 ml portions of concentrated nitric acid was added to each tube, using a dispenser. Each tube was mixed thoroughly using a whirlimixer, and then the test tubes were placed in the hot block at 50 °C and left overnight.

2-The test tubes were removed from the hot block and allowed to cool. 1 ml of 60 % Perchloric acid and 5 ml of hydrofluoric acid was added to each test tube. The next step was to replace the tubes in the hot block and heate in the following sequence of temperatures; 100 °C for 3 hours, then 150 °C for 3 hours and 190 °C overnight. Then they were removed to cool.

3-About 1ml of concentrated hydrochloric acid was added to each tube and mixed thoroughly. Then the tubes were heated on the hot block at 50 °C for 1 hour, after which they were removed and allowed to cool. Each was diluted with 9-ml of deionised water and mixed thoroughly.

4-The tubes were decanted into 15-ml centrifuge tubes and centrifuged for 15 minutes. Then they were ready for analysis, by ICP.

Appendix 4

Archaeological and Historical Background to the Sites

In this appendix, I have only dealt with the archaeological background of the sites of Saqqara, Dahshur and Luxor, where I took samples. However, the historical background, architecture and the decoration styles have been studied in the previous work of my M.Sc. "Mural Design in Ancient Egyptian Architecture "3100-1085 BC" (see Appendix 1). Whilst, I focused in particular on the archaeological and historical background as well as the decoration style of the Alexandrian sites of Mustafa Pasha and Anfushi, as it was my interest to carry out research on Græco-Roman wall paintings as this area has not been touched before. In addition, the Piddington Roman site was used, to differentiate the Egyptian and Græco-Roman wall plasters and mortars that used in ancient and classical times.

4.1-Saqqara and Dahshur

The first site from which I took samples was excavated in 1995-96 north of Saqqara, to the north of the Tite pyramids, and north west of the archaeologist's offices (houses of antiquities). The archaeological evidence indicated that the tomb was dated to the late first dynasty. The architectural structure appeared to be different from others dated to the same period; it contained a big hall, rectangular in shape, about 17 m x 45 m, built of mud bricks. The external walls were covered with a plaster of mud mixed with straw.

In Dahshur, the mastaba/tomb excavated in 1996. It was dated to the 4th dynasty Dahshur mastaba was built of mud brick, being about 29 m x 14.5 m. The bricks used were about 30x 15x 10 cm. The tomb is to the Southeast of the Amen-Mhat III pyramid. The fragments of pottery found, being black as well as red inside, indicated the style of the 4th dynasty, the characteristic of which is Nile mud with a rough surface.

South of Saqqara, there are two tombs which were excavated in 1994-97. The first is dated to the 5th dynasty. This was indicated by the name of Ka-Nefr, found somewhere in the tomb, a name in use in the period of the Old Kingdom. This tomb is to the south of the tomb of Hor-Mo-Heb and to the north of the Idy tomb (the other new excavation). It was built of mud brick. Some architectural as well as archaeological evidence was found such as; ceramics (small statues), talismans and a stone chest, dated to the New Kingdom, indicating its use in the 18th dynasty. The tomb was for family use. In 1997, another site excavated was the tomb of Idy dated to the late 6th dynasty (the excavation of Tabate El-Gash). This tomb was also built of mud brick and consisted of 8 rooms of varied size, the walls being covered with a layer of mud plaster and painted white. One of the rooms was decorated in a rough style of painting. The chamber was covered in stone and decorated with scenes of daily life.

This data was abstracted from of unpublished reports written by the Egyptian archaeologists.

4.2-Luxor

The site of Sheikh Abd El-Qurna is to the north of Qurnet Murai and Deir El-Medina, wedged between the cultivated areas and the cliffs of the western desert. The tombs of the nobles at Sheikh Abd El-Qurna are varied in size and detail. At this site, tomb number 56 of Userhēt (plan 2 A), was dated to the 18th dynasty (Amenhotep II). The second, tomb number 55, the tomb of Ramose (plan 2 B) was dated to the 18th dynasty (Amenhotep III-Amenhotep IV).

The site of Drah Abu El-Naga extends for over 1km north of the Bay of Deir El-Bahri, modern occupation begin quite heavy near the southern end. The area has suffered heavily from spoliation and unscientific excavation. Some of the earliest and latest tombs built at Thebes during the New Kingdom were found on the site of Drah Abu El-Naga. The two tombs from this site used in this study are; number 255 the tomb of Roy (plan 2 D) dated to the 18th dynasty and the tomb of Shuroy number 13 (plan 2 E), dated to Ramesside times (19th dynasty).

The areas of Khokha and Asasif are between Sheikh ABD El-Qurna and the bay of Deir El-Bahri, near rise the hill of Khokha, where some of the greatest 18th dynasty tombs are found and a few of small rock-tomb burials of the Old Kingdom. Many tombs of the later period (the 22nd–26th dynasty) are found on the Asasif site with a few dating to the 11th dynasty. At the site of Khokha, there is the tomb of Kheruef, number 192 (plan 2 C), dated to the 18th dynasty (Amenhotep III-Amenhotep IV). *The Epigraphic Survey et al.* (1980) have published the work in full. The tomb from Asasif is number 414, the tomb of Ankh-Hor (plan 3 E), dated to the 26th dynasty (Pasmatik II-Apries). The tomb has been excavated and restored by the Austrian Archaeological institute.

In the Valley of the Kings, where the rulers of the New Kingdom were buried, is the site at the rear of the Bay of Deir El-Bahri. The three tombs are dated to different dynasties; tomb number 14 of Queen Tausert (wife of Sethos II) (plan 3 C), tomb number 15 of Sethos II (plan 3 A), both dated to the 19th dynasty and the tomb of Ramesses III number 11 (plan 3 B), dated to the 20th dynasty.

The Deir El-Medina site is located in a little valley behind the hill of Qurnet Murai at the southern end of the Thebean necropolis. This site was the worker's village of the Deir El-Medina with its cemeteries and temples. The tombs are arranged in terraces on the hills to the north and west of the village (Ramesses III-Ramesses IV, Ramesside period). The tomb of Inherkhau from Deir El-Medina is number 359 (plan 3 F), dated to the 20th dynasty (Murnane 1983: 70 & 280-298). See also Porter and Moss part 1& 2 where are included all the plans of the tombs as well as the topographical bibliography of the ancient hieroglyphic texts, reliefs and paintings, with more details on the distributions of the decorative scenes (Porter and Moss 1960: 21-28, 107-114, 339, 421 and 1964: 518-526).

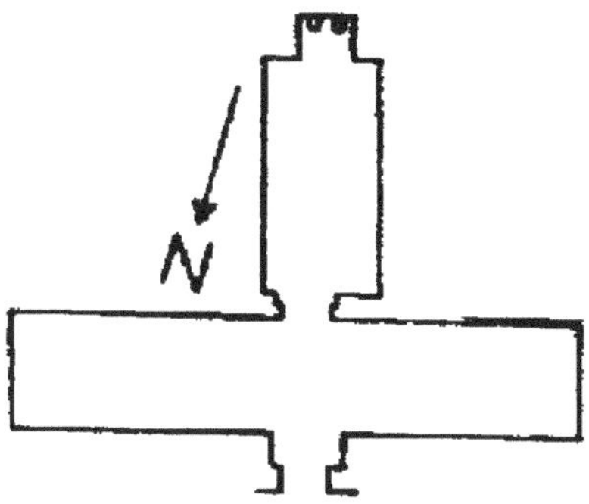

(A) T.56

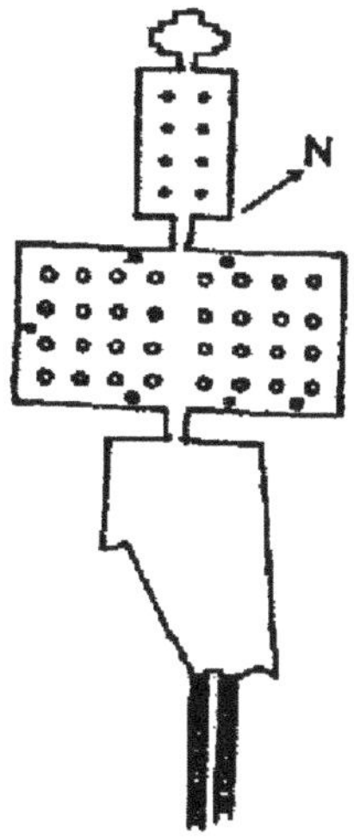

(B) T.55

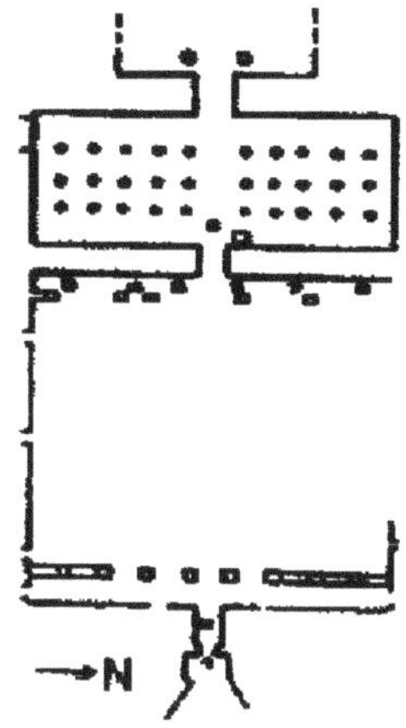

(C) T.192

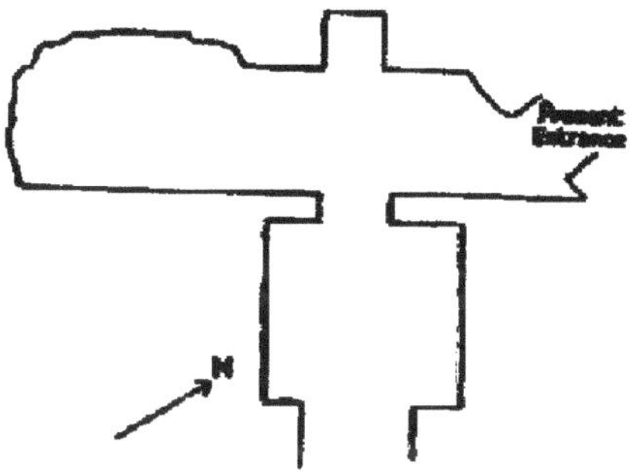

(D) T.255

(E) T.13

Plans 2 A-E

A, is the plan of the tomb of Userhēt No 56, Sheikh Abd El-Qurna, Luxor
B, is the plan of the tomb of Ramose No 55, shows about 32 columns in the broad hall, which is about 25 metres wide, Sheikh Abd El-Qurna, Luxor.
C, the plan of the tomb of Kheruef No 192, Khokha; these tombs are dated to the 18th dynasty, Luxor
D, is the plan of the tomb of Roy No 255, Dra Abu El-Naga, Luxor
E, Shuroy tomb No 13, Dra Abu El-Naga, Luxor
(After Porter & Moss 1960).
(See Appendices 5.4-5.6 for sampled areas).

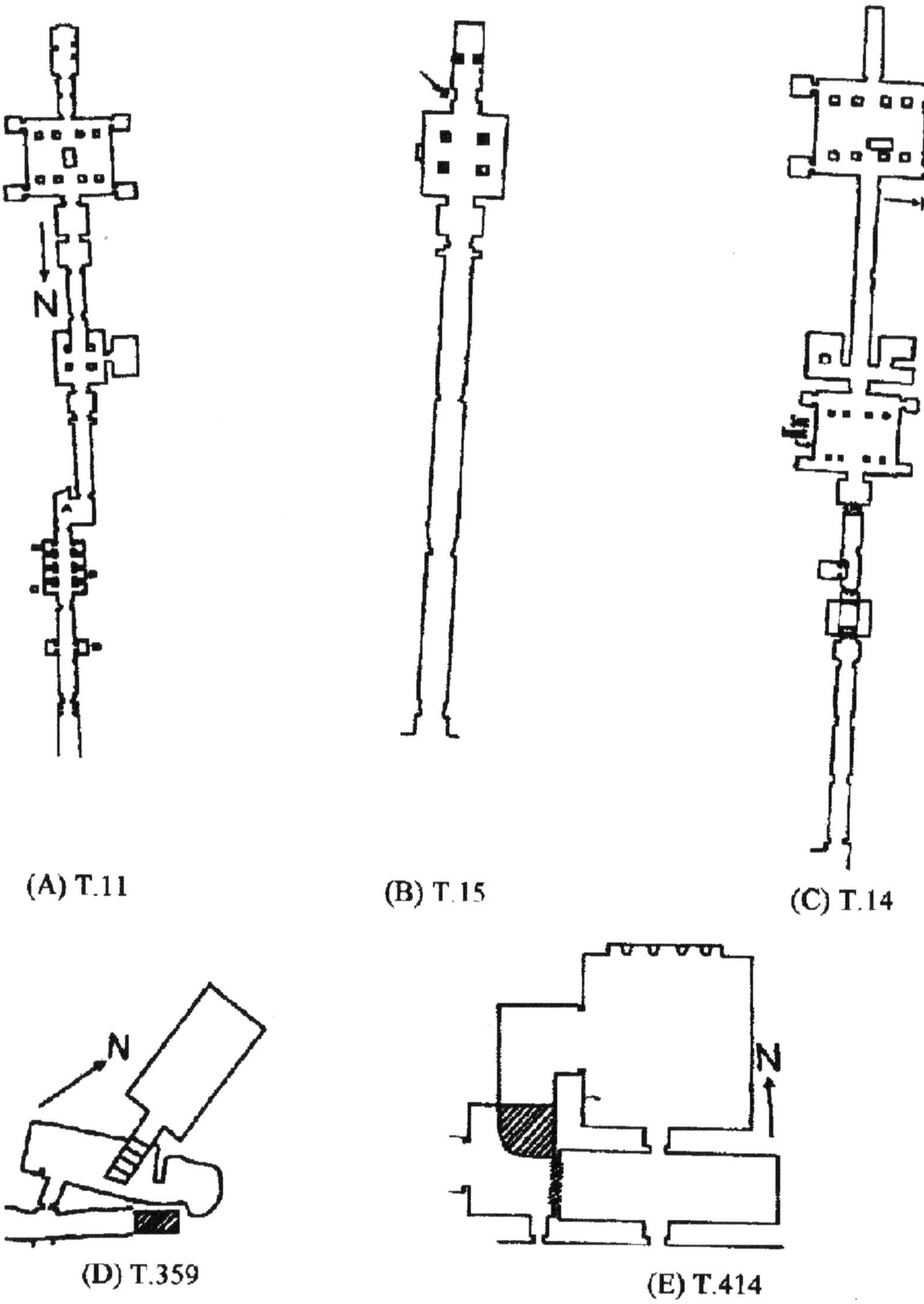

Plans 3 A-E

The different architectural designs of the New Kingdom tombs.
A, the tomb of Queen Tausert No.14, The valley of Kings, Luxor
B, the tomb of Sethos II No. 15, 19th dynasty, The valley of Kings, Luxor
C, the tomb of Ramesses III No.11, The valley of Kings, Luxor.
D, the tomb of Inherkhau No.359, Deir El-Medina, Luxor.
E, the tomb of Ankh-Hour No.414, Asasif, Luxor
(After Porter and Moss 1964).
(See Appendices 5.6-5.7 for sampled areas).

4.3-Alexandria

In 332 BC Alexander the Great conquered Egypt, founding Alexandria in 331 BC. His conquest of the old empires to the East and Southwest opened up vast new areas to the influence of Greek culture. After his death, his generals divided his empire into a series of independent Kingdoms and these lands were gradually "Hellenised". The term "Hellenistic" is used to describe the period between Alexander's death and the fall of Egypt to Rome. The Macedonian general Ptolemy took control of Egypt and declared himself king in 305 BC. He achieved a remarkable marriage between the conservative Egyptian world and his own lively Hellenistic culture. The Egyptian and Greek settlers had lived separate from and unaffected by each other, but during this period Egyptian culture was absorbed by the Greeks and, through them has passed into the mainstream of western civilization. The Ptolemies were tolerant in their attitude to Egyptian culture in general and Egyptian religions in particular (Ruffle 1977: 100). (See also Bowman 1986)

The third century BC was the zenith of Ptolemic rule, when Alexandria was famous for its intellectual and scientific achievements. The Kings, who were the patrons of Hellenic culture, established, as the centre of that culture, the museum and library. These were the essential features of Hellenism in Egypt as a whole and became Alexandria's claims to fame in Antiquity under reign of the second and third Ptolemy. The social, intellectual and artistic life of the Alexandrian was Greek, although it is known that at the beginning of the Roman period Alexandria did not start a senate until the Ptolemies changed their policy (Noshy 1937: 6, 7).

4.3.1-Architectural Style in Alexandrian Tombs

The styles of the tombs were the same as the cemeteries in Greek times and have been found in many parts of Egypt. There were two types: Firstly, pit tombs were rectangular or irregular in shape, cut in rock or excavated in soil, and covered with slabs or earth, varying in size and depth according to the number of people buried. The second type was the underground tomb, of which there are two types: *Kline* tombs for the "Aristocracy" and *loculus* tombs, found at Alexandria and Fayum for the middle classes, necessary because of the growing population and limited space. The style of tomb which had been changed to eliminate the *Klinai* in favour of *loculi*, starting at Shataby, found its culmination at Max, where the single *loculus* tomb lost out in favour of mass *loculus* tomb and the re-use the old ones. *Loculi* were usually closed with slabs, decorated with false doors, executed in painting and occasionally in relief. Noshy, in his study of the architecture in Alexandria, noted developments compared with places outside Egypt. He also noted the similarity between the plan of the *Kline* tombs and Greek Hellenistic houses and the Greek character of tomb architecture. The only exception was the use of the Egyptian fillet bands. The introduction of Egyptian-bands into two Doric half-columns and Greek dentils into a few Egyptian *naiskoi* and *loculus* doors as well as the decoration of some *loculus* slabs with an architecture half-Greek and half Egyptian was as a fashion of the architecture style. By the second century this was an indication of a certain de-Helleniziation of the Alexandrian Greeks. Noshy mentioned that Alexandria was the home of the original motifs of the first and second Greek style. The *loculus* tombs are decorated in a Greek style, but these are a minor attempt at fusion in the decoration of some late *loculus* doors (Noshy 1937: 20, 21, 39 & 40). It was noticed that in the architecture style of the Mustafa Pasha tombs, all the rooms are around a court whereas in Anfushi they are on one axis (plan 4 & 5). These tombs were selected for sampling.

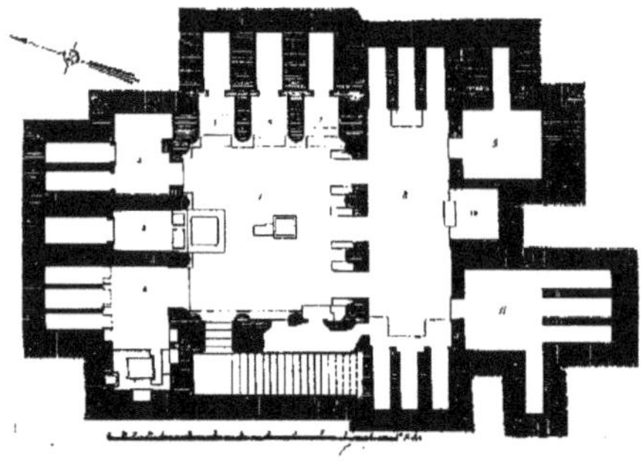

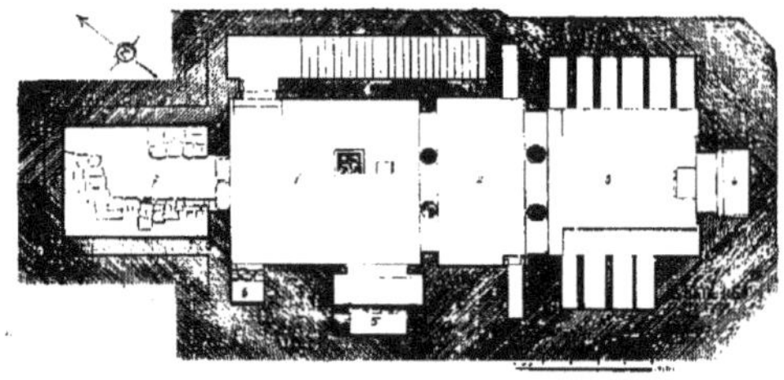

4 A 4 B

Plan 4 A & B shows the architectural style of the Mustafa Pasha tombs 1 (A) and 2 (B), where samples were collected (after Adriani 1936). (See Appendix 5.8 for sampled areas).

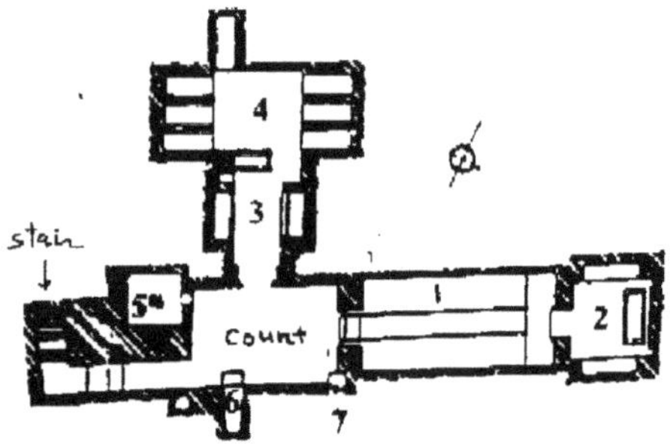

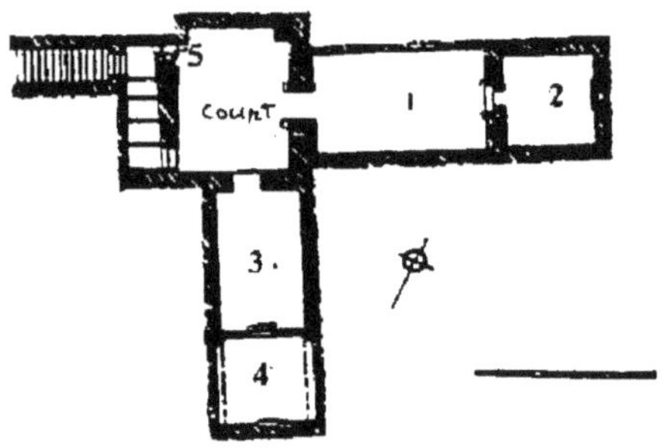

5 A 5 B

Plan 5 A & B, these showing the architectural style of Anfushi tombs 1 (A) and 2 (B) (after Hassen 1963). (See Appendices 5.5.9-5.11 for sampled areas).

4.3.2-The Decoration Style in Alexandria in Classical Times

Alexandria is one of the few sites where examples of Hellenistic tombs dated to the second century BC survive. The decorated wall paintings found by Rostovtzeff in southern Russia were important to complete a series of monuments from the 4th century BC to the fourth century AD, because these depict the evolution of mural painting in the east (Rostovtzeff 1919: 147). However, in Alexandria, as everywhere, the best paintings were lost, and those remaining are seldom of sufficient quality to attract attention. As a result, information about Hellenistic painting can not help us to know more about Alexandrian painting (Brown 1957: 1).

Vase painting and decoration were varied in comparison with wall painting. The most important ones are the red figure, black glazed and white ground vase-painting techniques which reflect the scheme of decoration and colours which give an outline of the Hellenistic civilization. It could be said that the style of wall painting is similar to the schemes of drawing, motives and composition used on vases. Painting on vases was more complicated than those on walls because the painting has to be applied to a small three dymantional structure. However, Adriani states that the decoration styles used in the Alexandrian tomb sites of Mustafa Pasha gave important evidence about the decoration style (the zonal style) used in the Greek times, which is different from those used in the Anfushi tombs. The tombs of Mustafa Pasha and Sidi Gaber were decorated in the zonal style, but so were the tombs at Shataby and El Hadara. Adriani stated that these examples were applied to the same divisions of the wall and were painted in the scheme of the zonal style. Unfortunately most of the painting was lost when the tombs were excavated, but the remaining parts give an idea of the decorative style used.

The zonal style was not only used in Alexandria, but also spread out to other areas over a long period of time. There were many changes in the details of the schemes of the decoration, as well as in the variety of colours used for painting below the intermediate band, which had previously been painted in red, following the traditional Aegean style. Although the development of the zonal style was followed in Alexandria, there was not enough evidence to assume that the First Pompeii style originated in Alexandria, or that the transformation from one style to another was found only in Alexandria and not its neighbourhood (Adriani 1936: 120-130).

The oldest tombs in Alexandria were decorated or adorned with the zonal style, while the more recent tombs in Mafrosa and Anfushi were decorated in the First Pompeiian style. The 3rd Century BC. was the date of the zonal style, as well as the formation of the First Pompeiian style, but it should be borne in mind that there was a transformation period during which the old style was replaced by the new style in the 3rd Century BC. Adriani said that it had been believed that the zonal style was rarely employed in Alexandria (Adriani 1936: 120). However, Mustafa Pasha is purely Hellenistic in decoration, and Anfushi tomb 2 has the Egyptian iconography and style, found on the walls of the first landing of the staircase, leading to the chamber room (photograph 5 A). It still survives as two intermediary bands of decoration and one on the lunette of the wall facing the second ramp (Brown 1957: 52, 53).

Adriani reported that most art historians had ignored the zonal style, so was difficult to claim it as being in existence prior to that of first Pompeiian style, but, from his point of view, he considered that the six tombs of Mustafa Pasha, Shataby and El-Hadara were painted in the zonal style, as all these tombs had shown editions of the zonal style, which pre-dated the first Pompeiian period because the zonal style was not known in Pompeii. Some examples from Agua have shown that the zonal style was applied for decoration. Similar examples have been found in Etruria and south Italy. The majority of the cases show that the edges were divided into horizontal polychrome areas of different size. Some have been painted with motifs and others with figures, but the majority were monochrome (Adriani 1936: 121).

4.3.2.1-The Decoration Scheme in the Mustafa Pasha Tombs 1 & 2

The Mustafa Pasha tombs were representative of the last stage in the development of the zonal style and gave a clear idea of how the transformation from the late stage of the zonal style to the First Pompeian style occurred. Adriani said that it can be reasonably certain that the First and Second Pompeian styles were founded on the zonal style (Adriani 1936: 12, 13 & 127).

Adriani described the scheme of decoration and the colours used in the tombs of Mustafa Pasha. He illustrated the imitation marble style and coloured stones used, as well as the comparing those in Alexandria, reporting the similarities and differences between the elements of decorative art of the tombs of Sidi Gaber and Mustafa Pasha. The 2nd tomb of Mustafa Pasha and the tomb of Sidi Gaber have the same characteristic decorative style. The only difference was the use of the red colour applied on the base of the wall (*dado/ socolo*). He assumed that, as the chamber rooms were small, they were painted in the zonal style, which was different from the traditional style. *Stucco* was applied in a very thin layer to cover the mortar walls and the edges of the doorframes. The scheme of decoration and the colours used for painting, e.g. in rooms 2 and 4, were red, white ivory and yellow edges, with a black band between them. They were always painted *dedo* (see glossary page xvi) in a red colour with the separating band in a brown colour. A golden yellow was applied on top and the internal and external rooms were painted in blue (Adriani 1936: 131, 115, 116 & 20-31).

A picture still remains on the lintel of room 8, protected by glass. The colours tend to be lighter and cracks have appeared on the surface as well. Adriani and Brown gave a description of the decoration, which was in much better than it is now. The drawing was executed in the indistinguished *fresco* technique on a thick layer of plaster on layers of *stucco* applied to the stone. The colours used were; yellow, blue, black, and brown, also red brown tending to violet or red violet. However Adriani states that there was a wide brown band around the picture, like a shadow, and the colour was actually blue, but had turned brownish (Adriani 1936: 38 & Brown 1957: 52).

Unfortunately that picture is now very faded, the colours are very faint so that they could not be determined. In addition, the painting was only in the tomb on the horizontal stone above the lintel, which leads from the courtyard to a burial chamber room 8. Adriani mentioned and described the five niches that have been found in the tomb of Mustafa Pasha, which were decorated in blue and pink colours and the clothes on the figures which were painted in pink and violet colours. However, none of the niches were found on the site. He mentions also traces of the violet colour noticed on the ceiling (Adriani 1936: 20, 24),

Brown reported that after the foundation of Alexandria, these styles were brought from Greece where they had been used in the 4th century BC and continued in use with the funeral stela. The popular style of the production was differentiated in the Hellenistic period in other places, compared with the motifs used in Alexandrian stela decoration, which were isolated and abstract, with more friezes and repeated symbols which became more noticeable in Roman times. These stelae were found in the Alexandrian catacombs and elsewhere. In the second century BC, the decoration style of the stelae developed to serving a symbolic rather than realistic function which represented the fourth style of *loculus* slabs and continued to be the dominant style with modification of the colouristic motives. By the third century BC the Post-Proxile Lan style with small scale, lively and fine decoration were found in Mustafa Pasha in both richer and poorer tombs. These tended to be more naturalistic and dynamic in style in the early part of that century and were called the first Pergamene style. These have significant connections with the development of the late antique and early Christian style (Brown 1957: 86-94).

4.3.2.2-The Decoration Scheme in the Anfushi Tombs 1 & 2

Noshy stated that the Anfushi tombs were covered with *stucco* and painted in the style of architectural decoration. This was a development of the second style (Noshy 1937: 34). It is interesting to notice that in Anfushi tomb 2, one could determine by naked eye the layers of the paintings, which indicated re-decoration of the tomb, which was very common at that time. Brown noted that the decoration, which followed, included rectangles containing figured scenes in the second style. There are three pictures, which are Egyptian in iconography and style, painted on the two contiguous walls (layers) at the first landing of the staircase, which leads to tomb chambers. The decoration is over the intermediary band of decoration and on one of the lunette of the walls facing the second ramp (Brown 1957: 53). In Anfushi tomb 2 the decoration combines Greek in the first-second style and Egyptian styles, the date of the tomb was considered to be 200 BC as in the decoration of hypogeum. On the other hand, Noshy has discussed that point and concluded that the graffiti on the right of room 1 shows a warship, which has been identified as a Roman *naviturita* (Noshy 1937: 26-27).

The Egyptian paintings or Egyptianizing style existed in the first decoration of the tomb in the second century BC. Egyptian decoration did not take place before that period, as there were no parallels in Ptolemaic times. Anfushi was therefore a Macedonian tomb with origins in Greek architecture and decoration, which was requisitioned in the Roman times by a second proprietor who gave it an Egyptian touch (Noshy 1937: 28 & Brown 1957: 59). The Egyptian and Greek style at the cemetery of Tuna El-Geble provides another example of wealthier burials near the Ibis catacomb, where there grew up a small town of house-tombs for the rich Greeks of Herompolis, for example the tomb of Petosiris. The interior walls were decorated with well preserved painted relief in Egyptian conventions, engaged in the traditional activities of Egyptian tomb-paintings, but strongly influence by Greek styles (Ruffle 1977: 159). However, Gabra and Droton suggested that it is largely native Egyptian and much earlier than the Egyptianizing tombs of the early centuries AD. Ling (1991) said that Egyptianizing elements appeared in second style painting after the battle of Actium during the Augustan period. This may imply the arrival of Alexandrian artists in Roman, appearing to suggest that an Alexandrian painter was responsible for some of the decorations in the house of Augutus (1991: 212).

There is scattered evidence of the style of painting which developed in late Hellenistic times, occurring in Alexandria, for example in the ceiling painting of Anfushi tomb 2, which includes rectangles containing figured scenes. Brown has given a description of the decoration scheme of the ceiling, which is a continuation of the naturalistic lively style, becoming at once more spirited and different to the third style of the third century BC, being painted in black, yellow, light blue, red, white, and brown-black. The development also included vase paintings (Brown 1957: 53). Some say that the *gesso* technique used to cover the walls was mixed with marble powder, as it was very commonly used for sculpture and kept the colour for the longest time (Hassen 1963: 119).

It can be concluded that, richer tombs had a decoration original in motive, but Greek in style, which points to Alexandria as the home of the first as well as second styles, whereas poorer tombs were decorated in a Greek style. At a later date some Loculus slabs were decorated with doors (Noshy 1937: 143). However, these tombs as at Anfushi, were re-used in Roman times, as mentioned above, and two different decoration styles were used in tomb 2, indicated by the re-plastering as well as the re-decoration of the walls, as shown on Photograph 67. Some evidence has also shown that the kind of the mortar used was applied by Romans, as that kind of mortar was not used previously then (see Chapter 6.2, 2.2.6).

4.4-Piddington Roman Villa was first built as a villa in the Hadrianic period during the first quarter of the second century, burnt down about AD 200 and rebuilt. Then the villa had a long history. The main feature of the villa (plan 6) was a long corridor in the front, floored with herringbone tiling in two colours, red and yellow, which was called *"opus spilatum"*. It had worn through to the pitched limestone foundations. The whole settlement covered some eight acres, there were two round houses, one of the Belgic period, distinguished by a circular timber slotted gully and two opposed door ways, and the other of the later fourth century. The fire may have started in the bathhouse between two ranges. The south wing was a separate building; only in the late second century was it joined to the main villa. The whole wing was substantially reconstructed and widened after a disastrous fire in the late 2nd century.

The villa continued in use and was rebuilt through the fourth century. There was only one heated room (room 26). The northern side was not a full wing, but a short wing consisting of a single very large room or hall with no internal divisions.

A recent discovery from this area has shown some wall plasters were painted deep red, supporting short stubby pillars, which had been plastered and then painted in three colours. In the courtyard there were piles of tessera waiting to be laid down in new mosaic floors and in the small suite of baths the final plastering had been prepared and polished but never painted (see Friendship-Taylor *et al.* 1982, 1989 and 1996 and also Rollo 1994, in which there is more information and description of Piddington excavations).

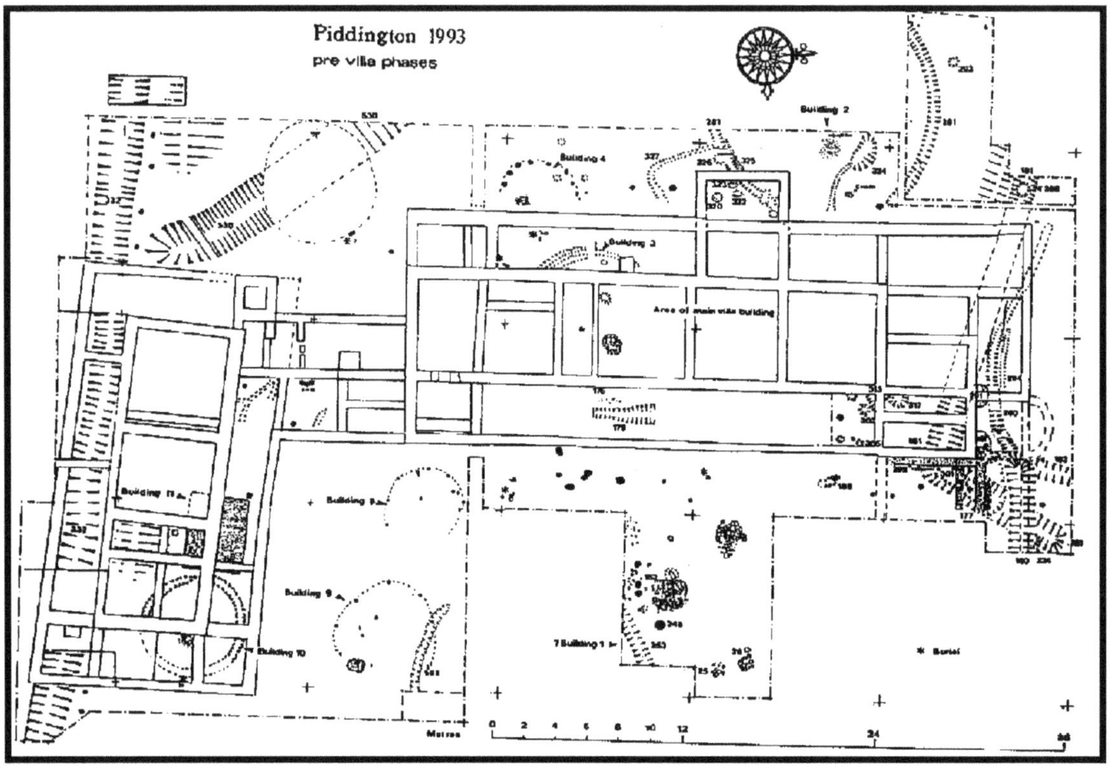

Plan 6, showing the pre villa of Piddington (from Rollo 1994)

4.4.1-Decoration style

The Romans were masters of painted wall decoration, creating spatial illusions to enhance the interiors of their houses. Ling mentioned that wall painting did not appear in isolation, but formed part of a decoration ensemble, which included work in mosaic and other media (*Stucco*, marble veneer) (see Appendix 2). The Romano-British craftsmen used two main stages, the application of the plaster and the application of the paint (Ling 1991: 47). Davey and Ling (1981) explain the techniques used in Roman Britain and Ling (1991) for Roman painting reported the styles of the decoration and techniques as well. The techniques used depended on the plaster having a good adherence to the underlying surface, for example by pecking the surface as found (see photograph 56 A).

Romans decorated their buildings (houses, villas, temples and tombs) with wall paintings applied both on internal and external surfaces, but the latter have rarely survived for obvious reasons. The techniques used were *frescoes* as stated in Chapter 2 (see 4.8.1-3).

Appendix 5-20

Description of Samples, Analytical Techniques & the Results of Chemical and Physical Analysis, in order from the Earliest to Latest

From the Sites of; Saqqara, Dahshur, Luxor, Alexandria in Egypt & Piddington Roman Villa in Britain

Catalogues of Samples

Appendix 5.1

Saqqara

Sample name	Sample area	Description
		First Dynasty (E E)
S.104	Different areas of the wall	Mud brick
S.100 S.101 S.102 S.103		Mud plaster (coarse plaster), covering the mud brick, samples were collected from the remains of the Tomb of Her-Nit. Samples were hard and solid.
S.102/1		Layer covering the mud plaster of sample (S.102), see photograph 7 B (part 1) and X-RPD, Appendix 16.15.

		First Dynasty (N E 1995-96)
S.142 (S.142/2 L)	Different areas of the wall	Clay plaster "*hib*" (powdered limestone and clay), covered with a very thin layer of gypsum/lime (X-RDS). Sample (S.142) consisted of two layers, some analyses were carried out on the second layer (S.142/2L). The plasters were very soft and fragile.
S.143/1		
S.143/2		

Catalogues of Samples

Appendix 5.2

Dahshur

Sample name	Sample area	Description
		Dahshur the 4th Dynasty Mastaba (N E)
S.52 S.54	From the Mastaba Dahshur	Limestone.
S.50 S.51 S.55/1 S.56	West side East wall, west side of the Mastaba Right entrance of the east side of Mastaba	Mud brick, very soft and fragile.
S.53 S.55/2 S.57	East side East wall, west side of the Mastaba	Clay plaster, very soft and powdery. Investigation of the structure and composition of the clay plaster/mortar has shown that the mixture consisted of; quartzite, quartz sand, powdered limestone, small pieces of chalk/lime and some other minerals.
S.55/3 S.57/3		Gypsum plaster (see photograph 8, part 1), this layer was found to cover clay plaster (sample S.53 and S.57), as a finishing layer. There were black spots on the surface.

Catalogues of Samples

Appendix 5.3

Saqqara

Sample name	Sample area	Description
Tomb of Ka-Nefr 5th Dynasty (N E)		
S.9 S.145	Different areas	Mud bricks, the supporting structure for the rendering, consisting of; quartz, mud, dark straw, a piece of flint and charcoal. The mixture was soft and fragile.
S.49 S.134/3 S.134/5 S.144		Mud plasters (soft and fragile), covering the mud bricks. The mixture contained mud, straw, quartz and limestone, and charcoal. Samples (S.134/5), covered with a gypsum plaster.
S.10 S.46 S.47 S.48 S.146/1		Clay plasters (soft and fragile), all samples show the same mixture; straw powdered limestone, quartzite, and a few pieces of chalk/lime, in addition to limestone and charcoal.
S.134/2	Different areas	Gypsum plaster consisted of many layers (found without any paint).
S.46/1 S.146/2		Very fine layer of plaster covering the clay plaster (sample S.46, photograph 9, part 1) (See X-RDS) Consisted of many layers, could be very thin gypsum layer covering S.146/1 (see X-RDS).
Idy Tomb Late 6th Dynasty (N E)		
S.139	Different areas	Mud brick.
S.137 S.138		Mud plaster/mortar the same above, with mollusc inside and some seeds inside sample S.137.
S.140		Clay plaster, same mixture as previous samples.

Catalogues of Samples

Appendix 5.4

Luxor

Sample name	Sample area	Description
Sheikh Abd El-Qurna T. 56 18th Dynasty		
S.36 S.66	Inside the tomb Internal room	Limestone, soft.
S.107		
S.37 S.105	Outside the tomb	
S.106	Court	Mud plaster/mortar covered with whitish layer, the plaster was very soft and fragile.
S.67	East wall	Salt in form of fibres/needles, the surface had many holes.

T. 55 18th Dynasty		
S.110/2		Limestone, very fine grain, soft.
S.113	The remaining wall on the right side of the entrance.	Mud brick, hard and solid.
S.111	North wall	Clay plaster.
S.110/1		Salt in rounded shape and fibres/needles, some were in form of yellowish/brownish fossil dust.
S.112	West wall	
S.76	First column on the right side	Finishing plaster, hard and pinkish.

Catalogues of Samples

Appendix 5.5

Luxor (continued)

Sample name	Sample area	Description
		Khokha T. 192 18th Dynasty
S.27/3	Room to the right of the entrance	Limestone, the stone was slightly pinkish and solid, showed slight bands of iron oxide, as confirmed by micro-chemical tests.
S.30		
S.124	West wall to the left room	Mortar filler used between the mud brick.
S.129	Corridor	Mud brick, the samples were soft and friable as well.
S.131	The final wall in the tomb	
S.27/1-2	Room to the right of the entrance	Small fragments of plaster.
S.28	Third room	
S.29	Second wide hall on the right	
S.31 & 32	Second wide hall on the right	
S.33	Second entrance on the right	
S.34	Right wall to the entrance	Clay plaster, very soft and fragile.
S.125	Court	
S.126		
S.127	Different areas of the left side of the entrance	
S.128		
S.130		

Catalogues of Samples

Appendix 5.6

Luxor (continued)

Sample name	Sample area	Description
		Drah Abu El-Naga T. 255 18th Dynasty
S.83/1-2	Ceiling of the north side of the internal room	Limestone.
S.122		Mud plaster, hard and solid.
S.121	Right side of the entrance	Clay plaster.
S.120	North side	Finishing plaster.
S.114/1-4	Right entrance	Pigments; black, blue and yellow.
		T. 13 19th Dynasty
S.118	South side of the last room	Limestone.
S.115		Mud plaster.
S.116	Ceiling of the second room	Clay plaster.
S.119/1-3 S.119/y	Un-known	Small fragments (119/1painted layer in red, 119/2, finishing plaster and 119/3 clay plaster) 119/y painted in yellow.
		The Valley of the Kings T. 14 19th Dynasty
S.RT	The Valley Site	Limestone.
S.61		
S.3	To right the of the entrance	Clay plaster.
S.86	Ceiling of the long hall	
S.88	Left side of the hall	Salt.

Appendix 5.7

Catalogues of Samples

Luxor (continued)

Sample name	Sample area	Description
		T. 15 19th Dynasty
S.4	Left side of chamber room	Clay plaster, very soft and powdery.
		T. 11 20th Dynasty
S.59	Ceiling of chamber room	Limestone.
S.64		Clay plaster.
		Deir El-Medina T. 359 20th Dynasty
S.1	Small room on the right of the chamber room	Limestone.
S.8		
S.6	Different areas	Mud plaster covering the stone, covered with traces of a whitish layer Coloured plaster (finishing plaster).
S.40		
S.5		
		Asasif T. 414 26th Dynasty
S.13	Chamber room	Limestone.
S.14	Right entrance	
S.84	Left of the column room	
S.16	Left wall of the entrance	Mortar filler for leveling of irregular surface of the rock.
S.15	South wall of the entrance	Mud plaster.
S.17	East wall to the entrance	
S.18	Column hall	
S.19		
S.20	Corridor	
S.85	East wall to the chapel room	

Catalogues of Samples

Appendix 5.8

Alexandrian Sites

Sample name	Sample area	Description
		Mustafa Pasha Tomb 1 (MP T. 1)
S.1	Room 2	plaster/mortar painted in red-brown.
S.2	Room 5	plaster/mortar, fossil shell, quartz, mica and some other minerals, the plaster was covered with a coloured *intonaco* layer in yellow.
S.3	Room 6	Plaster from the edge of the frame door, painted in yellow.
S.4	Room 6	Plaster (first layer) with traces of red colour on top.
S.5	Room 6	Mortar covered with coloured *intonaco* in red (see photograph 20 B, part 1).
S.6	Room 6	Plaster, second layer.
S.7	Room 7	Pigment.
S.8	Room 8	Pigment.
S.9	Room 8	Plaster painted in bright red.
S.10	Room 8	Pigment.
		Mustafa Pasha Tomb 2 (MP T. 2)
S.11	Room 2	Pigments (green, yellow and red).
S.12	Room 2	Lime plaster, painted in bright red and covered with a very thin of calcite film, microorganisms were noticed on the painted layer and between the plaster as well (see photograph 3 B, detail).
S.13	Room 3	Two small pieces of lime plaster/mortar painted in red and yellow and another showing traces of red colour.
S.14	Site	Limestone.

Appendix 5.9

Catalogues of Samples

Alexandrian Sites (continued)

Sample name	Sample area	Description
Anfushi T. 1 (A T. 1)		
S.8	Right entrance	Salt (wall faced to the right entrance to tomb 1).
S.5	Room 1	Salt (left wall). Pigments.
5/1		
S.5/2		
S.5/3		
S.1	Room 2	Mortar and plaster.
S.2	Court left wall right wall	Calcite salt.
S.3		Plaster.
S.4/1 A-5A		Small mortar/plaster without paint.
S.7		Salt.
S.8		Salt.

Catalogues of Samples

Appendix 5.10

Alexandrian Sites (continued

Anfushi T. 2, (A T. 2)

Sample name	Sample area	Description
S.1	The entrance	Plaster.
S.2		Plaster.
S.4		Salt right entrance (1998).
S.6		Salt right side (from middle wall) (1998).
S.7		Salt from the top right of the Egyptian wall painting (1998) (see photograph 5A, part 1).
S.9	Court	Salt (right side) to room 1.
S.10/1		Salt left side from the entrance to the court.
S.10/2		Salt right side from the entrance to the court.
S.10/3		Salt right side (east wall).
S.10/4		Salt left side.
S.10/5		Salt left side.
S.15/5		Salt right side
S.14/2		Limestone, left side court.
S.17/1	Room 1	Coloured plaster.
S.17/2		Pigment (brownish).
S.17/3		Yellow.
S.17/4		Blue/green.
S.18/1		Salt right side.
S.18/2		Salt with traces of yellow on top.

Catalogues of Samples

Appendix 5.11

Alexandrian Sites (continued)

Sample name	Sample area	Description
Anfushi T. 2 (A T. 2)		
S.11	Room 2	Plaster, right side.
S.12		Salt left side.
S.12/2		Plaster.
S.13		Right side.
S.15/1		Pigment, brown.
S.15/2		Green/black.
S.15/3		Coloured plaster.
S.15/4		Blue.
S.15/5	Room 2	Pigment, brown.
S.16		Plaster (*gesso*).
S.4	Room 3	Salt from the entrance to room 3 (1998).
S.5		
S.6		
S.7		
Anfushi T. 3 (A T. 3)		
S.19	One room	Limestone.
S.20		Coloured plaster (left side).
S.21		Coloured plaster (right side).
S.22		Limestone (light wall).
S.23		Limestone (left wall).

Catalogues of Samples

Appendix 5.12

Libya and Cyprus

(These samples supplied from Dr. G. Morgan' collections)

Sample name	Sample area	Description
S.1 L	Libya Tolmeta beach	Limestone.
S.2 *C	Cyprus Queens Bay	Limestone.

Catalogues of Samples

Appendix 5.13

Fragments from Piddington

Sample name	Sample area	Description
S.12	Room 21 -DD/6 (AA H/F 95) -Box 70 (AA L2 H/F 95) -Box 80, recorded 96-97 (bag 8)	Fragments different in size and decoration scheme; some painted in red coloured *intonaco* and burnished; some others had paint in different colours, green, brown and black on white *intonaco*. Fragments consisted of two layers in different thicknesses between 10-60 mm. A few fragments were of three layers in different thicknesses, but the *intonaco* layer was between 0.1-0.5 mm.
S.13	Room 26 -Y/26 L5 M/P 90 -Bag 63 A, 1990	The fragments consist of one layer, between 7-10 mm in thickness and some of two layers with no paint on the first layer; red *intonaco* layer covered the plaster and was burnished. The plaster was red in colour (burnt). Whilst some of the group of fragments were painted a brownish colour and painted over in black with white panel between, others were painted in red; these fragments seem to be detached from the other layer, as the back appeared to be on a level.
S.1 S.2 S.3	Room 53 -Box 78	All fragments were painted in a cream colour and covered with a very thick deep yellow layer of calcite, as fossil dust (ancient dust). Plasters consisted of two layers; the first layer from bottom about 7 mm, and the second layer about 10 mm, which was covered with an *intonaco* layer of 0.1 mm in thickness. Some fragments were painted in yellow ochre with a red panel.

Appendix 5.14

Fragments from Piddington (continued)

Sample name	Sample area	Description
S.10	Room 53 -N15, N16	The thickness of this fragment was different compared with the previous fragments; the first layer (from bottom) between 35.45 mm; the second layer about 16 mm was covered in very thick layers of calcite, with very thin *intonaco* layer 0.1 mm and paint in a cream colour was splashed with green colours; very deteriorated surface (photograph 55 A).
S.14	Room 53 -GG/002-003 -HH/001 L3 (U) -66/001-002 -Boxes 75, 76, 1994-96	Different fragments; some were wall plasters and others from floor and window mouldings so that the thickness of the mortars were different and the marks on the back indicated that the mortars were mouldings and rough surface was noticed as well. The plasters consisted of two layers and some of three layers in different thickness and decoration schemes, with one panel and others in different panels; dark reddish/brownish, white on pink and green, in addition to Egyptian blue. Some fragments were coloured in red *intonaco,* in thickness about 0.2-1mm and splashed colours were found on the painted layer.
S.15	Room 55 -99/001 16 LUD M/F 94, box 71 A	These fragments have shown the same structures as above. The plasters were in two layers; the thickness of the first layer from bottom between 15-20 mm, and the second layer about 7 mm, painted in different panels; of yellow ochre, red and green/grey or yellow onto a cream background.

Catalogues of Samples

Appendix 5.15

Fragments from Piddington

Sample name	Sample area	Description
S.8 S.9 S.16	Room 56 JJ/003 810 JJ/005007 L4-L7 JJ/005.006 L4 L-7	The plasters consisted of two layers in different thicknesses, the first layer from bottom about 10-22 mm, was painted in different panels; black, purple onto red. The second layer about 4 mm, was painted in cream. Some fragments were painted in yellow and red panels, the first layer from bottom about 20-28 mm, and the second 7mm painted in a cream colour on *intonaco* layer about 0.1 mm. Egyptian blue (coarse) was found on some samples and a coloured *intonaco* layer was noticed on those fragments 1-3 mm in thickness.
S.4 S.5	Room 57 JJ/2-3 F 810 Box 71	The fragments consisted of two layers in different thicknesses, the first layer from bottom, about 20-38 mm, and the second layer between 12-20 mm, which was covered with an *intonaco* layer of about 0.1 mm, and painted in a cream colour with a wide red panels in some cases; the first layer from bottom was about 10-30 mm and the second 7mm, painted in a cream colour.
S.6 S.7	Box 80 F 10	The fragments appeared in different coloured schemes, with green and black panels; some were in green colour on red and others painted in yellow and brown. The plaster consisted of two layers; the first from bottom between 17-28 mm and some about 20-25 mm and second layer 7 mm covered with an *intonaco* layer of 0.1 mm in thickness. The painted layers were covered with a calcite film, indicating the use of *fresco* techniques.

Examination of Cross-Sections: Measurements of Plasters/Mortars

Appendix 6.1

Luxor

Sample name	Painted Layer mm	Finishing Plaster mm	Clay Plaster/Mortar mm	Mud Plaster mm
T. 192				
S.29	0.1-0.3	0.2-0.3	13	-
S.31	0.1	0.4-0.5	-	-
S.32	-	0.3	7-9	-
T. 13				
S.119	Very thin	0.1-0.5	15-17	-
S.119/Y		0.1-0.3		-
T. 359				
S.40	-	-	-	0.5-11
T. 414				
S.18	-	-	-	0.3-0.5

Examination of Cross-Sections: Measurements of Plasters/Mortars

Appendix 6.2

Alexandrian Sites

Sample name	Painted Layer mm	*Intonaco* Layer mm	Plaster/Mortar First Layer mm	Plaster/Mortar Second Layer mm	Plaster/mortar Third Layer mm
MP T. 1					
S.1	-	-	0.2-0.5	0.1	0.1
S.1	Very thin	0.1-0.15	0.25	-	-
S.2	Very thin	0.1-0.3	0.1-0.3	-	-
S.3	Very thin	0.1-0.15	0.35-0.5	-	-
S.5	-	2	8-10	-	-
S.8	Very thin	-	2	-	-
MP T. 2					
S.12	-		0.1-0.2	0.2-0.3	-
S.13	-	0.1	0.3	0.25-0.4	-

Appendix 6.3

Alexandrian sites (continued)

Sample name	Painted Layer mm	*Intonaco* Layer mm	Plaster/Mortar First Layer mm	Plaster/Mortar Second Layer mm	Plaster/Mortar Third Layer mm
A T. 1					
S.1A	-	-	0.3-8	-	-
S.1B	-	-	0.1-0.25	0.1-0.15	-
S.4/A1	-	0.2	15	-	-
S.4/A2	-	-	0.15-0.3	0.5-9	-
S.4/A3	-	-	7-13	-	-
S.4/A4	-	-	0.2-0.4	0.2-0.4	-
S.4/A5	-	-	0.2-7	-	-
A T. 2					
S.2	0.5	0.1	0.3-0.35	-	-
S.11	0.1	-	0.2	0.15-0.2	-
S.13	0.1	-	0.5-0.2	-	-
S.15/2	-	0.2-0.5	0.2	-	-
S.15/3	-	-	0.3	-	-
S.16	0.1	0.5	0.3-0.35	-	-
S.17/1	0.1	0.2	0.2	-	-
A T. 3					
S.21	-	0.1-0.2	0.3-0.4	-	-

Examination of Cross-Sections: Measurements of Plasters/Mortars

Appendix 6.4

Piddington

Sample name	Painted Layer mm	*Intonaco* Layer mm	Plaster/Mortar First Layer mm	Plaster/Mortar Second Layer mm
S.1	0.5	0.1	-	-
S.2	0.1-0.3	0.1-1.5	-	-
S.3	0.5	0.5	-	-
S.4	0.5	1	-	-
S.5	0.1	0.2-0.4	-	-
S.6	0.1	0.2-0.4	-	-
S.7	0.5	0.1-0.2	-	-
S.8	0.3	0.1-0.3	-	-
S.9	0.1	0.1	-	-
S.10	0.1	0.1-1.3	9	7-9

Catalogues of Analytical Techniques Used in The Examination and Analysis of the Samples

Appendix 7.1

Saqqara and Dahshur

Chemical Analysis				Physical Methods					
ST	**SA**	**SS**	**M $CaCO_3$**	**PCS**	**X-RD**	**X-RPD**	**ICP**	**PLM**	**SEM**
S.102/1	S.104	S.104							
	S.100	S.100	S.143/1			S.101	S.104	S.143/1	S.101
	S.103	S.101				S.102	S.101	S.55	S.102
		S.102	S.52			S.102/1	S.102	S.52	S.142
	S.143/1	S.143/1	S.54			S.103	S.103	S.54	S.55/3
	S.143/1	S.54	S.50			S.142/2L	S.143/1	S.50	S.55/5
S.103		S.50	S.55/1			S.143/1			
	S.50	S.51	S.56			S.143/2	S.54		
	S.51	S.55/1	S.53				S.50		
	S.55/1	S.56	S.57			S.52	S.55/1		
	S.56	S.53				S.55/3	S.56		
S.55/3	S.53	S.57			S.53/3	S.57/3	S.53		
S.57/3	S.57	S.55/2					S.57		
		S.55/3 & S.57/3							

ST: Spot Test **SA**: Standard Analysis (qualitative & quantitative analysis) **SS:** Soluble Salts
M $CaCO_3$: Measurements of Calcium Carbonate **PCS:** Polished Cross Section (identification of composition)
X-RD: X-Ray Diffraction (identification of single crystal) **X-RPD:** X-Ray Powder Diffraction (identification of minerals)
ICP: Induction Coupled Plasma Spectroscopy (metal ions) **PLM:** Polarised Light Microscopy (internal structure)
SEM: Scanning Electron Microscope **-EDS:** Energy Dispersive Spectrometer (surface structure)

Catalogues of Analytical Techniques Used in The Examination and Analysis of the Samples

Appendix 7.2

Saqqara and Dahshur (continued)

Chemical Analysis				Physical Methods					
ST	**SA**	**SS**	**M $CaCO_3$**	**PCS**	**X-RD**	**X-RPD**	**ICP**	**PLM**	**SEM**
	S.134/3	S.9	S Stone		S.144	S.46			S.9
	S.145	S.145	S.145			S.134/2	S.135		S.134/2
	S.49	S.49	S.10		one piece of the mixture		S.136		S.145
	S.134/5	S.134/3	S.46			S.140	S.140		S.140
	S.144	S.144	S.48						
	S.46/1	S.10	S.146						
	S.46/2	S.46	S.140		S/146/2				
S.134/2	S.47	S.47							
S.146/2	S.48	S.48							
	S.146/1	S.146/1			S.140				
	S.139	S.139							
	S.137	S.137							
	S.138	S.138							
	S.140	S.140							

ST: Spot Test
SA: Standard Analysis (qualitative & quantitative analysis)
SS: Soluble Salts
M $CaCO_3$: Measurements of Calcium Carbonate
PCS: Polished Cross Section (identification of composition)
X-RD: X-Ray Diffraction (identification of single crystal)
X-RPD: X-Ray Powder Diffraction (identification of minerals)
ICP: Induction Coupled Plasma Spectroscopy (metal ions)
PLM: Polarised Light Microscopy (internal structure)
SEM: Scanning Electron Microscope **-EDS:** Energy Dispersive Spectrometer (surface structure)

Catalogues of Analytical Techniques Used in The Examination and Analysis of the Samples

Appendix 7.3

Luxor

Chemical Analysis				Physical Methods					
ST	**SA**	**SS**	**M $CaCO_3$**	**PCS**	**X-RD**	**X-RPD**	**ICP**	**PLM**	**SEM**
S.113	S.106	S.37	S.36	S.29	S.67	S.36	S.36	S.111	S.107
	S.129	S.105	S.105	S.31		S.105	S.105		S.67
S.76		S.107	S.106	S.32	S.110/1		S.106	S.27/3	
	S.6	S.67	S.66		S.110/2	S.111	S.66	S.129	S.110/1
S.32	S.40			S.119	S.112		S.67	S.131	S.112
		S.113	S.113	S.119/Y		S.27/3		S.33/2	S.76
	S.15	S.111	S.111		S.88	S.129	S.113	S.127	
	S.16	S.110/1		S.40		S.131	S.111	S.27/1	S.32
	S.18	S.112	S.27/3			S.124	S.110/1	S.27/2	
	S.20	S.27/3	S.27/1	S.18		S.33/2	S.122		S.83/1
		S.129	S.27/2			S.127	S.27/3		
		S.131	S.29			S.27/1	S.129	S.121	S.88
		S.28	S.32/2				S.131	S.120	
		S.29	S.33/2			S.83/1	S.27/1		S.8
		S.31	S.125			S.121	S.27/2		
		S.33	S.126			S.120	S.32/2		S.18
		S.33/2	S.31			S.118	S.33/2		
		S.124	S.124			S.119/2	S.127		

ST: Spot Test
SA: Standard Analysis (qualitative & quantitative analysis)
SS: Soluble Salts
M $CaCO_3$: Measurements of Calcium Carbonate
PCS: Polished Cross Section (identification of composition)
X-RD: X-Ray Diffraction (identification of single crystal)
X-RPD: X-Ray Powder Diffraction (identification of minerals)
ICP: Induction Coupled Plasma Spectroscopy (metal ions)
PLM: Polarised Light Microscopy (internal structure)
SEM: Scanning Electron Microscope **-EDS:** Energy Dispersive Spectrometer (surface structure)

Catalogues of Analytical Techniques Used in The Examination and Analysis of the Samples

Appendix 7.4

Luxor (continued)

Chemical Analysis				Physical Methods					
ST	SA	SS	M $CaCO_3$	PCS	X-RD	X-RPD	ICP	PLM	SEM
	S.111	S.83/1	S.83/1			S.86	S.83/1		
	S.113	S.122	S.121			S.3	S.122		
		S.121	S.120				S.121		
	S.34	S.120				S.4	S.120		
	S.126		S.118						
	S.127 & S.128	S.118	S.115			S.59	S.118		
	S.130	S.115	S.116				S.115		
		S.116	S.119/3			S.1	S.116		
	S.116	S.119/3	S.119			S.8	S.119/3		
	S.119/3	S.119					S.119		
			S.RT			S.13			
	S.121	S.RT	S.61			S.14	S.RT		
		S.61	S.3			S.16	S.61		
	S.3	S.3	S.86			S.18	S.3		
	S.86	S.86					S.86		
		S.88	S.4				S.88		

ST: Spot Test **SA**: Standard Analysis (qualitative & quantitative analysis) **SS:** Soluble Salts
M $CaCO_3$: Measurements of Calcium Carbonate **PCS:** Polished Cross Section (identification of composition)
X-RD: X-Ray Diffraction (identification of single crystal) **X-RPD:** X-Ray Powder Diffraction (identification of minerals)
ICP: Induction Coupled Plasma Spectroscopy (metal ions) **PLM:** Polarised Light Microscopy (internal structure)
SEM: Scanning Electron Microscope **-EDS:** Energy Dispersive Spectrometer (surface structure)

Catalogues of Analytical Techniques Used in The Examination and Analysis of the Samples

Appendix 7.5

Luxor (continued)

Chemical Analysis				Physical Methods					
ST	SA	SS	M $CaCO_3$	PCS	X-RD	X-RPD	ICP	PLM	SEM
	S.4	S.4	S.59				S.4		
	S.64		S.64						
		S.59	S.1				S.59		
		S.64	S.8				S.64		
	S.74		S.6						
	S.79	S.1	S.40				S.1		
		S.8	S.14				S.8		
	S.87	S.6	S.15				S.6		
		S.40	S.16				S.40		
	S.109		S.18				S.14		
		S.15	S.19				S.15		
		S.16	S.20				S.16		
		S.18	S.74				S.18		
		S.19	S.79				S.19		
		S.20	S.76				S.20		
		S.14	S.87				S.87		

ST: Spot Test
SA: Standard Analysis (qualitative & quantitative analysis)
SS: Soluble Salts
M $CaCO_3$: Measurements of Calcium Carbonate
PCS: Polished Cross Section (identification of composition)
X-RD: X-Ray Diffraction (identification of single crystal)
X-RPD: X-Ray Powder Diffraction (identification of minerals)
ICP: Induction Coupled Plasma Spectroscopy (metal ions)
PLM: Polarised Light Microscopy (internal structure)
SEM: Scanning Electron Microscope **-EDS:** Energy Dispersive Spectrometer (surface structure)

Catalogues of Analytical Techniques Used in The Examination and Analysis of the Samples

Appendix 7.6

Alexandrian Sites

Chemical Methods				Physical Methods				
ST	**S A**	**SS**	**M $CaCO_3$**	**PCS**	**X-RD**	**ICP**	**PLM**	**SEM**
Mustafa Pasha T. 1 (MP T. 1)								
On all Samples	S.14	S.4		S.1	S.5			
		S.5		S.2	S.6	S.4	S.4	S.2
		S.14		S.3	S.7	S.5	S.14	S.5
				S.4	S.8	S.14		S.6
				S.5				S.9
				S.8				
Mustafa Pasha T. 2 (MP T. 1)								
On all Samples				S.12	S.11			S.13
				S.13	S.13			

ST: Spot Test
SA: Standard Analysis (qualitative & quantitative analysis)
SS: Soluble Salts
M $CaCO_3$: Measurements of Calcium Carbonate
PCS: Polished Cross Section (identification of composition)
X-RD: X-Ray Diffraction (identification of single crystal)
X-RPD: X-Ray Powder Diffraction (identification of minerals)
ICP: Induction Coupled Plasma Spectroscopy (metal ions)
PLM: Polarised Light Microscopy (internal structure)
SEM: Scanning Electron Microscope **-EDS:** Energy Dispersive Spectrometer (surface structure)

Catalogues of Analytical Techniques Used in The Examination and Analysis of the Samples

Appendix 7.7

Alexandrian Sites (continued)

Chemical Analysis				Physical Methods					
ST	SA	SS	M $CaCO_3$	PCS	X-RD	X-RPD	ICP	PLM	SEM
	Anfushi T. 2 (AT. 2)								
On all Sample	S.14/2	S.9		S.11	S.3 /98	S.12/2	S.9		S.1
		S.10/1		S.13	S.5		S.10/1		S.2
		S.10/2		S.15/2	S.6		S.10/2		S.15/2
		S.10/3		S.15/3	S.7		S.10/3		S.16
		S.10/4		S.17/1	S.9		S.10/4		S.18/2
		S.10/5			S.10/1		S.10/5		
		S.12/2			S.10/2		S.12/2		
		S.14/1			S.10/3		S.14/1		
					S.10/4				
					S.1				
					S.13				
					S.15/2				
					S.15/4				
					S.16				
					S.17/2				

ST: Spot Test **SA**: Standard Analysis (qualitative & quantitative analysis) **SS:** Soluble Salts
M $CaCO_3$: Measurements of Calcium Carbonate **PCS:** Polished Cross Section (identification of composition)
X-RD: X-Ray Diffraction (identification of single crystal) **X-RPD:** X-Ray Powder Diffraction (identification of minerals)
ICP: Induction Coupled Plasma Spectroscopy (metal ions) **PLM:** Polarised Light Microscopy (internal structure)
SEM: Scanning Electron Microscope **-EDS:** Energy Dispersive Spectrometer (surface structure)

Appendix 7.8

Alexandrian Sites (continued)

Chemical Analysis				Physical Methods				
ST	**SA**	**SS**	**M $CaCO_3$**	**PCS**	**X-RD**	**X-RPD**	**ICP**	**PLM**
Anfushi T. 3 (AT. 3)								
On all Sampl es		S.19		S.21	S.20	S.19	S.19	
		S.22			S.21	S.22	S.21	
		S.23				S.23	S.22	
							S.23	
							S.1 L	
							S.2 *C	

ST: Spot Test **SA**: Standard Analysis (qualitative & quantitative analysis) **SS:** Soluble Salts
M $CaCO_3$: Measurements of Calcium Carbonate **PCS:** Polished Cross Section (identification of composition)
X-RD: X-Ray Diffraction (identification of single crystal) **X-RPD:** X-Ray Powder Diffraction (identification of minerals)
ICP: Induction Coupled Plasma Spectroscopy (metal ions) **PLM:** Polarised Light Microscopy (internal structure)
SEM: Scanning Electron Microscope **-EDS:** Energy Dispersive Spectrometer (surface structure)

Catalogues of Analytical Techniques Used in The Examination and Analysis of the Samples

Appendix 7.9

Piddington

Chemical Analysis				Physical Methods				
ST	**SA**	**SS**	**M $CaCO_3$**	**PCS**	**X-RD**	**ICP**	**PLM**	**SEM**
	S.12 A small fragments	S.4	calcite film	S.1			S.11	S.1A
	S.12 B second layer	S.10	S.1A& B	S.2, second layer	S.10 green	S.1A	S.1	S.1B
	S.13 different fragments		S.2	S.3 first layer	S.15 grey	S.1B	S.2	S.2B
	S.3 first layer		S.3	S.10 L1	S.8 yellow	S. 2B	S.3	S.3B
	S.3 second layer		S.10L1	S.10.L2	S.8 grey	S.10	S.10	S.8 first layer
	S.10L1 first layer		S.7	S.4 first layer	S.8 pink 2	S.4	S.9	
On all Samples	S.10L2 second layer			S.5 first layer	S.8 purple	S.7	S.8	
	S.14 A& B window moulding			S.6 first layer	S.4 dark red		S.5	
	S.14 C floor				S.6 pink 1			
	S.14 D different fragments				S.6 pink 3			
	S.14 E 1 first layer Three layers of plaster							
	S.8 first layer							
	S.4 first & second layer							
	S.7 first & second layer							

ST: Spot Test **SA**: Standard Analysis (qualitative & quantitative analysis) **SS:** Soluble Salts
M $CaCO_3$: Measurements of Calcium Carbonate **PCS:** Polished Cross Section (identification of composition)
X-RD: X-Ray Diffraction (identification of single crystal) **X-RPD:** X-Ray Powder Diffraction (identification of minerals)
ICP: Induction Coupled Plasma Spectroscopy (metal ions) **PLM:** Polarised Light Microscopy (internal structure)
SEM: Scanning Electron Microscope **-EDS:** Energy Dispersive Spectrometer (surface structure)

Appendix 8.1

Chemical analysis of Samples from Saqqara and Dahshur

Sample name	Original weight/ g	Solvent	Residue	% Residue	% Sol.
First Dynasty (E E)					
S.104	50	Water	49.02	98.04	1.96
Mastaba of Dahshur 4th Dynasty (N E)					
S.50	35.231	HCl	31.713	90.01	9.99
S.51	45.78		41.85	91.42	8.58
S.56	40.18		37.802	94.08	5.95
Ka-Nefr Tomb 5th Dynasty (N E)					
S.9	62.78	HCl	58.33	92.91	7.09
S.145	109.41		99.27	90.73	9.27
Idy Tomb late 6th Dynasty (N E)					
S.139	32.325	HCl	31.432	97.24	2.76

Appendix 8.2

The Percentages of Sieving Weights of Samples from Saqqara and Dahshur

Site name	**First Dynasty (E E)**	**Dahshur 4th Dynasty**			**Tomb Ka-Nefr 5th Dynasty**	
Sample name	**S.104/W**	**S.50**	**S.51**	**S.56**	**S.9**	**S.145**
Weight	50.000	35.231	45.780	40.180	62.780	109.410
Residue	49.020	31.713	41.85	37.802	58.330	99.270
% Acid Soluble	**1.960**	**9.99**	**8.58**	**5.92**	**7.090**	**9.270**
Sieving Weight Percentages						
Mesh Size mm				**% Residue**		
16.00	0.00	0.00	0.00	0.00	0.00	0.00
11.20	0.00	0.00	0.00	0.00	6.76	0.00
8.00	0.00	0.00	0.00	0.00	0.00	0.41
5.60	0.00	0.00	0.00	0.00	0.00	0.38
4.00	0.00	0.00	0.00	0.00	0.00	0.83
2.80	0.00	0.00	0.00	0.00	0.00	0.23
2.00	0.00	0.20	0.00	0.00	0.00	0.42
0.85	0.10	0.77	0.04	0.39	0.05	1.75
0.43	6.51	12.40	7.01	18.82	1.82	8.54
0.25	22.22	12.99	15.74	23.23	3.51	14.25
0.18	17.58	7.11	4.78	11.35	5.14	8.45
0.15	22.44	10.19	5.05	13.08	16.66	10.31
<0.15	31.15	56.34	67.83	33.13	66.06	54.43
	98.040	**90.01**	**91.42**	**94.08**	**92.910**	**90.730**

The Results of Standard Analysis of the Mud Brick and Mortar Filler

Appendix 8.3

The Percentages of Sediment of Mud Brick and Their Identifications

First Dynasty (Emery excavation)

Sample name	Acid soluble	Gravel > 2 mm	Sand	Silt < 0.15 μm	Residue Identification
S.104	2.08	-	61.5	38.5	Silt/mud; red, pink, smoky and transparent quartz sand, charcoal and mica.
Dahshur 4th Dynasty					
S.50	9.99	-	43.66	56.34	Mud/silt, various colours of quartz sand, quartzite, chert, ironstone, feldspar, charcoal, green olivine and mica. And sandstone, and brick in S.56.
S.51	8.58	-	32.62	67.38	
S.56	5.95	-	66.87	33.13	
Ka-Nefr Tomb 5th Dynasty					
S.9	7.09	6.76	27.18	66.06	Mud/silt, quartz sand, pyroclastic pebbles of volcanic source, ironstone, and brick, charcoal green olivine, and feldspar.
S.145	9.27	1.85	43.72	54.43	The same as above with quartzite, straw, olive charcoal, sandstone, burnt mud, gypsum crystal, cemented materials, chert, mica.
Idy Tomb late 6th Dynasty					
S.139	2.76	1.31	16.97	81.72	Mud/ silt, straw, quartz sand, charcoal, pebble, sandstone chert, green olivine and mica.

The Results of Standard Analysis of the Mud Brick and Mortar Filler

Appendix 8.4

Chemical analysis of Mud Brick Samples from Luxor

Sample name	Original weight/ g	Solvent	Residue	% Residue	% Sol.
T. 55					
S.113	49.88	HCl	43.936	88.08	11.92
T. 192					
S.129	49.28 G	HCl	39.872	80.91	19.09

Appendix 8.5

The Percentages of Sieving weights of Mud Brick

Site name	**T. 55**	**T. 192**
Sample name	**S.113**	**S.129**
Weight	49.880	49.280
Residue	43.936	39.872
% Acid Soluble	**11.920**	**19.09**

Sieving Weight Percentages

Mesh Size mm

	% Residue	
16.00	0.00	0.00
11.20	0.44	0.00
8.00	1.98	0.00
5.60	0.56	1.30
4.00	0.33	0.46
2.80	0.77	0.23
2.00	1.08	0.36
0.85	2.50	0.28
0.43	8.17	0.69
0.25	4.49	1.11
0.18	2.49	0.99
0.15	4.68	2.81
< 0.15	72.53	91.86
	88.080	**80.91**

The Results of Standard Analysis of the Mud Brick and Mortar Filler

Appendix 8.6

The Percentages of Sediment of Mud Brick and Their Identifications

Sample name	Acid soluble	Gravel > 2 mm	Sand	Silt < 0.15 μm	Residue Identification
T. 55					
S.113	11.92	4.062	23.407	72. 531	Mud/silt, straw, quartzite, quartz sand varying in colour, brick, chert, silica, charcoal, marl, ironstone, green olivine, mica and one piece of wheat seed.
T. 192					
S.129	19.09	1.994	6.147	91.859	Mud/silt, quartz sand, quartzite, straw, sandstone, charcoal, ironstone, gypsum, mica and brick.

The Results of Standard Analysis of the Mud Brick and Mortar Filler

Appendix 8.7

The Distributions of Particle Sizes of Mud Brick from Saqqara & Dahshur

W= Washed sample

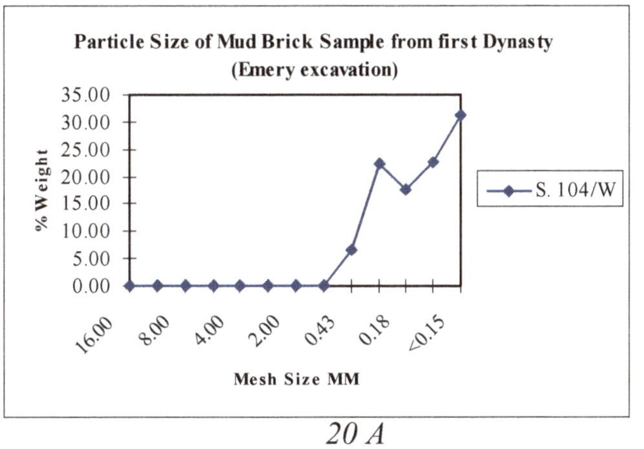

20 A

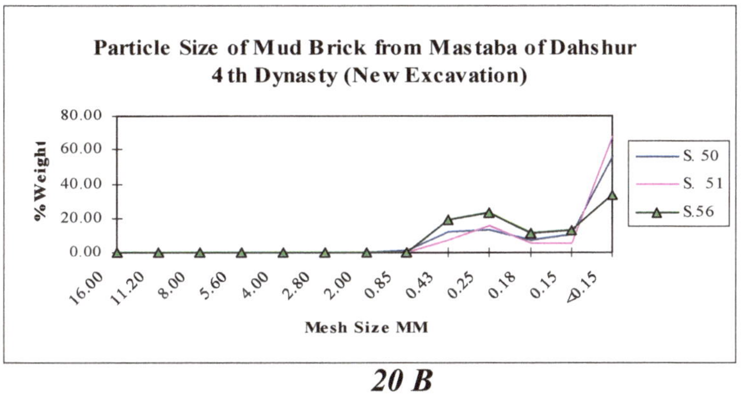

20 B

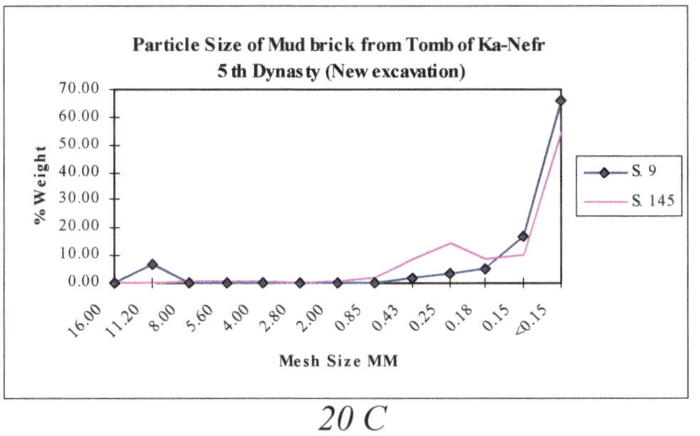

20 C

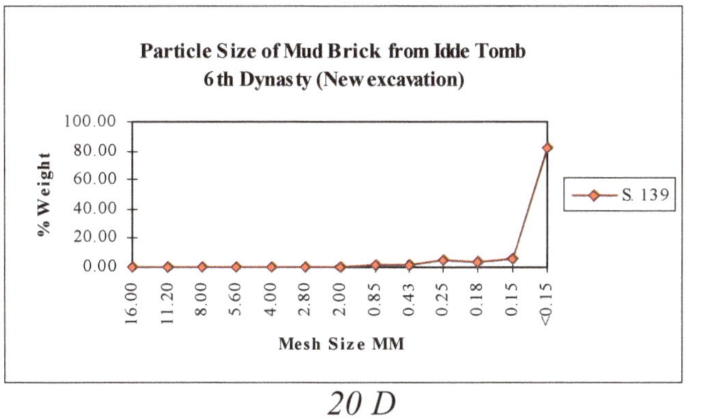

20 D

Graphs 20 A-D, show the distribution of aggregate particle sizes in mud bricks analysed from different areas of Saqqara and Dahshur, graph A is very different to the others as a washed sample. 1 B shows that the mixture of the mud brick was nearly the same, with a little difference due to the percentage of the sand as stated in Appendix 8.6. It was possible to note such differences between the same kinds of material by using different method of analysis. Graph C, is slightly differentdue to the percentage of the gravel and mud/silt. D shows a very high percentage of slit/mud.

The Results of Standard Analysis of the Mud Brick and Mortar Filler Appendix 8.8

The Distributions of Particle Sizes of Mud Brick from Luxor

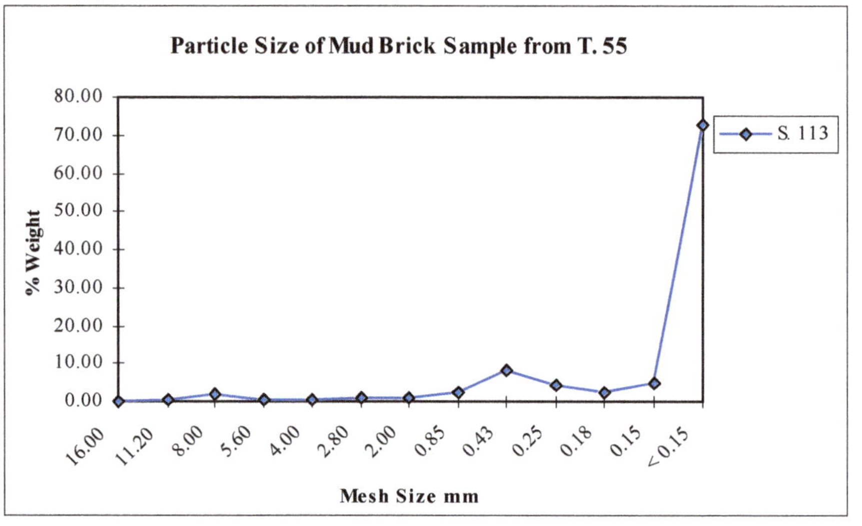

21 A

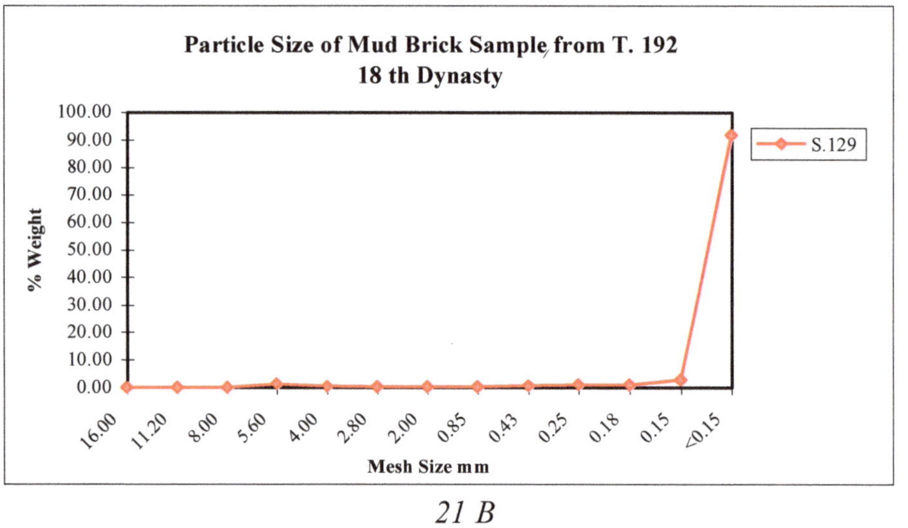

21 B

Graphs 21 A and B show the distribution of the aggregate particle sizes in mud brick, from T. 55 and T. 192. The differences shown on the graphs are due to the percentage of sand being higher in sample T. 55. The percentage of silt is about 91.86%, indicating that the mud brick had finer particles, showing similarities to those from Dahshur (see graph 1 D).

The Results of Standard Analysis of the Mud Brick and Mortar Filler

Appendix 8.9

Chemical Analysis of Samples of Mortar Filler/Joining Mortar from Luxor

Sample name	Original weight/g	Solvent	Residue	% Residue	% Sol.
Khokha T. 192 18th dynasty					
S.124	35.718	HCl	21.610	60.50	39.50
Asasif T. 414 26th dynasty					
S.16	60.59	Water	59.307	97.88	2.12

Appendix 8.10

The Percentages of Sieving Weights of Samples of Mortar filler

Site name	**T. 192**	**T. 414**
Sample name	**S.124**	**S.16/w**
Weight	35.718	60.590
Residue	21.610	59.307
% Acid Soluble	**39.500**	**2.120**

Sieving Weight Percentages

Mesh Size mm

	% Residue	
16.00	0.00	0.00
11.20	0.00	0.00
8.00	5.36	0.00
5.60	0.00	0.20
4.00	0.41	0.61
2.80	0.04	7.20
2.00	0.01	6.69
0.85	0.34	13.33
0.43	0.46	10.53
0.25	3.08	3.88
0.18	1.50	1.28
0.15	3.03	1.81
< 0.15	85.77	54.47
	60.500	**83.290**

Appendix 8.11

The Percentages of Sediment of Mortar Filler and Their Identifications

Sample name	Acid soluble	Gravel > 2 mm	Sand	Silt < 0.15 μm	Residue Identification
				T. 192	
S.124	39.50	5.817	8.418	85.765	Silt/clay, quartz sand, straw, quartzite, limestone, gypsum, burnt mud, ironstone, charcoal, brick and mica.
				T. 414	
S.16	2.12	8.007	37.519	54.474	Silt/clay, limestone, gypsum, ferruginous fossil, quartz, gypsum (crystal), charcoal & some seeds.

The Results of Standard Analysis of Mud Plasters/Mortars

Appendix 9.1

Chemical Analysis of Samples of Mud Plasters/Mortars from Saqqara and Dahshur

Sample name	Original weight/g	Solvent	Residue	% Residue	% Sol.
First Dynasty (E E)					
S.100	25.344	HCl	24.12	95.17	4.83
S.103	5.323	Water	5.196	97.61	2.39
Ka-Nefr Tomb 5th Dynasty (N E)					
S.49	58.69	HCl	54.30	92.52	7.48
S.134/3	46.25		39.77	85.99	14.01
S.134/5	21.370		19.141	89.57	10.43
S.144	43.11		38.179	88.56	11.44
Idy Tomb late 6th Dynasty (N E)					
S.137	15.721	HCl	14.170	90.13	9.87
S.138	21.769		19.975	91.76	8.24

Appendix 9.2

The Percentages of Sieving Weights of Samples of Mud Plasters/Mortars from Saqqara and Dahshur

	First Dynasty (E E)		Tomb Ka-Nefr 5th Dynasty				Idy Tomb 6th Dynasty	
Sample name	**S.100**	**S.103/W**	**S.49**	**S.134/3**	**S.134/5**	**S.144**	**S.137**	**S.138**
Weight	25.344	5.323	58.690	46.250	21.370	43.110	15.721	21.769
Residue	24.120	5.196	54.300	39.770	19.141	38.179	14.170	19.975
% Acid Soluble	**4.830**	**2.390**	**7.480**	**14.010**	**10.430**	**11.440**	**9.870**	**8.240**
Sieving Weight Percentage								
Mesh Size mm				**% Residue**				
16.00	0.00	0.00	0.00	0.00	0.00	0.00	0.00	0.00
11.20	0.00	0.00	0.00	0.00	0.00	0.00	0.00	0.00
8.00	0.00	0.00	0.00	1.51	0.00	0.00	0.00	0.00
5.60	0.00	0.00	0.00	0.29	0.00	0.00	0.00	0.00
4.00	0.00	0.00	0.00	0.83	0.00	0.00	0.00	0.00
2.80	0.00	0.00	0.00	0.77	0.00	0.00	0.00	0.00
2.00	0.62	0.00	0.00	0.63	0.59	0.70	1.33	1.91
0.85	0.62	0.04	0.15	2.37	1.59	0.44	1.35	2.03
0.43	2.82	2.29	4.47	10.73	9.76	2.90	1.58	2.04
0.25	9.58	7.97	13.72	19.66	15.74	11.10	2.67	2.63
0.18	3.65	3.87	10.90	7.39	8.51	9.64	1.82	1.64
0.15	2.70	3.10	17.25	12.13	10.21	13.41	3.64	4.31
<0.15	80.01	82.73	53.51	43.69	53.61	61.81	87.61	85.44
	95.170	**97.610**	**53.510**	**85.990**	**89.570**	**88.560**	**90.130**	**91.760**

The Results of Standard Analysis of Mud Plasters/Mortars

Appendix 9.3

The Percentages of Sediment of Mud Plasters/Mortars and Their Identifications

Sample name	Acid soluble	Gravel > 2 mm	Sand	Silt < 0.15 μm	Residue Identification
First Dynasty (Emery Excavation)					
S.100	4.83	-	19.99	80.01	Silt/mud, straw, quartz sand, charcoal, brick, a few seeds and mica.
S.103	2.39	-	17.27	82.73	Silt/mud; red, pink, smoky and transparent quartz sand, charcoal and mica.
Ka-Nefr Tomb 5th Dynasty					
S.49	7.48	-	46.49	53.51	Mud/silt, quartz sand, sandstone, brick, charcoal straw and mica. And quartzite, chert, ironstone, some seeds in S.134/1.
S.134/3	14.01	3.4	52.91	43.69	
S.134/5	10.43	-	46.39	53.61	The same with burnt mud, feldspar, green olivine and mica.
S.144	11.44	-	38.19	61.81	Mud/silt, quartz sand, straw, charcoal, ironstone, quartzite, some seeds and mica.
Idy Tomb Late 6th Dynasty					
S.137	9.87	-	12.39	87.61	Mud/silt, various colours of quartz sand, brick, straw charcoal, burnt mud and mica sandstone, and quartzite in S.138.
S.138	8.24	-	14.56	85.44	

The Results of Standard Analysis of Mud Plasters/Mortars

Appendix 9.4

Chemical Analysis of Samples of Mud Plasters/Mortars from Luxor

Sample name	Original weight/g	Solvent	Residue	% Residue	% Sol.
Sheikh Abd El-Qurna T. 56 18th Dynasty					
S.106	34.615	HCl	27.066	78.19	21.81
Deir El Medina T. 359 20th Dynasty					
S.6	42.10	HCl	36.564	86.85	13.15
S.40	43.71		33.458	76.55	23.45
Asasif T. 414 26th Dynasty					
S.15	83.89	HCl	69.867	83.28	16.72
S.18	43.49		35.704	82.10	17.90
S.20	39.44		32.849	83.29	16.71

Appendix 9.5

The Percentages of Sieving Weights of Samples of Mud Plasters/Mortars from Luxor

Site name	T. 56	T. 359		T. 414		
Sample name	**S.106**	**S.6**	**S.40**	**S.18**	**S.20**	**S.16/w**
Weight	34.615	42.100	43.710	43.490	39.440	60.590
Residue	27.066	36.564	33.458	35.704	32.704	59.307
% Acid Soluble	**21.810**	**13.150**	**23.450**	**17.900**	**16.710**	**2.120**
Sieving Weight Percentages Mesh Size mm		**% Residue**				
16.00	0.00	0.00	0.00	0.00	0.00	0.00
11.20	0.00	0.00	1.27	0.00	0.00	0.00
8.00	5.51	0.00	0.00	0.00	0.00	0.00
5.60	0.00	0.27	0.13	0.00	0.00	0.20
4.00	0.26	1.32	0.54	0.00	0.67	0.61
2.80	0.16	0.93	0.54	0.00	0.33	7.20
2.00	0.21	1.12	0.48	0.42	0.68	6.69
0.85	1.15	3.00	3.22	1.26	2.30	13.33
0.43	1.84	5.72	5.85	2.50	3.72	10.53
0.25	1.49	2.44	6.21	1.94	2.58	3.88
0.18	1.03	1.86	2.07	0.85	1.41	1.28
0.15	2.89	4.38	4.60	1.52	3.44	1.81
< 0.15	85.46	78.97	75.09	91.52	84.86	54.47
	78.190	**86.850**	**76.550**	**97.980**	**82.100**	**83.290**

The Results of Standard Analysis of Mud Plasters/Mortars

Appendix 9.6

The Percentages of Sediment of Mud Plasters/Mortars and their Identifications

Sample name	Acid soluble	Gravel > 2 mm	Sand	Silt < 0.15 μm	Residue Identification
T. 56					
S.106	21.81	5.93	8.61	85.46	Mud/silt, quartz sand of various colours, quartzite, chert, charcoal, straw, brick, ironstone, micaceous silt stone, some green olivine and feldspar, and mica in different colours.
T. 359					
S.6	13.15	2.52	18.515	78.985	Mud/silt, straw, brownish/reddish mud quartz sand, quartzite, ferruginous fossil, chert/flint gypsum, charcoal, sandstone, brick and mica & some seeds.
S.40	23.45	2.478	22.428	75.094	Mud, straw, mud, amorphous silica, quartz sand, quartzite, ferruginous fossil, gypsum crystals, charcoal, ironstone, brick, mica, shell fragments and a few seeds.
T. 414					
S.15	16.72	0.702	13.529	58.769	Mud, quartzite, straw, red, opaque and transparent quartz sand, marl, charcoal, ironstone, brick and mica.
S.18	17.90	-	8.483	91.517	Mud, quartz sand, quartzite, marl, straw, gypsum, charcoal, ironstone, mica and some seeds in S.18, feldspar & green olivine in S.20.
S.20	16.71	1.002	14.134	84.864	

Appendix 9.7

The Distributions of Particle Sizes of Mud Plasters/Mortars from Saqqara & Dahshur

W= Washed sample *22 A*

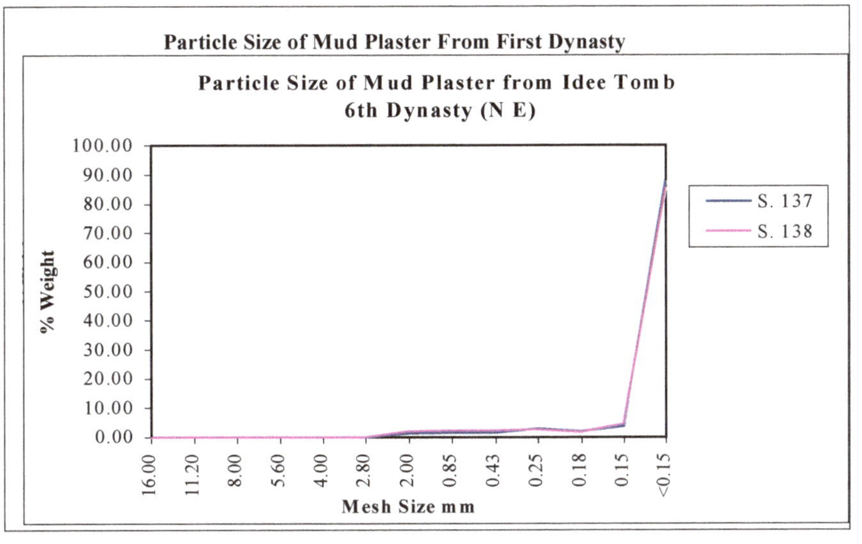

22 B

On graph 22 A & B, note that the sample S.100, which was treated with HCl, gave the same result as sample S.103 (although it was only washed) resulting in the loss of carbonated materials such as lime/chalk or even limestone. This was low, as the percentage dissolving was less than 5%. B shows more or less the same percentages of sediment types, however, the acid soluble was low, as stated in Appendix 9.3.

Appendix 9.8

The Distributions of Particle Sizes of Mud Plasters/Mortars from Luxor

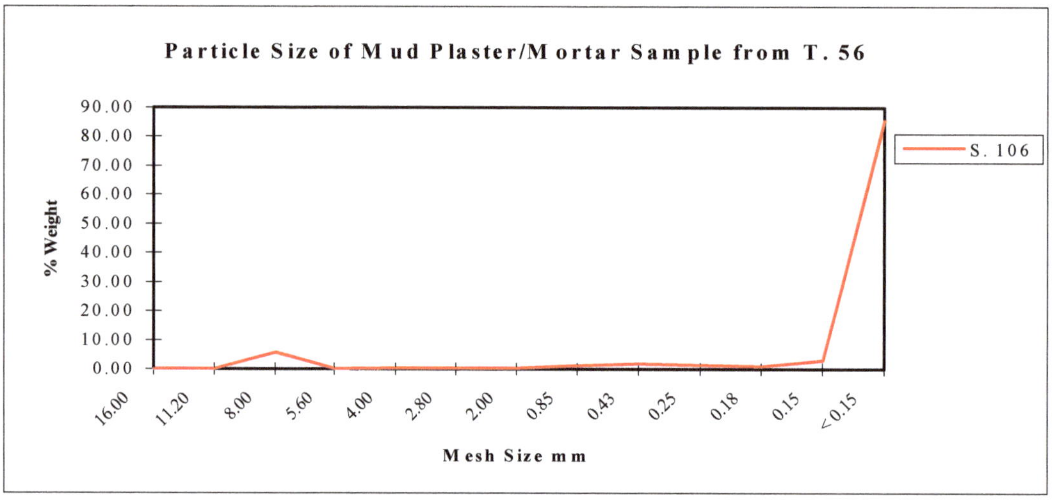

23 A

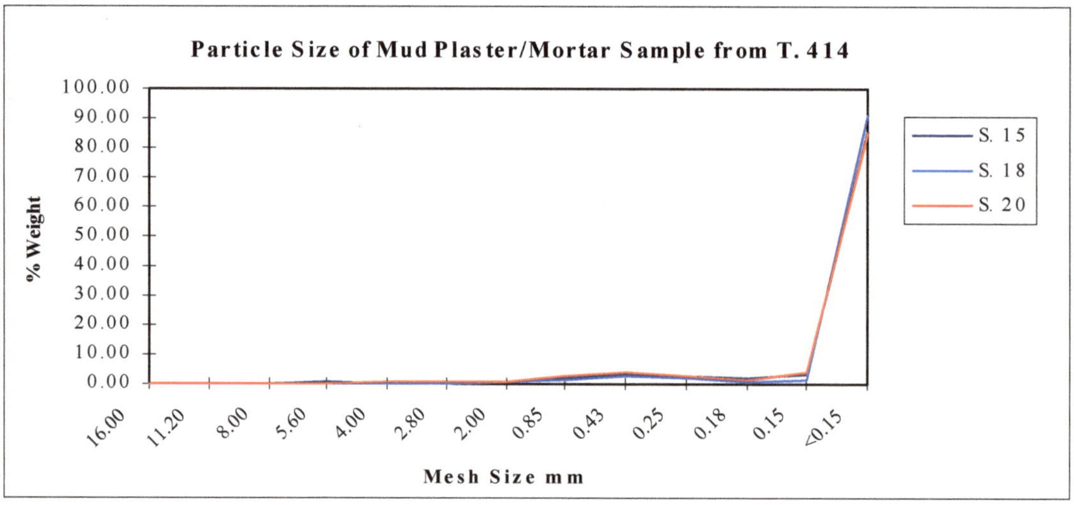

23 B

Graphs 23 A & B show the differences in the distribution of sediment type of gravel, sand and silt. A, shows a little difference due to the percentage of the gravel, 5.93% being higher then the others. In B, the percentages of silt were completely different from each other, however, the similarity in graphs due to the high percentages of silt being more than 75%, except S.15 about 58%, as stated in Appendix 9.6

The Results of Standard Analysis of the Clay Plasters/Mortars

Appendix 10.1

Chemical Analysis of Samples of Clay Plasters/Mortars from Saqqara and Dahshur

Sample name	Original weight/g	Solvent	Residue	% Residue	% Sol.
First Dynasty (N E)					
S.143/1	7.400	HCl	5.696	79.97	23.03
S.143/1	6.311	Water	6.180	97.92	2.08
Mastaba of Dahshur 4th Dynasty (N E)					
S.53	43.46		35.549	81.80	18.20
S.55/1	51.67	HCl	41.65	80.61	19.39
S.57	75.58		65.14	86.19	13.81
Ka-Nefr Tomb 5th Dynasty (N E)					
S.10	42.62		24.879	58.37	41.63
S.46	19.615		9.468	48.27	51.73
S.47	100	HCl	91.09	91.09	8.91
S.48	42.17		23.009	54.56	45.44
S.146/1	35.741		17.794	49.79	50.21
Idy Tomb late 6th Dynasty (N E)					
S.140	53.68	HCl	48.51	90.37	9.63

Appendix 10.2

The Percentages of Sieving Weights of Samples of Clay Plasters/Mortars from Saqqara and Dahshur

Site name		**First Dynasty (N E)**		**Dahshur 4th Dynasty**	
Sample name	**S.143/1A**	**S.143/1W**	**S.53**	**S.55/1**	**S.57**
Weight	7.400	6.311	43.460	51.670	75.580
Residue	5.696	6.180	35.549	41.65	65.14
% Acid Soluble	**23.030**	**2.080**	**18.20**	**19.39**	**13.81**
Sieving Weight Percentages					
Mesh Size mm					
		% Residue			
16.00	0.00	0.00	0.00	0.00	21.48
11.20	0.00	0.00	0.00	0.00	0.00
8.00	0.00	0.00	0.00	0.00	2.86
5.60	0.00	0.00	0.00	0.00	0.00
4.00	0.00	0.00	0.00	0.00	0.76
2.80	0.00	0.00	0.00	0.00	0.16
2.00	0.00	3.30	0.12	0.07	0.08
0.85	0.00	2.18	0.32	0.97	0.70
0.43	7.39	5.89	4.86	6.62	4.65
0.25	25.90	18.44	7.67	7.90	5.71
0.18	17.08	13.09	4.97	5.48	3.20
0.15	23.89	18.66	8.22	6.19	4.64
<0.15	25.74	38.50	73.84	72.77	55.76
	76.970	**97.920**	**81.70**	**80.61**	**86.19**

Appendix 10.3

The Percentages of Sieving Weights of Samples of Clay Plasters/Mortars from Saqqara and Dahshur

Site name		Tomb Ka-Nefr 5th Dynasty				Idy Tomb 6th Dynasty
Sample name	**S.10**	**S.46**	**S.47**	**S.48**	**S.146/1**	**S.140**
Weight	42.620	19.615	100.000	42.170	35.741	53.680
Residue	24.879	9.468	91.090	23.009	17.794	48.510
% Acid Soluble	**41.879**	**51.730**	**8.910**	**45.440**	**50.210**	**9.630**
Sieving Weight Percentages Mesh Size mm						
			% Residue			
16.00	0.00	0.00	0.00	0.00	0.00	0.00
11.20	0.00	0.00	0.00	0.00	0.00	0.00
8.00	0.00	0.00	1.60	0.00	0.00	1.61
5.60	0.27	0.00	0.16	0.00	0.00	0.98
4.00	0.44	0.00	0.16	0.00	0.00	1.97
2.80	0.24	0.00	0.75	0.00	0.00	1.80
2.00	0.63	0.00	0.34	0.04	0.15	0.84
0.85	1.39	0.23	1.48	0.13	0.28	1.72
0.43	4.02	2.28	10.91	2.02	2.45	3.44
0.25	9.75	5.11	16.23	5.13	5.70	3.06
0.18	4.88	4.38	8.92	4.13	4.05	1.75
0.15	5.72	7.82	12.66	7.31	7.34	9.03
<0.15	73.09	80.18	46.79	81.24	80.03	73.80
	58.370	**48.270**	**91.090**	**54.560**	**49.790**	**90.370**

The Results of Standard Analysis of the Clay Plasters/Mortars

Appendix 10.4

The Percentages of Sediment of Clay Plasters/Mortars and Their Identifications

Samples name	Acid soluble	Gravel > 2 mm	Sand	Silt < 0.15 µm	Residue Identification
					First Dynasty (N E)
S.143/1	23.03	-	74.26	25.74	Silt/clay, various quartz sand; red, smoky and transparent, feldspar, mica and amorphous silica.
S.143/1w	2.08	-	61.5	38.5	As above and gypsum crystal deposits were visible.
					Dahshur 4th Dynasty (N E)
S.53	18.20	-	26.16	73.84	Clay/silt, quartz sand, quartzite, chert, ironstone, sandstone, fossil shell, feldspar, green olivine and mica.
S.55/1	19.39	-	27.23	72.77	The same mixture as above with gypsum and charcoal.
S.57	13.81	25.26	18.98	55.76	The same as S.53 with quartz sand showing sand blasting as an environmental effect, chert, pebbles, brick, iron oxide and gypsum.

Appendix 10.5

The Percentages of Sediment of Clay Plasters/Mortars and Their Identifications

Samples name	Acid soluble	Gravel > 2 mm	Sand	Silt < 0.15 µm	Residue Identification
Ka-Nefr Tomb 5th Dynasty (N E)					
S.10	41.63	0.95	25.95	73.09	Clay/silt, straw, quartz sand, charcoal, feldspar and mica, chert and green olivine in S.46
S.46	51.73	-	19.82	80.18	
S.47	8.91	2.67	50.54	46.79	Clay, silt, quartz sand, quartzite, sandstone, ironstone, chert, pebble quartz round environment, charcoal, gypsum crystal, brick, feldspar, green olivine and mica.
S.48	45.44	-	18.76	81.24	Clay/silt, straw, quartz sand, charcoal, gypsum crystal.
S.146/1	50.21	-	19.97	80.03	Clay/silt and quartz sand, quartzite, charcoal and feldspar.
Idy Tomb Late 6th Dynasty (N E)					
S.140	9.63	6.36	19.84	73.8	Clay/silt, quartz sand, gypsum crystal deposit, sandstone, chart, quartzite, ironstone, charcoal, green olivine feldspar, some cemented materials and mica.

The Results of Standard Analysis of the Clay Plasters/Mortars

Appendix 10.6

Chemical Analysis of Samples of Clay Plasters/Mortars from Luxor

Sample name	Original weight/g	Solvent	Residue	% Residue	% Sol.
Sheikh Abd El-Qurna T. 55 18th Dynasty					
S.111	24.166	Water	23.682	98	2
Khokha T.192 18th Dynasty					
S.34	57.44	HCl	25.917	45.12	54.88
S.126	23.141		19.764	84.41	14.59
S.127&128	67.48		24.395	36.15	63.85
S.130	29.833		24.786	83.08	16.92
Drah Abu El-Naga T. 255 18th Dynasty					
S.121	46.22	HCl	13.027	28.18	71.82
T. 13 19th Dynasty					
S.116	41.68	HCl	21.884	52.50	47.50
S.119/3	43.80		12.431	28.38	71.62
Valley of the Kings T. 14 19th Dynasty					
S.3	60	HCl	29.005	48.34	51.66
S.86	62.73		25.302	40.33	59.67
T. 15 19th Dynasty					
S.4	32.682	HCl	14.768	45.19	54.81
T. 11 20th Dynasty					
S.64	45.65	HCl	7.381	16.17	83.83

Appendix 10.7

The percentages of Sieving Weights of Samples of Clay Plasters/Mortars from Luxor

Site name	T. 55	T. 255		T. 13
Sample name	**S.111**	**S.121**	**S.116**	**S.119/3**
Weight	24.166	46.220	41.680	43.800
Residue	23.682	13.027	21.880	12.431
% Acid Soluble	**2.000**	**71.82**	**47.500**	**71.620**
Sieving Weight Percentages				
Mesh Size mm				
			% Residue	
16.00	0.00	0.00	0.00	0.00
11.20	0.00	0.00	0.00	0.00
8.00	0.00	0.00	0.00	0.00
5.60	0.00	0.81	0.00	0.79
4.00	0.00	0.65	0.00	0.00
2.80	0.00	1.64	0.00	0.37
2.00	1.57	0.52	0.29	0.51
0.85	8.62	4.98	1.30	3.38
0.43	23.77	5.37	5.98	6.62
0.25	10.15	5.80	3.66	6.01
0.18	3.34	1.01	2.74	1.47
0.15	2.65	2.79	3.47	2.13
<0.15	50.51	76.42	82.57	78.72
	98.000	**28.180**	**52.500**	**28.380**

The Results of Standard Analysis of the Clay Plasters/Mortars

Appendix 10.8

The Percentages of Sieving Weights of Samples of Clay Plasters/Mortars from Luxor

Site name			**T. 192**			**T. 14**	**T. 15**	**T. 11**
Sample name	**S.34**	**S.126**	**S.127& 128**	**S.130**	**S.3**	**S.86**	**S.4**	**S.64**
Weight	57.440	23.141	67.480	29.833	60.000	62.730	32.682	45.650
Residue	25.917	19.764	24.395	24.786	29.005	25.302	14.768	7.381
% Acid Soluble	**54.880**	**14.590**	**63.850**	**16.920**	**51.660**	**59.670**	**54.810**	**83.830**
Sieving Weight Percentages Mesh Size mm								
					%Residue			
22.40	0.00	0.00	0.00	0.00	0.00	38.61	0.00	0.00
16.00	15.62	0.00	0.00	0.00	0.00	0.00	0.00	0.00
11.20	0.00	0.00	0.00	0.00	0.00	0.00	0.00	0.00
8.00	2.13	0.00	0.00	0.00	0.00	0.00	0.00	0.00
5.60	6.81	0.00	0.00	0.40	0.00	0.00	0.00	0.77
4.00	4.66	0.00	3.56	0.00	0.00	0.13	0.00	0.27
2.80	4.10	0.00	3.36	0.17	0.62	0.00	0.53	0.98
2.00	3.64	1.93	3.55	0.61	0.49	0.00	0.00	0.80
0.85	6.19	2.34	8.94	1.21	5.06	0.27	0.40	3.64
0.43	3.29	3.08	8.36	1.89	16.77	1.81	0.38	5.43
0.25	0.81	3.58	4.55	2.30	7.99	0.77	0.39	4.59
0.18	0.54	3.79	0.65	0.87	2.24	1.05	1.27	1.90
0.15	0.81	3.99	0.70	2.40	3.49	4.27	8.64	5.60
<0.15	51.41	81.29	66.28	90.14	63.40	53.10	88.46	76.02
	45.120	**85.410**	**36.150**	**83.080**	**48.340**	**40.330**	**45.190**	**16.170**

Appendix 10.9

The Percentages of Sediment of Clay Plasters/Mortars and Their Identifications

Samples name	Acid soluble	Gravel > 2 mm	Sand	Silt < 0.15 μm	Residue Identification
T. 55					
S.111	2	-	49.492	50.508	Clay/silt, limestone, quartz sand, straw, ferruginous fossil, iron oxide, charcoal, gypsum and mica.
T. 192					
S.34	54.88	33.326	15.267	51.407	Clay/silt, marl of many layers, quartz sand, straw, quartzite, gypsum, ironstone, brick, charcoal, mica and some black minerals, feldspar, ironstone in S.126, gypsum, amorphous silica, brick, (gypsum crystal) in S.127 & S.128 and green olivine was visible in S.130.
S.126	14.59	-	18.706	81.294	
S.127 & S.128	63.85	6.965	26.755	66.280	
S.130	16.92	0.569	9.287	90.144	
T. 255					
S.121	71.82	3.101	20.477	76.422	Silt/clay, mud limestone/amorphous silica, straw, quartz sand, ironstone, charcoal, brick, green olivine, chert, quartzite, mica and some seeds.
T. 13					
S.116	47.50	-	17.433	82.567	Silt/clay, straw, amorphous silica, reddish marl, quartz sand, gypsum (crystal), ironstone, ferruginous fossil, charcoal, mica and straw, quartzite in S.119/3.
S.119/3	71.62	0.788	20.49	78.722	

The Results of Standard Analysis of the Clay Plasters/Mortars

Appendix 10.10

The Percentages of Sediment of Clay Plasters/Mortars and Their Identifications

Samples name	Acid soluble	Gravel > 2 mm	Sand	Silt < 0.15 μm	Residue Identification
T. 14					
S.3	51.66	0.62	35.98	63.4	Clay/silt, marl in different colours, straw, chert, quartzite, ironstone, brick, sandstone, charcoal, feldspar and mica.
S.86	59.67	38.736	8.169	53.095	Clay/silt, burnt limestone, amorphous silica, quartzite, quartz sand, charcoal, iron oxide, green olivine and mica.
T. 15					
S.4	54.81	0.528	11.012	88.46	Clay/silt, amorphous silica, quartz sand, flint, straw, gypsum, ironstone, charcoal and mica.
T. 11					
S.64	83.83	2.012	21.962	76.02	Clay/silt, amorphous silica, straw, quartz sand, brick, ferruginous fossil, ironstone, charcoal, burnt limestone and mica.

Appendix 10.11

The Distributions of Particle Sizes of Clay Plasters/Mortars from Saqqara & Dahshur

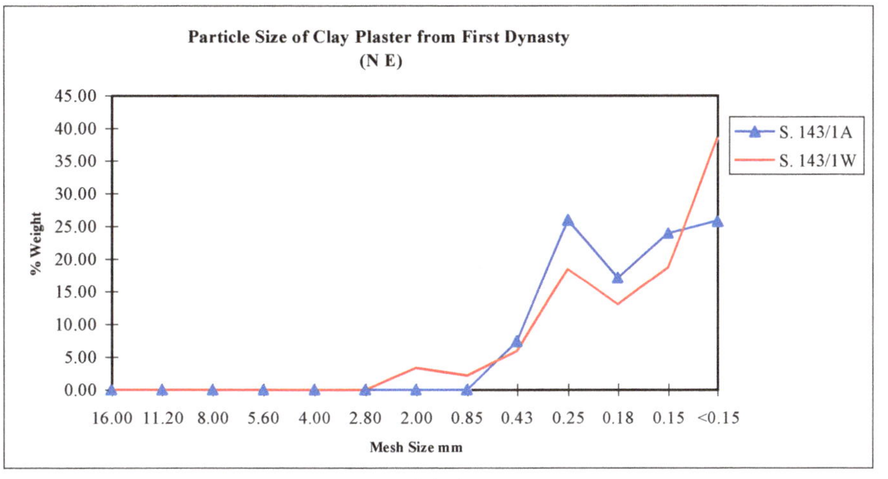

A= Acid used 24 A

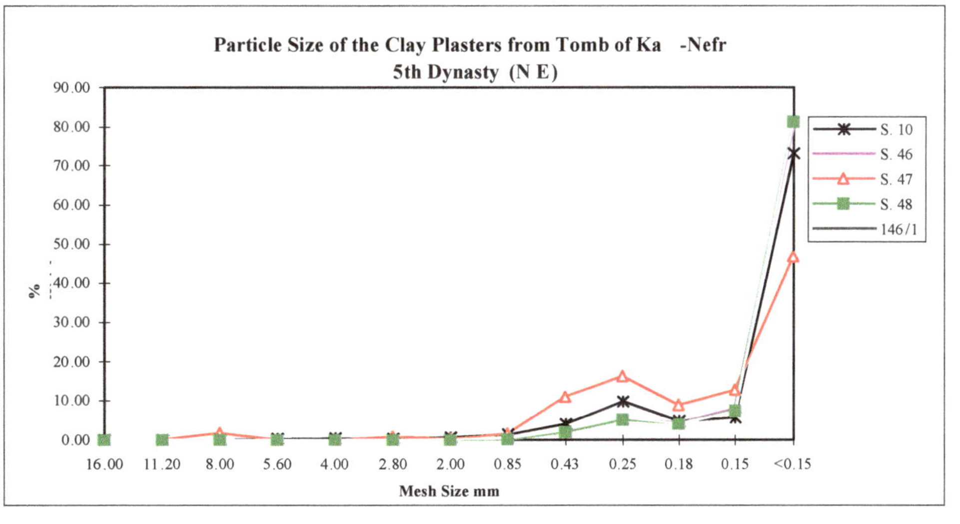

W= Washed sample 24 B

Note the only slight differences between the same sample treated with different methods, the acid soluble elements (powdered limestone) was less than 25%. The percentage of sand was higher in S.143/1A, where the silt was in S.143/w as shown on the graph. As shown on graph B, the samples are the same except S.47, where there is a different percentage of sand as stated, in Appendix 10.4, which was more than 50 %.

Appendix 10.11.1

The Distributions of Particle Sizes of Clay Plasters/Mortars from Saqqara

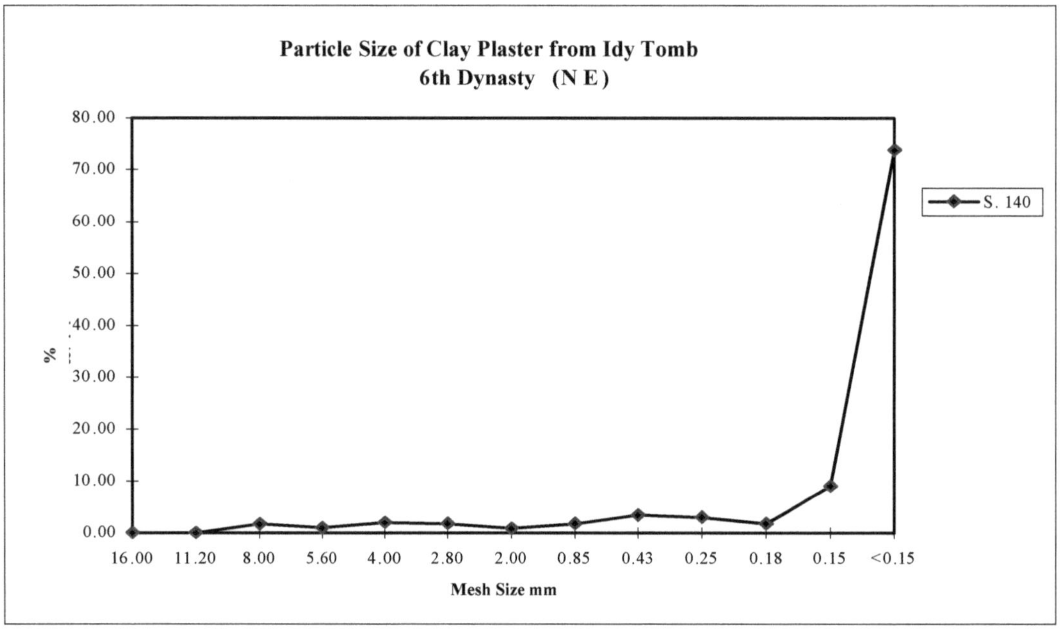

24 C

This graph shows the high percentages of the silt/clay used (see graph) and show the similarity in the distribution of the particle sizes used from the analysis of clay/mud plaster/mortar and mud brick, silt being the highest in all samples.

Appendix 10.12

The Distributions of Particle Sizes of Clay Plasters/Mortars from Luxor

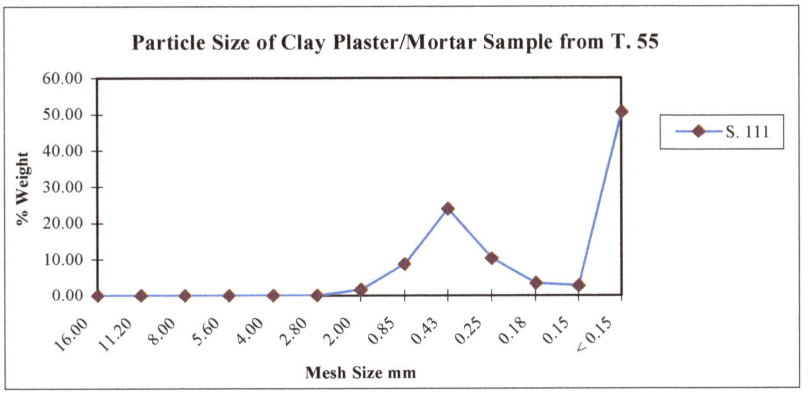

Graph 25, shows the distribution of the aggregate particle sizes of a washed sample from T.55, note the percentage of the sand, which was nearly the same as the silt.

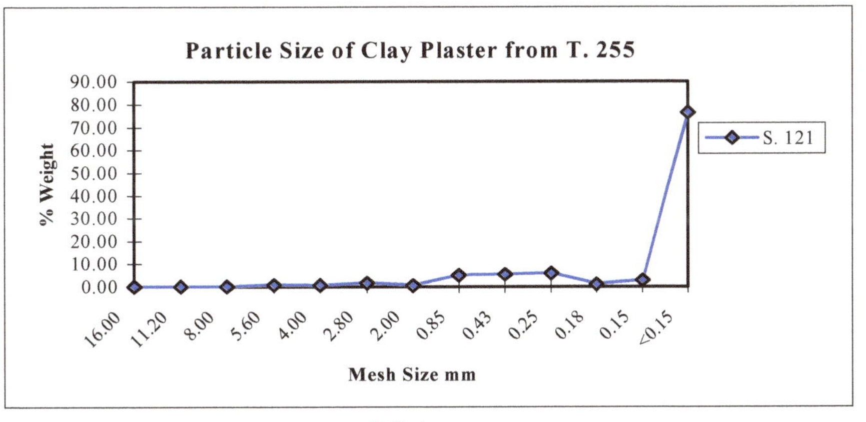

26 A

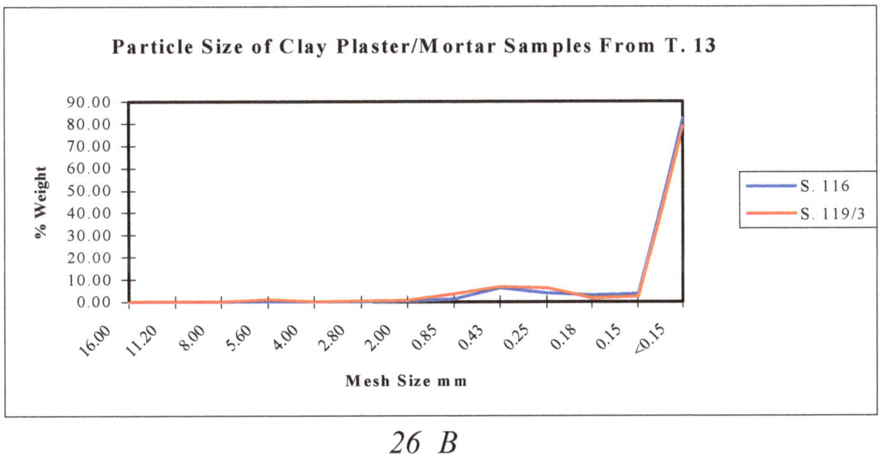

26 B

Graphs 26 A & B show the differences in the distribution of aggregate particle sizes of different samples from different tombs and areas. Note the similarity in graph A & B, although the samples were from a different area. B shows similar distributions of the aggregate particle sizes in samples S.121 and S.119/3.

Appendix 10.13

The Distributions of Particle Sizes of Clay Plasters/Mortars from Luxor

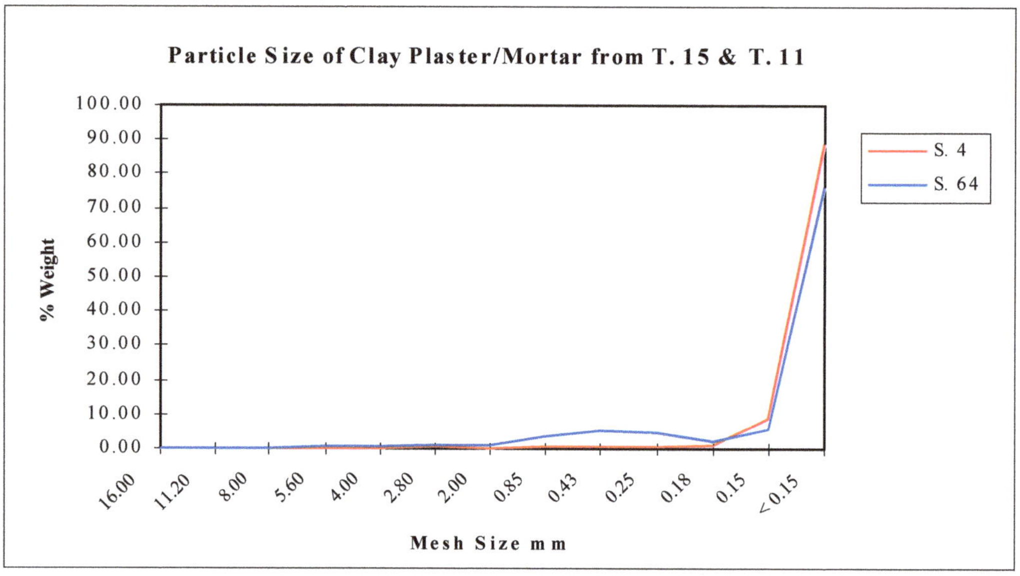

Graph 27 shows the distribution of the aggregate particle sizes of a sample from the Valley the of Kings, the differences are due to the variation of the sediment type used. It also showed the highest percentage of silt, found in sample S.4, and the sand was in different proportions in both S.4 and S.64. The highest percentage of acid soluble found in the samples S.64, being 83.83%, was due to the hight percentage of carbonate in the clay plaster/mortar .

Appendix 10.14

Chemical Analysis of Plaster Samples Used for Restoration of Tombs (255, 13 and T. 14)

Sample name	Original weight/g	Acid used	Residue	% Residue	% Sol.
T. 255 and T. 13					
S.74	14.672	HCl	8.357	56.96	43.04
S.79	46.69		25.630	47.30	52.70
T. 14					
S.87	14.603	HCl	6.577	45.04	54.96

Appendix 10.15

The Percentages of Sieving Weights of Plaster Samples Used for Restoration

Site name	Drah Abu El-Naga T. 255 & T.1 3		Valley of Kings T. 14
Sample name	**S.74**	**S.79**	**S. 87**
Weight	14.672	46.690	14.603
Residue	8.357	47.300	6.577
% Acid soluble	**43.040**	**52.700**	**54.960**
Sieving Weight Percentages			
		% Residue	
Mesh Size mm			
22.00	0.00	0.00	0.00
16.00	0.00	0.00	0.00
11.20	0.00	0.00	0.00
8.00	0.00	0.00	0.00
5.60	0.00	0.00	0.00
4.00	0.00	0.00	0.00
2.80	0.00	0.00	0.00
2.00	0.00	0.57	0.00
0.85	0.02	17.58	1.11
0.43	0.40	44.41	14.87
0.25	0.19	15.25	11.42
0.18	0.16	4.00	4.05
0.15	0.56	3.27	4.58
<0.15	98.98	14.93	63.97
	56.960	**52.700**	**45.040**

Appendix 10.16

The Percentages of Sediment of Plaster Samples Used for Restoration and their Identifications

Samples name	Acid soluble	Gravel > 2 mm	Sand	Silt < 0.15 μm	Residue Identification
T. 13					
S.74	43.04	-	1.328	98.672	Clay/silt, quartz sand, amorphous silica, ferruginous fossil, ironstone, iron oxide, burnt limestone, charcoal, flint, olivine green and mica., shell fragment and brick.
S.79	52.70	-	85.072	14.928	
T. 14					
S.87	54.96	-	36.036	63.971	Clay/silt, quartzite, quartz sand in various colour, burnt limestone, olivine green and mica.

The Results of Standard Analysis of Lime Plasters/Mortars

Appendix 11.1

Chemical Analysis of Lime Plasters/Mortars from Alexandrian Sites

Sample name	Original weight/g	Solvent	Residue	% Residue	% Sol.
MP T. 1					
S.14 Site	11.164	Acetic acid	3.175	28.44	71.56
A T. 1					
S.1/A	11.263	HCl	3.04	26.99	73.01
S.14 Site	8.548		5.468	63.97	36.03

Appendix 11.2

The Percentages of Sieving Weights of Lime Plasters/Mortars and Stone from Alexandrian Sites

Site name	**Mustafa Pasha**	**Anfushi**	
	Stone	**T. 1**	**Stone**
Sample name	**S.14**	**S.I**	**S.14/2**
Weight	11.164	11.263	8.548
Residue weight	3.175	3.040	5.468
% Acid Soluble	**71.560**	**73.010**	**36.030**
Sieving Weight Percentages			
Mesh Size mm		**% Residue**	
16.00	0.00	0.00	0.00
11.20	0.00	0.00	0.00
8.00	0.00	0.00	0.00
5.60	0.00	0.00	0.00
4.00	0.00	0.00	0.00
2.80	0.00	0.00	0.00
2.00	0.00	1.78	0.00
0.85	0.60	3.88	0.49
0.43	9.95	4.47	13.53
0.25	8.50	4.38	33.32
0.18	9.26	2.20	16.48
0.15	34.08	2.66	8.19
<0.15	39.59	80.79	18.80
	28.440	**26.990**	**63.960**

The Results of Standard Analysis of Lime Plasters/Mortars

Appendix 11.3

The Percentages of Sediment of Lime Plasters/Mortars & Stone and Their Identifications

Sample name	Acid Soluble	Gravel > 2 m	Sand	Silt < 0.15	Residue identification
Mustafa Pasha Site					
S.14	71.56	0.6	61.79	39.59	Mainly transparent quartz and basalt & mica.
Anfushi T. 1					
S.1/A	73.01	1.78	17.13	80.79	Brick, feldspar, quartz, dust brick, sandstone and mica (biotite).
Anfushi Site					
S.14	36.03	0.49	82.11	18.80	Quartz, red quartz, feldspar and black minerals.

Appendix 12.1

Chemical Analysis of Lime Plasters/Mortars-First Layers from Piddington

Sample name	Original weigh/g	Solvent	Residue	% Residue	% Sol
			R. 21 & R. 26		
S.12A	90.01	HCl	48.29	53.65	46.35
S 13A	79.65		34.49	43.30	56.70
			R. 53		
S.10L1	25.88	HCl	12.58	48.60	51.40
S.14E.1	50.33		31.51	62.61	37.39
S.14D	74.47		37.03	49.72	50.28
			R. 56		
S.3L1	9.23	HCl	5.41	58.62	41.38
			R. 57		
S.4L1	19.06	HCl	9.25	48.54	51.46
S.7AL1	19.31		12.21	63.21	36.79
S.7BL1	17.93		8.69	48.45	51.55

Appendix 12.2

The Percentages of Sieving Weights of the First Layers

Site name	R. 21	R. 26		R. 53		R. 56	Other area		R. 57	
Sample name	**S.12A**	**S.13A**	**S.10L1**	**S.14E1**	**S.14D**	**S.3L1**	**S.8 L1**	**S.4L1**	**S.7AL1**	**S.7BL1**
Weight	90.01	79.65	25.88	50.33	74.47	9.23	10.55	19.06	19.31	17.93
Residue	48.29	34.49	12.58	31.51	37.03	5.41	5.60	9.25	12.21	8.69
% Lime Element	**46.35**	**56.70**	**51.04**	**37.39**	**50.28**	**41.38**	**46.89**	**51.46**	**36.79**	**51.55**
Sieving Weight Percentages										
Mesh Size mm										
						% Residue				
16.00	0.00	0.00	0.00	0.00	0.00	0.00	0.00	0.00	0.00	0.00
11.20	0.00	0.00	0.00	0.00	0.00	0.00	0.00	0.00	0.00	0.00
8.00	1.72	0.00	0.00	0.00	0.00	0.00	0.00	7.00	0.00	0.00
5.60	11.22	0.00	12.80	6.59	7.26	0.00	0.00	13.80	5.70	4.70
4.00	9.46	0.55	9.70	8.16	8.45	2.10	0.00	4.60	7.80	3.50
2.80	8.14	2.06	7.20	4.08	9.45	10.60	4.90	10.70	9.40	12.10
2.00	7.37	1.48	4.60	2.49	7.35	4.70	7.10	5.80	5.40	5.20
0.85	10.44	3.48	13.00	6.48	10.59	11.40	13.70	10.00	10.20	12.80
0.43	10.33	10.44	11.80	10.17	11.53	26.70	18.40	10.60	14.10	15.60
0.25	6.67	19.02	7.00	12.89	7.70	17.60	10.10	6.60	7.40	8.60
0.18	3.42	11.48	3.40	8.37	3.92	6.20	6.50	3.30	4.70	4.40
0.15	4.39	12.29	4.20	10.10	5.18	6.50	10.30	14.60	8.60	8.20
< 0.15	24.48	37.90	25.20	30.83	28.00	13.20	28.40	35.10	18.80	24.50
	53.65	**43.30**	**48.60**	**62.61**	**49.72**	**58.62**	**53.11**	**48.54**	**63.21**	**48.45**

Appendix 12.3

The Percentages of the Sediment of the First Layers and Their Identifications

Sample name	Acid Soluble	Gravel > 2 m	Sand	Silt < 0.15	Residue Identification
R. 21 & 26					
S.12A	46.35	40	35	25	Silt, mainly brick, quartz sand, iron and sandstone, quartzite, burnt flint, silica, coal & mica.
S.13A	56.70	6	56	38	Silt, iron stone, sandstone, quartzite quartz sand, burnt flint & mica.
R. 53					
S.10L1	51.40	35	40	25	Silt, brick (black & white), quartz sand, ironstone, flint & silica
S.14E1	50.28	33	39	28	Silt, brick, quartz sand, quartzite, ironstone, flint, coal & mica. Charcoal, burnt flint and sandstone in S.14 D.
S.14D	37.39	21	48	31	
R. 56					
S.3L1	41.38	17	68	16	Silt, brick, quartz sand, ironstone, quartzite & charcoal.
Another Area					
S.8L1	46.89	12	63	27	Silt, brick, quartzite, quartz sand, flint, charcoal, ironstone & mica.
R. 57					
S.4L1	51.46	20	45	35	Silt, brick (black & red), quartz sand, ironstone, silica and charcoal.
S.7AL1	36.79	25	45	23	Silt, brick, flint, quartz sand, quartzite, ironstone, silica and mica.
S.7BL1	51.55	25	50	25	

Appendix 12.4

Chemical Analysis of Lime Plasters/Mortars-Second Layers from Piddington

Sample name	Original weight/g	Solvent	Residue	% Residue	% Sol
R. 21 & 26					
S.12B	33.85	HCl	20.38	60.21	39.79
S.13B	48.37		20.78		57.04
R. 53					
S.10L2	25.86	HCl	17.35	67.08	32.92
S.14E.2	105.65		53.88	5I.00	49.00
S.14E.3	30.40		15.53	51.08	48.92
R. 56					
S.3L2	13.80	HCl	7.58	54.94	45.06
S.8	10.55		5.60		46.89
R. 57					
S.4L2	24.86	HCl	13.45	54.11	45.89
S.7AL2	20.30		10.82	53.28	46.72
S.7BL2	16.20		9.71	59.95	40.05

Appendix 12.5

The Percentages of Sieving Weights of the Second Layers

Third Layer

Site name	R. 21	R. 26	R. 53		R. 56		R. 57		R. 53
Sample name	**S.12B**	**S.13B**	**S.14E2**	**S.10L2**	**S.3L2**	**S.4L2**	**S.7AL2**	**S.7BL2**	**S.14E3**
Weight	33.85	48.37	105.65	25.86	13.80	24.86	20.30	16.20	30.40
Residue	20.38	20.78	53.88	17.35	7.58	13.45	10.82	9.71	15.53
% Lime Element	**39.79**	**57.04**	**49.00**	**32.92**	**45.06**	**45.89**	**46.72**	**40.05**	**48.92**
Sieving Weight Percentages									
					% Residue				
Mesh Size mm									
16.00	0.00	0.00	0.00	0.00	0.00	0.00	0.00	0.00	0.00
11.20	0.00	0.00	0.00	0.00	0.00	0.00	0.00	0.00	0.00
8.00	0.00	0.00	0.00	0.00	0.00	0.00	0.00	8.00	0.00
5.60	13.54	0.00	5.51	0.00	6.10	1.60	9.70	4.70	4.55
4.00	7.21	0.48	8.43	1.90	4.80	2.80	7.00	10.00	6.44
2.80	9.67	1.40	7.46	5.40	1.90	5.10	8.70	4.70	7.21
2.00	7.61	1.68	5.98	4.80	3.00	2.80	4.60	4.00	7.63
0.85	8.93	3.85	9.89	11.60	9.90	6.80	10.10	9.50	11.01
0.43	9.57	11.74	9.91	17.70	22.80	13.60	10.30	10.70	12.01
0.25	9.22	19.78	11.12	14.90	14.70	14.40	9.70	8.20	10.56
0.18	4.91	12.20	5.25	6.70	5.40	7.40	5.30	4.60	5.89
0.15	5.64	10.88	6.98	8.40	5.60	8.10	6.60	5.50	7.67
< 0.15	22.96	37.99	29.70	30.30	15.10	36.20	28.20	20.70	30.56
	60.21	**42.96**	**51.00**	**67.08**	**54.94**	**54.11**	**53.28**	**59.95**	**51.08**

Appendix 12.6

The Percentages of the Sediment of the Second layers and Their Identifications

Sample name	Acid Soluble	Gravel > 2 m	Sand	Silt < 0.15	Residue Identification
R. 21 & R. 26					
S.12B	39.79	38	38	24	Silt, brick, quartz sand, iron & sandstone, flint, burnt flint, quartzite, silica, mica and coal & charcoal in S.13 B.
S.3B	57.04	4	58	38	
R. 53					
S.10L2	32.92	11	59	30	Silt, brick, quartz sand, quartzite, flint, ironstone, sandstone, burnt flint, silica, charcoal, coal & mica. Flint and fossil shell in S.10 L2.
S.14E2	49.00	38	48	20	
R. 56					
S.3L2	45.06	16	58	26	Silt, brick, quartz sand, ironstone, quartzite, flint & sandstone.

R. 57

Sample name	Acid Soluble	Gravel > 2 m	Sand	Silt < 0.15	Residue Identification
S.4L2	45.89	12	50	38	Silt, brick, quartz sand, quartzite, flint, ironstone and mica.
S.7AL2	46.72	30	42	28	
S.7BL2	40.05	32	39	27	

Third layer

R. 53

Sample name	Acid Soluble	Gravel > 2 m	Sand	Silt < 0.15	Residue Identification
S.14E3	48.92	23	47	30	Silt, brick, quartz sand, quartzite, flint, ironstone, sandstone, burnt flint, silica, charcoal, coal & mica.

Appendix 12.7

The Distributions of Particle Sizes of Lime Plasters/Mortars of First Layers

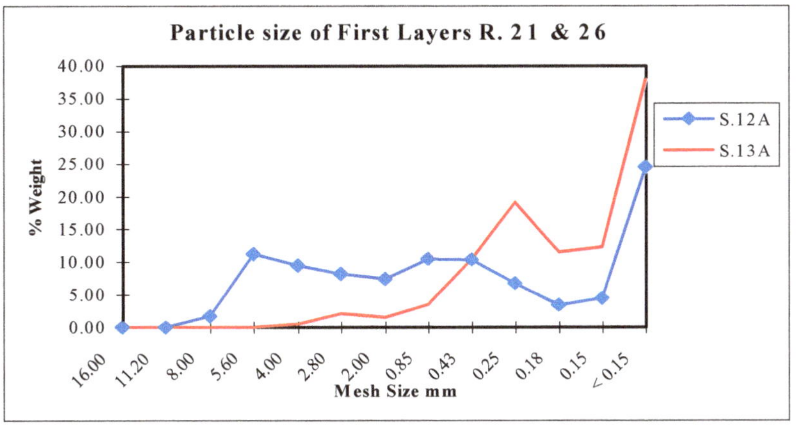

28 A

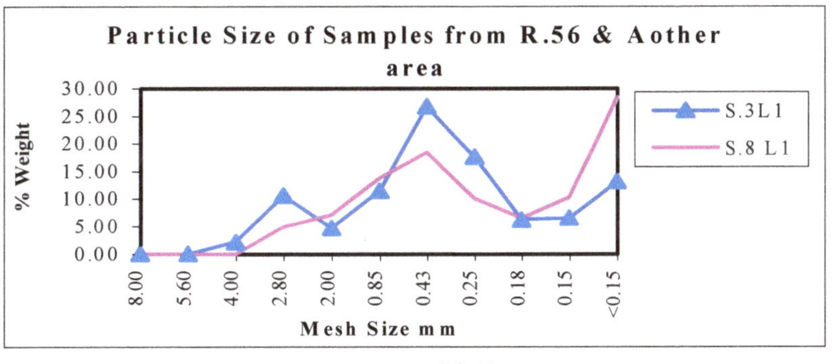

28 B

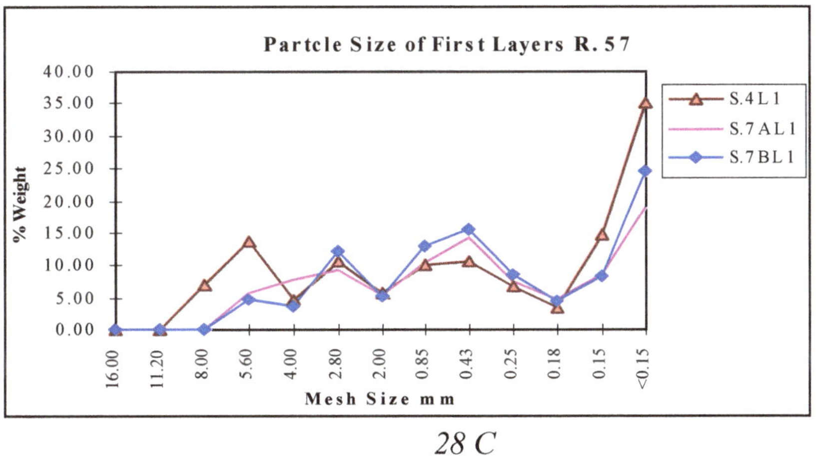

28 C

Graphs 28 A-C show the distribution of the aggregate of particle size of the first layers of the plasters/mortars from different rooms. It is possible to note the differences in graphs due to the percentages of the mixture used as reported in Appendix 12.1 & 12.4.

Appendix 12.8

The Distributions of Particle Sizes of Lime Plasters/Mortars of Second Layers

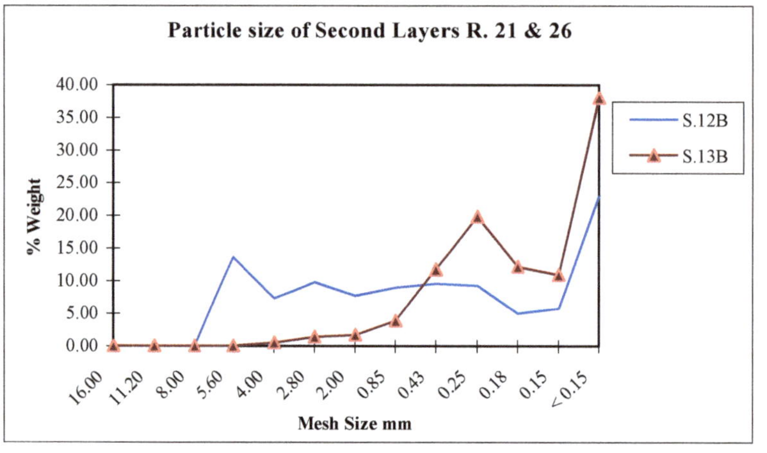

29 A

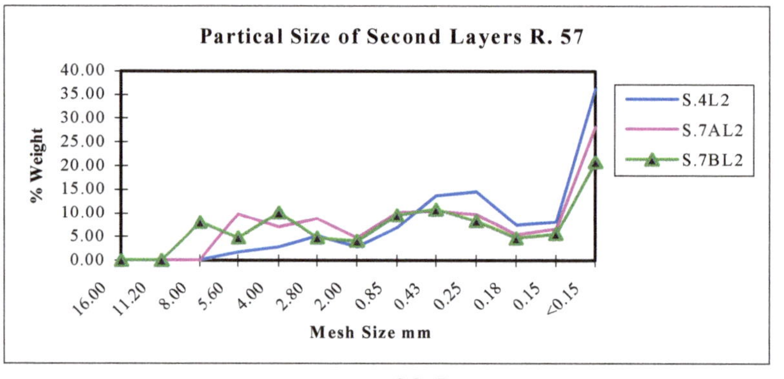

29 B

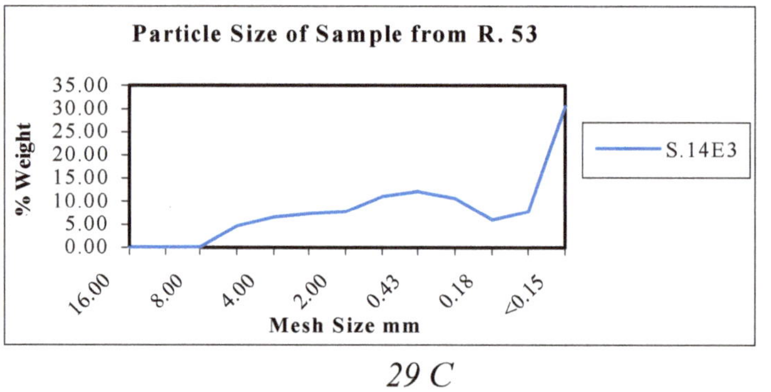

29 C

Graphs 29 A & B show the distribution of the aggregate of particle size of the second layers of the plasters/mortars. Graph A, showed more or less the same distribution of the aggregate particle size as graph 9 A, however, others were different as seen in the graphs due to the percentages of sediment types found in samples as stated Appendix 12.3 & 12.6. C shows different distribution of aggregate of the particle size of the third layer.

The Results of Chemical Analysis of Lime Plasters/Mortars

Appendix 12.9

Chemical Analysis of Mortar Mouldings

Sample name	Sample analyzed	Original weigh/g	Residue	% Solution
		R. 53		
S.14A	Window moulding	48.23	15.45	67.96
S.14B		64.57	32.74	49.29
S.14C	Floor	105.17	64.49	38.68
		R. 57		
S.17	Door moulding	81.44	49.83	38.81

Appendix 12.10

The Percentages of Sieving Weights of Mortar Mouldings

Site name	**R. 53**			**R. 57**
Material analysis	**Window Moulding**		**Floor**	**Door**
Sample name	**S.14A**	**S.14B**	**S.14C**	**S.17**
Weight	48.23	64.57	105.17	81.44
Residue	15.45	32.74	64.49	49.83
% Acid Soluble	**67.96**	**49.29**	**38.68**	**38.81**
Sieving Weight Percentages				
Mesh Size mm		**% Residue**		
16.00	0.00	0.00	33.52	0.00
11.20	0.00	0.00	18.36	0.00
8.00	0.00	1.70	11.10	2.31
5.60	0.00	12.96	12.96	9.31
4.00	5.95	12.69	6.31	11.70
2.80	3.95	7.62	2.56	8.21
2.00	5.88	6.96	1.71	9.07
0.85	0.84	9.88	2.00	14.10
0.43	25.42	10.49	1.27	11.09
0.25	17.62	5.95	0.98	6.32
0.18	15.31	3.38	0.51	3.03
0.15	5.14	3.95	0.71	3.92
< 0.15	19.71	24.08	6.92	20.53
	32.04	**50.71**	**61.3**	**61.19**

Appendix 12.11

The Percentages of the Sediment Types of Mortar Mouldings and Their Identifications

<table>
<tr><th>Sample name</th><th>Acid Soluble</th><th>Gravel > 2 m</th><th>Sand</th><th>Silt < 0.15</th><th>Residue Identification</th></tr>
<tr><td colspan="6">Window Moulding R. 53</td></tr>
<tr><td>S.14A</td><td>67.96</td><td>16</td><td>64</td><td>20</td><td rowspan="4">Silt, brick, quartz sand, quartzite, iron and sandstone, flint, burnt flint, silica, coal & mica.</td></tr>
<tr><td>S.14B</td><td>49.29</td><td>42</td><td>34</td><td>24</td></tr>
<tr><td colspan="5">Floor</td></tr>
<tr><td>S.14C</td><td>38.68</td><td>88</td><td>5</td><td>7</td></tr>
<tr><td colspan="6">Door Moulding R. 57</td></tr>
<tr><td>S.17</td><td>38.81</td><td>41</td><td>39</td><td>20</td><td>Silt, brick, quartzite, quartz sand, iron and sandstone, charcoal, coal & flint.</td></tr>
</table>

Appendix 12.12

The Distributions of Particle Sizes of Mortar Mouldings

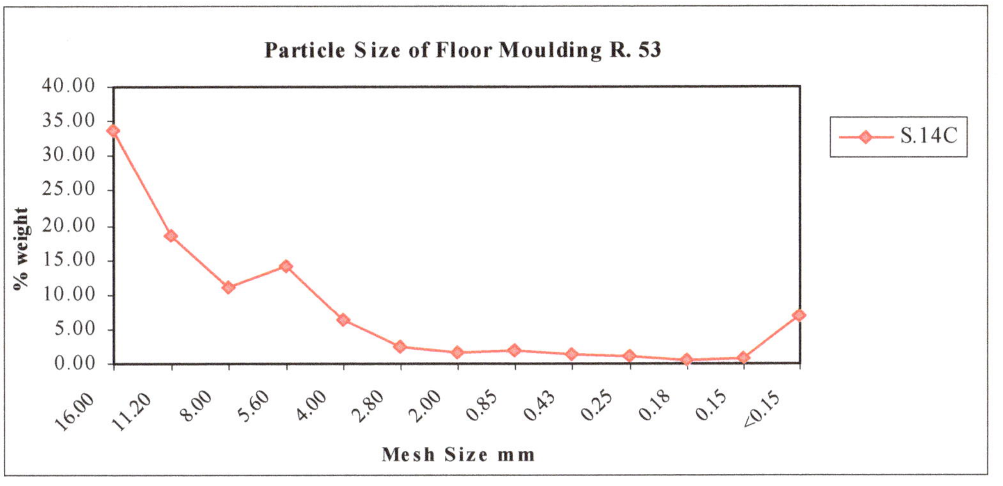

30 A

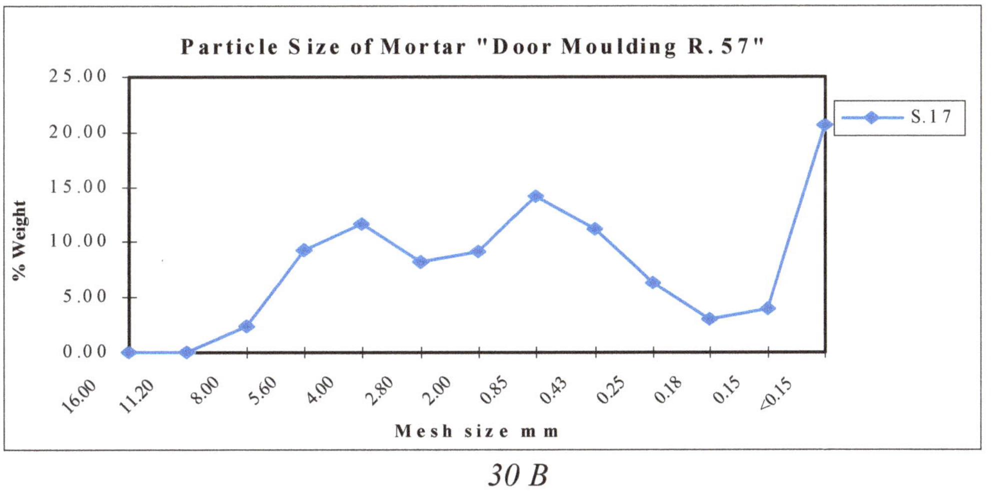

30 B

Graphs 30 A and B, show different distribution of the aggregates particle size of the door and floor. As shown that the high percentage of gravel found in sample S.14 C. However, samples S.17 showed different percentages of the sediment type used as showing in graph.

The Results of Analysis of Mud Alluvium

Appendix 13.1

Washing Sample of Mud Alluvium from Luxor

Sample name	Original weight/ g	Solvent	Residue	% Residue	% Sol.
S.109	64.20	Water	63.688	99.20	0.08

Appendix 13.2

The Percentages of Sieving Weights of Nile Mud

Sample name	**S.109**
Weight	64.200
Residue	63.688
% Acid soluble	**0.080**

Sieving Weight Percentages
Mesh Size mm

	% Residue
8.00	0.00
5.60	0.00
4.00	0.18
2.80	0.35
2.00	0.23
0.85	0.60
0.43	1.43
0.25	0.82
0.18	0.52
0.15	9.58
< 0.15	86.32
	99.20

Appendix 13.3

The Percentages of Sediment of Nile Mud and their Identifications

Sample name	Acid Soluble	Gravel > 2 m	Sand	Silt < 0.15	Residue Identification
S.109	0.08	0.528	13.153	86.319	Mud/silt, quartzite, quartz sand, limestone, brick, shell fragments, charcoal, lime/chalk, gypsum (crystal) straw and mica.

- The precence of charcoal suggest the precence of human activity

The Results of Chemical Analysis of Lime Plasters/Mortars

Appendix 13.4

Corrected Results

Alexandria Sample

First Analysis | Second Analysis

Sample name	Original weight/g	Residue	% Residue	% Sol.		Residue Weight/g	Residue	% Residue	% Sol.
S.1/A AT1	11.164	4.448	68.35	31.65		4.448	3.040	26.99	73.01

Piddington Samples

S.7A1	19.31	10.782	44.32	55.68		19.31	12.21	36.79	63.21
S.10L2	25.86	14.973	42.11	57.86		25.86	17.35	32.92	67.07

Results of Measurements of Water-Soluble Salts

Appendix 14.1

Saqqara and Dahshur

Sample name	Original Weight/g	Dry weight after analysis NaCl, $CaSO_4$ & KNO_3	% Residue	% Dissolved		Dry weight after analysis carbonates	% Residue	% Dissolved
First Dynasty (E E)								
S.104		1.959	97.95	2.05		1.932		
S.100		1.878	93.9	6.1		1.841	98.03	1.97
S.101	2	1.888	94.4	5.6		1.801	95.39	4.61
S.102		1.997	99.85	0.15		1.815		
S.103		1.868	93.4	6.6		1.827	90.81	9.11
First Dynasty (N E)								
S.143/1	2	1.910	95.5	4.5		1.745	91.36	8.64
Mastaba of Dahshur 4th Dynasty (N E)								
S.54		1.952	97.6	2.4		1.468	75.20	24.80
S.50		1.925	96.25	3.75		1.873	97.30	2.70
S.51		1.887	94.35	5.65		1.820	96.45	3.55
S.55/1		1.935	96.75	3.25		1.811	93.59	6.41
S.56	2	1.912	95.6	4.4		1.814	94.87	5.13
S.53		1.942	97.1	2.9		1.789	92.12	7.88
S.55/2		1.918	95.9	4.1		1.875	97.76	2.24
S.57		1.929	96.45	3.55		1.830	94.87	5.13
S.55/3		1.936	96.8	3.2		1.672	86.36	13.64
S.57/3		1.933	96.65	3.35		1.678	86.81	13.19

Results of Measurements of Water-Soluble Salts

Appendix 14.2

Saqqara

Sample name	Original Weight/g	Dry weight after analysisNaCl, $CaSO_4$ & KNO_3	% Residue	% Dissolved		Dry weight after analysis carbonates	% Residue	% Dissolved
Ka- Nefr Tomb 5th Dynasty (N E)								
S.9	2	1.922	96.1	3.9		1.855	96.51	3.49
S.145		1.887	94.35	5.65		1.848	97.93	2.07
S.49		1.863	95.65	6.85		1.794	96.30	3.70
S.134/3		1.917	95.85	4.15		1.836	95.77	4.23
S.144		1.871	93.35	6.45		1.761	94.12	5.88
S.10		1.896	94.8	5.2		1.822	96.10	3.90
S.46		1.938	96.9	3.1		1.858	95.87	4.13
S.47		1.863	93.15	6.85		1.811	98.96	1.04
S.48		1.913	93.15	4.35		1.848	96.60	3.40
S.146/1		1.933	96.65	3.35		1.790	92.60	7.40
Idy Tomb late 6th Dynasty (N E)								
S.139	2	1.880	94	6		1.825	97.07	2.93
S.137		1.868	93.45	6.6		1.768	94.65	5.35
S.138		1.930	96.5	3.5		1.875	97.15	2.85
S.140		1.933	96.65	3.35		1.801	93.17	6.83

Results of Measurements of Water-Soluble Salts

Appendix 14.3

Luxor

Sample name	Original Weight/g	Dry weight after analysis NaCl, $CaSO_4$ & KNO_3	% Residue	% Dissolved		Dry weight after analysis carbonates	% Residue	% Dissolved
Sheikh Abd El-Qurna T. 56 18th Dynasty								
S.37	2	1.970	98.5	1.5		1.819	92.34	7.66
S.105		1.964	98.2	1.8		1.846	93.99	6.01
S.107		1.969	98.45	1.55		1.800	91.42	8.58
S.67		0.037	1.85	98.15		0.009	24.32	75.68
T. 55 18th Dynasty								
S.113	2	1.984	99.2	0.8		1.837	92.59	7.41
S.111		1.936	96.8	3.2		1.802	93.08	6.92
S.110/1		0.266	13.3	86.7		0.097	0.266	63.52
S.112		0.270	13.5	86.5		0.033	12.22	87.78
Khokha T. 192 18th Dynasty								
S.27/3	2	1.960	98	2		1.839	93.83	6.17
S.129		1.919	95.95	4.05		1.863	97.08	2.92
S.131		1.984	99.2	0.8		1.742	87.80	12.20
S.28		1.974	98.7	1.3		1.533	77.66	22.34
S.29		1.985	99.25	0.75		1.903	95.87	4.13
S.31		1.940	97	3		1.494	77.01	22.99
S.33		1.977	98.85	1.15		1.849	93.53	6.47
S.33/2		1.958	97.9	2.1		1.754	89.58	10.42

Results of Measurements of Water-Soluble Salts

Appendix 14.4

Luxor (continued)

Sample name	Original Weight/g	Dry weight after analysis NaCl, $CaSO_4$ & KNO_3	% Residue	% Dissolved		Dry weight after analysis carbonates	% Residue	% Dissolved
Drah Abu El-Naga T. 255 18th Dynasty								
S.83/1	2	1.963	98.15	1.85		1.809	92.75	7.85
S.122		1.927	96.35	3.65		1.910	99.12	0.88
S.121		1.909	95.45	4.55		1.855	96.51	3.49
S.120		1.948	97.4	2.6		1.631	83.73	16.27
T. 13 19th Dynasty								
S.118	2	1.989	99.45	0.55		1.839	92.46	7.54
S.115		1.938	96.9	3.1		1.807	93.24	6.76
S.116		1.924	96.2	3.8		1.762	91.58	8.42
S.119/3		1.946	97.3	2.7		1.855	95.32	4.68
S.119		1.954	97.7	2.3		1.808	92.53	7.47
The Valley of the Kings T. 14 19th Dynasty								
S.RT	2	1.978	98.9	1.1		1.693	85.59	6.78
S.61		1.975	98.75	1.25		1.841	93.22	14.41
S.3		1.948	97.4	2.6		1.777	91.22	8.78
S.86		1.925	95.25	3.75		1.783	92.62	7.38
S.88		0.840	42	58		0.598	71.19	28.81
T.1 5 19th Dynasty								
S.4	2	1.955	97.75	2.25		1.815	92.84	7.16

Results of Measurements of Water-Soluble Salts

Appendix 14.5

Luxor (continued)

Sample name	Original Weight/g	Dry weight after analysis NaCl, $CaSO_4$ & KNO_3	% Residue	% Dissolved		Dry weight after analysis carbonates	% Residue	% Dissolved
T. 11 20th Dynasty								
S.59		1.947	97.35	2.65		1.790	91.94	8.06
S.64	2	1.933	96.65	3.35		1.693	87.58	12.42
Deir El-Medina T. 359 20th Dynasty								
S.1		1.956	97.8	2.2		1.818	92.94	7.05
S.8		1.950	97.5	2.5		1.797	92.15	7.85
S.6	2	1.900	95	5		1.799	94.68	5.32
S.40		1.882	94.1	5.9		1.762	93.62	6.38
Asasif T. 414 26th Dynasty								
S.15		1.925	96.25	3.75		1.738	90.29	9.71
S.16		1.957	97.85	2.15		1.787	91.71	8.74
S.18		1.936	96.8	3.2		1.800	93.26	6.74
S.19	2	1.934	96.7	3.3		1.728	89.35	10.65
S.20		1.935	96.75	3.25		1.782	92.09	7.91
S.14		1.979	98.95	1.05		1.840	92.98	7.02

Appendix 14.6

Alexandrian Sites

Sample name	Original Weight/g	Dry weight after analysis NaCl, $CaSO_4$ & KNO_3	% Residue	% Dissolved
Mustafa Pasha T. 1				
S.4	2	1.9723	98.62	1.38
S.5		1.9748	98.70	1.26
S.14	1.386	1.3529	97.61	2.39
Anfushi T. 1				
S.4/A1	1	0.9499	94.99	5.01
S.4/A2		0.9455	94.55	5.45
S.4/A3		0.8663	68.63	13.37
S.4/A4		0.8672	86.72	13.28
S.4/A5		0.8423	82.23	15.77
S.5/1		0.3482	34.82	65.18
S.6		0.4382	43.82	56.18
S.7		0.558	5.50	94.42
S.8		0.0999	9.99	90.01
Anfushi T. 2				
S.1/2	1	0.8247	82.42	17.53
S.9		0.0438	4.38	95.42
S.10/1		0.956	9.56	90.44
S.10/2		0.0381	3.81	96.19
S.10/3		0.0969	9.67	90.33
S.10/4		0.0695	6.95	93.05
S.10/5		0.0589	5.89	94.11
S.12/2		0.0988	9.88	90.12

Appendix 14.7

Alexandrian Sites (continued)

Sample name	Original Weight/g	Dry weight after analysis NaCl, $CaSO_4$ & KNO_3	% Residue	% Dissolved
Anfushi T. 2, 1998				
S.4	1.132	1.013	89.48	10.51
S.5	2	0.216	10.51	89.2
S.6		0.286	14.3	85.7
S.7		1.611	80.55	19.45
Anfushi T. 3				
S.19	2	1.9802	99.01	0.99
S.22		1.8127	90.63	9.37
S.23		1.7291	86.46	13.54

Results of Measurements of Calcium Carbonate ($CaCO_3$)

Appendix 15.1

Saqqara and Dahsur

Sample name	Original weight/g	Weight of $CaCO_3$ in sample	Ml	% $CaCO_3$	Residue	% Residue	% Dissolving	Weight of $CaCO_3$/g "Standard	Ml	Weight of $CaCO_3$
First Dynasty (N E)										
S.143/1	0.35	0.06363	16	18.2	0.2599	74.25	25.75	0.35	88	0.0039772
Mastaba of Dahshur 4th Dynasty (N E)										
S.52	0.35	0.2262	59	64.8	0.0072	2.05	97.95	0.35	91	0.003846
S.54		0.140759	37	40.2	0.1714	48.97	51.03			
S.50		0.01538	4	4.4	0.3099	88.54	11.45			
S.55/1		0.073075	19	20.9	0.2729	77.97	22.03			
S.53		0.08076	21	23.1	0.2723	77.8	22.2			
S.56		0.06153	16	17.6	0.3099	88.54	11.45			
S.57		0.0846	22	24.2	0.2489	71.11	28.89			
Ka-Nefr Tomb 5th Dynasty (N E)										
S.145	0.35	0.04056	10.5	11.6	0.2843	81.22	18.78	0.35	88	0.0039772
S.10		0.12329	31	35.2	0.1839	52.54	47.46			
S.46		0.13522	34	38.6	0.1525	43.57	56.43			
S.48		0.19488	49	55.7	0.1299	37.11	62.89			
S.146		0.16704	42	47.7	0.1580	45.14	54.86			
Idy Tomb late 6th Dynasty										
S. Stone	0.35	0.26249	66	75	0.0394	11.25	88.75	0.35	88	0.0039772
S.140		0.00795	2	2.3	0.2972	84.91	15.09			

Results of Measurements of Calcium Carbonate ($CaCO_3$)

Appendix 15.2

Luxor

Sample name	Original weight/g	Weight of $CaCO_3$ in sample	Ml	% $CaCO_3$	Residue	% Residue	% Dissolving	Weight of $CaCO_3$/g "Standard	Ml	Weight of $CaCO_3$
T. 56										
S.36		0.2333282	58	66.67	0.0488	13.94	86.06			
S.105		0.2172366	54	62.07	0.1191	34.03	75.97			
S.106	0.35	0.03339	8.3	9.54	0.2764	78.97	31.03	0.35	87	0.0040229
S.66		0.2103976	52.3	60.11	0.1182	33.77	66.23			
T. 55										
S.113	0.35	0.0297088	73	8.49	0.3075	16.2	83.8			
S.111		0.2401123	59	68.60	0.567	87.86	12.14	0.35	86	0.0040697
S.76		0.0529061	13	15.12	0.0358	10.23	89.77			
T. 192										
S.27/3		0.248883	64	71.11	0.0414	11.83	88.17			
S.27/1		0.1399997	36	40	0.1833	52.37	47.63			
S.27/2		0.2255504	58	64.44	0.0619	17.69	82.31			
S.29		0.3033264	78	86.66	0.0440	12.57	87.43			
S.32/2		0.2877712	74	82.22	0.0484	13.83	86.17	0.35	90	0.003888
S.33/2	0.35	0.2877712	74	82.22	0.0712	20.34	79.66			
S.125		0.2877712	74	82.22	0.0829	23.69	76.31			
S.126		0.241106	62	68.89	0.0921	26.31	73.69			
S.31		0.0538454	14	15.38	0.1573	44.94	55.06		91	0.0038461
S.124		0.136108	35	38.89	0.1992	56.91	43.09		90	0.003888

Results of Measurements of Calcium Carbonate ($CaCO_3$)

Appendix 15.3

Luxor (continued)

Sample name	Original weight/g	Weight of $CaCO_3$ in sample	Ml	% $CaCO_3$	Residue	% Residue	% Dissolving	Weight of $CaCO_3$/g "Standard	Ml	Weight of $CaCO_3$
T. 255										
S.83/1	0.35	0.3230724	84	92.31	0.0164	4.68	95.32	0.35	91	0.0038461
S.121		0.230766	60	65.93	0.0869	24.83	75.17			
S.120		0.1076908	28	30.77	0.0319	9.11	90.89			
T. 13										
S.118	0.35	0.3329195	78	95.12	0.0325	9.29	90.71	0.35	82	0.0042682
S.115		0.0524988	12.3	14.99	0.2753	78.66	21.34			
S.116		0.149387	35	42.68	0.1753	50.09	49.91			
S.119/3		0.2390192	56	68.29	0.0877	25.06	74.94			
S.119		0.2185318	51.2	62.44	0.0655	18.71	81.29			
T. 14										
S.RT	0.35	0.2808093	69	80.23	0.0454	12.97	87.03	0.35	86	0.0040697
S.61		0.2970881	73	84.88	0.0358	10.23	89.77			
S.3		0.1912759	47	54.65	0.0693	19.8	89.2			
S.86		0.1587183	39	45.35	0.1036	29.6	70.4			
T. 15										
S.4	0.35	0.1621228	40.3	46.32	0.1500	42.86	57.14	0.35	87	0.0040229
T. 11										
S.59	0.35	0.281603	70	80.46	0.0492	14.06	85.94	0.35	87	0.0040229
S.64		0.213137	53	60.92	0.0977	27.91	82.09			

Results of Measurements of Calcium Carbonate ($CaCO_3$)

Appendix 15.4

Luxor (continued)

Sample name	Original weight/g	Weight of $CaCO_3$ in sample	MI	% $CaCO_3$	Residue	% Residue	% Dissolving	Weight of $CaCO_3$/g "Standard	MI	Weight of $CaCO_3$
T. 359										
S.1	0.35	0.0379512	9	10.84	0.2800	80	20	0.35	83	0.0042168
S.8		0.0442764	10.5	12.65	0.2746	78.46	21.54			
S.6		0.0295179	7	8.43	0.2818	80.51	19.49			
S.40		0.047307	12.3	13.52	0.2459	70.26	29.74		91	0.0038461
T. 414										
S.14	0.35	0.2358666	62	67.39	0.0760	21.71	78.29	0.35	92	0.0038043
S.15		0.0456516	12	13.04	0.2012	57.49	42.51			
S.16		0.02404317	63.2	68.69	0.0575	16.43	83.51			
S.18		0.076848	20.2	21.96	0.2581	73.74	26.26			
S.19		0.0798903	21	22.83	0.2538	72.51	27.49			
S.20		0.0654339	17.2	18.70	0.2574	73.54	36.46			

Results of Measurements of Calcium Carbonate ($CaCO_3$)

Appendix 15.5

Piddington

Sample name	Original weight/g	Weight of $CaCO_3$ in sample	Ml	% $CaCO_3$	Residue	% Residue	% Dissolving	Weight of $CaCO_3$/g "Standard	Ml	Weight of $CaCO_3$
S.1A	0.354	0.319385	80.3	91.24	0.0205	5.9	94.91	0.354	88	0.003981
S.1B	0.308	0.3330943	77	87.9	0.0393	11.2	98.98	0.354	81	0.0043259
S.2B	0.215	0.1799865	47	83.71	0.0026	1.2	99.98	0.354	92.7	0.003825
S.4	0.274	0.2320625	62.5	85	0.0171	6.23	93.77	0.35	93.8	0.003713
S.7	0.134	0.1110824	29.8	85.18	0.157	12	98	0.355	94	0.003726
S.10	0.332	0.2332191	64.5	70.62	0.049	14.93	89.07	0.32	88.51	0.0036158

The Identification by X-Ray Powder Diffraction (X-RPD)

Appendix 16.1

(See Chapter 6, 3.1.3, Part I)

Limestone Minerals from Dahshur

Sample name	X-Ray Powder Diffraction X-RPD	Identification JCPDS/File no
	Mastaba of Dahshur 4th Dynasty (N E)	
S.52	calcium carbonate/calcite ($CaCO_3$)	5-0586
	potassium sodium magnesium iron aluminum silicate hydrate/Illite 2M1 (K-Na-Mg-Fe-Al-Si-OH_2O)	9-0334D
	calcium sulphate hydrate/gypsum ($CaSO_4$ $2H_2O$)	33-0311

Limestone Minerals from Luxor

Sample name	X-Ray Powder Diffraction X-RPD	Identification JCPDS/File no
	T. 56	
S.105	calcium carbonate/calcite ($CaCO_3$)	5-0586
	silica/quartz (SiO_2)	3-1161
	iron oxide/hæmatite (Fe_2O_3)	3-06641
	potassium barium aluminum silicate/orthoclase, barian (K, Ba, Na) (Si, Ai)$_4$O$_8$	9-00021
	potassium aluminum silicate hydroxide/muscovite 2 Ml ($KA_{12}(Si_3Al)O_{10}(OH,F)_2$	6-02631
	calcium sulphate hydrate/gypsum ($CaSO_4.2H_2O$)	3-0311
	T. 192	
S.27/3	calcium carbonate/calcite ($CaCO_3$)	5-0586
	silica/quartz (SiO_2)	3-1161
	calcium iron magnesium carbonate/ankerite (Ca (Fe, Mg) $(CO_3)_2$)	3-02821

The Identification by X-Ray Powder Diffraction (X-RPD)

Appendix 16.2

Limestone Minerals from Luxor (continued)

Sample name	**X-Ray Powder Diffraction X-RPD**	**Identification JCPDS/File no**
	T. 255	
S.83/1	calcium carbonate/calcite ($CaCO_3$) calcium magnesium iron carbonate/dolomite, ferroan (Ca (Mgo.67FeO_3) $(CO_3)_2$) magnesium silicate hydrate/sepiolite ($Mg_4Si_6O15(OH)_2.6H_2O$)	5-0586 4-05171 3-05951
	T. 13	
S.118	calcium carbonate/calcite ($CaCO_3$) calcium iron magnesium carbonate/ankerite (Ca (Fe, Mg) $(CO_3)_2$)	5-0586 3-02821
	T. 11	
S.59	calcium carbonate/calcite ($CaCO_3$) silica/quartz (SiO_2) iron oxide/hæmatite (Fe_2O_3) potassium barium aluminum silicate/orthoclase, barian (K, Ba, Na) $(Si, Ai)_4O_8$ magnesium silicate hydrate/sepiolite ($Mg_4Si_6O15(OH)_2.6H_2O$) potassium aluminum silicate hydroxide/muscovite 2 Ml $(KA_{12}(Si_3Al)O_{10}(OH,F)_2$	5-0586 3-1161 3-06641 9-00021 3-05951 7-00421
	T. 359	
S.1	calcium carbonate/calcite ($CaCO_3$) silica/quartz (SiO_2) aluminum silicate hydroxide/dickite 2Ml ($Al_2Si_2O_5(OH)_4$	5-0586 3-1161 0-04461

Appendix 16.3

Limestone Minerals from Luxor (continued)

Sample name	X-Ray Powder Diffraction X-RPD	Identification JCPDS/File no
	T. 359	
S.8	calcium carbonate/calcite ($CaCO_3$) calcium sulphate/anhydrite ($CaSO_4$) silica/quartz (SiO_2) aluminum silicate hydroxide/kaolinite-1A($Al_2Si_2O_5(OH)_4$ sodium magnesium aluminum silicate hydroxide hydrate/montmorillonite (Na x (Al, Mg)$_2Si_4O_{10}(OH)_2.ZH_2O$ calcium aluminum silicate/anorthite, ordered ($CaAl_2Si_2O_8$)	5-0586 6-0226D 3-1161 4-01641 2-0204D 2-03011
	T. 414	
S.13	calcium sulphate/anhydrite ($CaSO_4$)	6-0226D
S.14	calcium carbonate/calcite ($CaCO_3$) calcium sulphate/anhydrite ($CaSO_4$) silica/quartz (SiO_2) aluminum silicate hydroxide/kaolinite-1A($Al_2Si_2O_5(OH)_4$	5-0586 6-0226D 3-1161 4-01641

The Identification by X-Ray Powder Diffraction (X-RPD)

Appendix 16.4

Minerals in the Mortar Filler/Joining from Luxor

Sample name	X-Ray Powder Diffraction X-RPD	Identification JCPDS/File no
	T. 192	
S.124	calcium carbonate/calcite ($CaCO_3$)	5-0586
	silica/quartz (SiO_2)	3-1161
	calcium sulphate/anhydrite ($CaSO_4$)	6-0226D
	calcium sulphate hydrate/gypsum ($CaSO_4.2H_2O$)	1-0816
	calcium magnesium iron carbonate/dolomite, ferroan (Ca (MgO.67Fe.33) $(CO_3)_2$	4-05171
	potassium aluminum silicate hydroxide/muscovite 2Ml (KA_{12} $(Si_3Al)O_{10}(OH,F)_2$	6-02631
	potassium sodium aluminum silicate/anorthoclase, disordered (Na, K) $AlSi_3O)_8$	7-04781
	T. 414	
S.16	calcium carbonate/calcite ($CaCO_3$)	5-0586
	calcium sulphate/anhydrite ($CaSO_4$)	6-0226D
	silica/quartz (SiO_2)	3-1161
	potassium sodium aluminum silicate/sanidine (K, Na) $AlSi_3O_8$)	9-1227
	aluminum silicate hydroxide/kaolinite-1A ($Al_2Si_2O_5$ $(OH)_4$)	4-01641

Appendix 16.5

Minerals in the Mud brick from Luxor

Sample name	X-Ray Powder Diffraction X-RPD	Identification JCPDS/File no
	T. 192	
S.129	sodium calcium aluminum silicate/albite, calcian, disordered (Na, Ca) (Si, Al)$_4$ O_8	0-0548C
	silica/quartz (SiO_2)	3-1161
	calcium carbonate/calcite ($CaCO_3$)	5-0586
	aluminum silicate hydroxide/kaolinite-1A ($Al_2Si_2O_5$ $(OH)_4$	4-01641
	potassium barium aluminum silicate/orthoclase, barian (K, Ba, Na) (Si, Al)$_4O_8$	7-00021
S.131	silica/quartz (SiO_2)	3-1161
	sodium calcium aluminum silicate/albite, calcian, disordered (Na, Ca) (Si, Al)$_4$ O_8	0-0572C
	potassium barium aluminum silicate/orthoclase, barian (K, Ba, Na) (Si, Al)$_4O_8$	7-00021
	calcium carbonate/calcite ($CaCO_3$)	5-0586
	aluminum silicate hydroxide/kaolinite-1A ($Al_2Si_2O_5$ $(OH)_4$	4-01641

The Identification by X-Ray Powder Diffraction (X-RPD)

Appendix 16.6

Minerals in the Mud Plaster/Mortar from Saqqara

Sample name	X-Ray Powder Diffraction X-RPD	Identification JCPDS/File no
	First Dynasty (E E)	
S.101	silicon oxide/quartz (SiO_2) potassium aluminum silicate hydroxide/muscovite-2M1 (KAl_2 (Si_3A1) O_{10} (OH, F)$_2$. aluminum silicate hydroxide/montmorillonite (AI_2O_3.4 SiO_2.H_2O.xH_2O. sodium aluminum silicate/ albite ordered ($NaAlSi_3O_3$).	33-1161 6-0263I 3-0016D 20-0572C
S.102	silicon oxide/quartz (SiO_2) sodium magnesium aluminum silicate hydroxide hydrate/ montmorillonite-15A (NaO_3 (AI , Mg) $_2$ Si_4 O_{10} $(OH)_2$.xH_2O) sodium aluminum silicate/albite (Na $AlSi_3O_3$). iron magnesium aluminum silicate/montmorillonite (Si-Al-Fe-Mg-O)	33-1161 29-1498 20-0572C 3-0009D
S.103	silicon oxide/quartz (low) (SiO_2) potassium aluminum silicate hydroxide/muscovite-1M (synthetic) (KA12 Si_3A1O $_{10}$ $(OH)_2$. calcium magnesium aluminum silicate hydroxide hydrate/montmorillonite-15A (Ca O_2 (Al, Mg)$_2$ Si $_4$O $_{10}$(OH) $_2$.xH_2O) sodium aluminum silicate/albite, calcian, disordered (Na, Ca) (Si, Al) $_4$ O_8)	33-1161 7-00251 13-0135

The Identification by X-Ray Powder Diffraction (X-RPD)

Appendix 16.7

Minerals in the Mud Plaster/Mortar from Luxor

Sample name	X-Ray Powder Diffraction X-RPD	Identification JCPDS/File no
	T. 414	
S.18	calcium sulphate/anhydrite ($CaSO_4$) calcium carbonate/calcite ($CaCO_3$) silica/quartz (SiO_2) iron oxide/hæmitite (Fe_2O_3) potassium sodium magnesium aluminum silicate hydroxide hydrate/ferrierite (Na, K) $2MGAl_3Si_{15}O_{36}$ (OH).H_2O magnesium phosphate/farringtonite (Mg_3 $(PO_4)_2$)	7-1496 5-0586 3-1161 3-06641 1-0429D 3-0876

Appendix 16.8

Minerals in the Clay Plaster/Mortar from Saqqara

Sample name	X-Ray Powder Diffraction X-RPD	Identification JCPDS/File no
	First Dynasty (N E)	
S.142/2 L	aluminum phosphate/berlinite ($AlPO_4$) potassium aluminum silicate/ sanidine, disordered ($KAISi_3O_8$)	10-04231 25-0618
	Ka-Nefr Tomb 5th Dynasty (N E)	
S.46	calcium carbonate/calcite ($CaCO_3$) silicon oxide/quartz (SiO_2) potassium sodium aluminum silicate/anorthoclase (Na O.7IKO.2)AI_3O_8) magnesium carbonate hydroxide hydrate/hydromagnesite (Mg_4 $(OH)_2$ $(CO_3)_3.3H_2O$) sodium aluminum silicate/albite, ordered ($NaALSi_3O_8$) cerium carbonate hydrate/lanthanite-(Ce) (Ce2 $(CO_3)3.6H_2O$	5-0586 33-1161 10-0361D 8-0179D 20-0554C 30-0295D
	Idy Tomb Late 6th Dynasty (N E)	
S.140	silicon oxide/quartz (SiO_2) silicon oxide /tridymite-2OH (SiO_2) calcium sulphate hydrate/gypsum ($CaSo_4$. $2H_2O$) potassium aluminum silicate hydrate/ muscovite-1M ($KA1_2Si_3AlO_{10}$ $(OH)_2$. sodium aluminum silicate/albite, ordered (Na $ALSi_3O_8$) sodium magnesium aluminium silicate hydroxide hydrate/montmorillonite-18A ($NaO._3$ (AI Mg) $2Si_4OIOOH_2$. xH_2O	33-1161 14-02601 36-0432D 7-00251 9-0466 12-0219

Appendix 16.9

Minerals in the Clay Plaster/Mortar from Luxor

Sample name	X-Ray Powder Diffraction X-RPD	Identification JCPDS/File no
	T. 55	
S.111	calcium carbonate/calcite ($CaCO_3$) silica/quartz (SiO_2) calcium iron magnesium carbonate/ankerite (Ca (Fe, Mg) $(CO_3)_2$	5-0586 3-1161 3-02821
	T. 192	
S.33/2	calcium carbonate/calcite ($CaCO_3$) calcium sulphate/anhydrite ($CaSO_4$) silica/quartz (SiO_2) calcium aluminum silicate hydroxide/muscovite-1M ($KAl_2Si_3AlO_{10}$ $(OH)_2$)	5-0586 7-1496 3-1161 7-00251
S.127	calcium carbonate/calcite ($CaCO_3$) silica/quartz (SiO_2) magnesium aluminum silicate hydroxide/clinochlore-llb (Mg_5Al $(Si_3Al)O_{10}$ $(OH)_8$) barium aluminum silicate/celsian ($BaAl_2Si_2O_8$)	5-0586 3-1161 6-1211D 8-0153D

The Identification by X-Ray Powder Diffraction (X-RPD)

Appendix 16.10

Minerals in the Clay Plaster/Mortar from Luxor (continued)

Sample name	X-Ray Powder Diffraction X-RPD	Identification JCPDS/File no
	T. 14	
S.3	calcium carbonate/calcite ($CaCO_3$) calcium sulphate/anhydrite ($CaSO_4$) silica/quartz (SiO_2) calcium magnesium iron carbonate/dolomite, ferroan (Ca (MgO.67Fe.33) $(CO_3)_2$) calcium sulphate hydrate/gypsum ($CaSO_4.2H_2O$) magnesium silicate hydrate/sepiolite (Mg_4Si_6O15 $(OH)_2.6H_2O$)	5-0586 7-1496 3-1161 4-05171 3-0311 3-05951
S.86	calcium carbonate/calcite ($CaCO_3$) calcium sulphate/anhydrite ($CaSO_4$) silica/quartz (SiO_2) potassium sodium aluminum silicate/anorthoclase, disordered (Na, K)$AlSi_3O_8$)	5-0586 7-1496 3-1161 9-04781
	T. 15	
S.4	calcium carbonate/calcite ($CaCO_3$) silica/quartz (SiO_2) iron oxide/hæmitite (Fe_2O_3) potassium barium aluminum silicate/orthoclase, barian (K, Ba) (Si, Al)$_4O_8$) calcium sulphate hydrate/gypsum ($CaSO_4.2H_2O$)	5-0586 3-1161 3-06641 9-0003 3-0311

The Identification by X-Ray Powder Diffraction (X-RPD)

Appendix 16.11

Minerals of the Finishing Plaster from Saqqara

Sample name	X-Ray Powder Diffraction X-RPD	Identification JCPDS/File no
	First Dynasty (E E)	
S.102/1	calcium carbonate/calcite ($CaCO_3$) silicon oxide/quartz (SiO_2) calcium magnesium iron carbonate/dolomite, ferroan (Ca (Mg O. 67Fe O.33) (CO_3) $_2$ sodium calcium aluminum silicate/anorthite, sodium, ordered (Ca, Na) (Si, AI) $_4O_8$ potassium aluminum silicate hydroxide/illite-1M (KO.7A12 (Si, Al) $_4$ O_{10} (OH) $_2$	5-0588 33-1161 34-0517I 20-0528 C 29-1496
	First Dynasty (N E)	
S.143/1	calcium sulphate hydrate/gypsum ($CaSO_4$. $2H_2O$) silicon oxide/quartz (SiO_2) calcium carbonate/calcite ($CaCO_3$) potassium sodium aluminum silicate/anorthoclase, disordered (Na, K) $AlSi_3O$.	6-0046D 33-1161 5-0586 9-04781
	Mastaba of Dahsur 4th Dynasty (N E)	
S.53/3	calcium sulphate hydrate/gypsum ($CaSO_4.2H_2O$) silicon oxide/quartz (low) (SiO_2) potassium sodium aluminum silicate/anorthoclase, disordered (Na, K) $AlSi_3O_8$)	33-0311 5-0490D 9-04781
S.57/3	calcium magnesium carbonate/dolomite (Ca Mg (CO_3)$_2$) calcium sulphate hydrate/gypsum (Ca $SO_4.2H_2O$) silicon oxide/quartz (low) (SiO_2) potassium aluminum silicate hydroxide muscovite 2M1 (KA_{12} (Si3Al)O_{10} (OH,F)$_2$)	36-0426 33-0311 5-0490D 6-0263I

The Identification by X-Ray Powder Diffraction (X-RPD)

Appendix 16.12

Minerals of the Finishing Plaster from Saqqara (continued)

Sample name	X-Ray Powder Diffraction X-RPD	Identification JCPDS/File no
	Ka-Nefr Tomb 5th Dynasty (N E)	
S.134/2	calcium sulphate hydrate/gypsum ($CaSO_4.2H_2O$) calcium sulphate anhydrite ($CaSO_4$) silicon oxide/moganite [NR] (SiO_2)	33-0311 37-1496 38-0360

Appendix 16.13

Minerals of the Finishing Plaster from Luxor

Sample name	X-Ray Powder Diffraction X-RPD	Identification JCPDS/File no
	T. 192	
S.27/1	calcium sulphate/anhydrite ($CaSO_4$) silica/quartz (SiO_2) calcium carbonate/calcite ($CaCO_3$) potassium sodium aluminum silicate/anorthoclase, disordered (Na, K) $AlSi_3O_8$)	7-1496 3-1161 5-0586 9-04781
	T. 255	
S.120	calcium sulphate/anhydrite ($CaSO_4$) calcium carbonate/calcite ($CaCO_3$) calcium sulphate hydrate/gypsum ($CaSO_4.2H_2O$) silica/quartz (SiO_2)	7-1496 5-0586 3-0311 3-1161
S.121	calcium carbonate/calcite ($CaCO_3$) silica/quartz (SiO_2) calcium sulphate/anhydrite ($CaSO_4$)	5-0586 3-1161 7-1496
	T. 13	
S.119/2	calcium sulphate/anhydrite ($CaSO_4$) silica/quartz (SiO_2) calcium magnesium manganese carbonate/kutonhorite, magnesian (CaO.97MnO.5MgO.5) ($CO_3)_2$)	7-1496 3-1161 0-02251

ICP Results (Solution/S and Powder/P)

Appendix 17.1

(See Chapter 6, 3.2, Part I)
Saqqara

Metals PPM	First Dynasty (E E)					First Dynasty (N E)
	Sample name					
	S.104/S	S.100/S	S.101/P	S.103/S		S.143/1/S
Al	0.360	0.600	**59618.3**	0.470		1.070
Ba	<0.050	<0.050	299.2	0.100		0.070
Co	0.020	0.040	21.9	0.040		0.150
Cr	<0.01	0.010	77.8	0.010		0.060
Cu	0.070	0.130	42.3	0.070		0.010
Ca	**163.220**	**186.410**	**20049.3**	**141.560**		**394.230**
Fe	0.070	<0.020	**49383.5**	0.070		0.080
K	14.500	12.250	6450.5	12.700		11.110
Mn	0.120	0.120	27.0	0.060		14.930
Mg	15.490	24.500	**6868.9**	14.950		0.050
Na	**344.010**	**351.640**	2346.7	**418.800**		411.83
Ni	0.170	0.230	<6.1	0.180		0.370
Sr	1.390	1.230	1237.2	1.250		2.700
Ti	0.000	0.000	78.6	0.000		0.000
V	0.060	0.080	<2.1	0.070		0.110
Zn	0.010	0.040	<5.8	0.000		0.220
Be	0.000	0.000	0.9	0.000		0.000

ICP Results (Solution/S and Powder/P)

Appendix 17.2

Saqqara and Dahshur (continued)

Metals PPM	Dahshur 4th Dynasty (N E)							Idy Tomb 5th Dynasty (N E)		
	Sample name									
	S.54/P	S.50/P	S.55/1/P	S.56/P	S.53/P	S.57/P		S.135/P	S.136/P	S.140/P
Al	**14362.9**	**25497.2**	**26189.7**	**28751.3**	**23099.2**	**26637.6**		2301.2	**7754.8**	**51738.1**
Ba	182.4	203.1	223.9	184.9	233.6	219.0		5.0	23.7	298.8
Co	9.0	8.5	10.0	8.0	8.0	10.7		<3.9	<3.9	34.6
Cr	11.9	18.8	15.6	29.2	29.2	16.3		<1.2	5.6	17.8
Cu	18.5	35.9	34.4	33.4	33.4	41.3		3.4	7.5	73.2
Ca	**157661.5**	**25032.0**	**55234.2**	**15414.8**	**41987.7**	**52962.7**		**332098.9**	**242032.0**	**18631.2**
Fe	**14855.9**	**19442.7**	**17251.1**	**25270.6**	**25270.6**	**17489.0**		**734.2**	**4394.7**	**44454.5**
K	**6609.2**	**8572.3**	**9030.9**	**6527.6**	**8612.4**	**9168.9**		282.5	1107.0	8663.2
Mn	327.3	324.2	187.9	283.6	205.6	163.9		29.1	81.9	1212.4
Mg	7026.4	3728.0	3029.9	4895.3	2440.7	2994.7		6457.9	4305.6	8148.8
Na	**7498.1**	**9724.4**	**9454.7**	**13650**	**8643.2**	**12051.1**		**22889.9**	**5047.5**	**16021.3**
Ni	34.9	36.8	41.1	44.1	39.3	41.5		<6.1	<6.1	53.9
Sr	419.5	126.5	177.4	99.0	148.1	182.5		1154.4	778.6	246.6
Ti	1705.6	3287.2	2757.8	3602.7	2899.2	2811.2		71.3	309.6	4969.0
V	31.2	52.5	51.1	63.1	43.2	48.2		<2.1	16.6	118.5
Zn	52.4	56.0	51.9	57.6	49.3	62.3		<5.8	8.8	59.1
Be	0.6	0.6	0.8	0.6	0.7	0.8		<0.1	0.1	0.9

ICP Results (Solution/S and Powder/P)

Appendix 17.3

Luxor

Metals PPM	T. 56 18th Dynasty					T. 55 18th Dynasty					
	Sample name										
	S.36/S	**S.105/S**	**S.106/S**	**S.66/S**	**S.67/S**	**S.113/S**	**S.111/S**	**S.110/1/S**	**S.110/1/P**	**S.110/2/P**	**S.112/P**
Al	1.520	0.380	1.900	0.950	<0.10	0.190	1.970	2.470	**4691**	575	641
Ba	0.010	0.170	0.130	0.070	0.340	0.110	0.060	0.130	149	6	8
Co	0.020	0.020	0.080	<0.010	<0.010	<0.01	0.070	0.060	13	1	5
Cr	<0.10	<0.10	<0.10	<0.10	<0.10	<0.10	<0.10	<0.10	21	3	7
Cu	0.580	0.150	0.360	0.140	<0.020	0.060	0.310	0.560	37	7	18
Ca	**601.450**	**104.590**	**291.170**	**21.990**	**41.330**	**59.180**	**697.720**	**936.490**	**126182**	**11433**	**35509**
Fe	<0.10	<0.10	<0.10	0.120	<0.10	<0.10	<0.10	<0.10	3490	468	688
K	<120	**17.760**	**27.520**	<120	<120	16.13	21.50	<120	1824	591	1118
Mn	1.160	0.020	0.670	0.020	0.130	0.00	0.030	0.010	63	2	11
Mg	10.540	4.31	**26.500**	1.110	0.540	**9.800**	**38.950**	**11.980**	4156	529	1514
Na	**1981.560**	**453.120**	**405.30**	**264.590**	**26386.860**	**26.970**	**687.900**	**25443.280**	**220360**	**296123**	**262950**
Ni	0.530	0.100	0.35	0.060	<0.050	<0.050	0.320	0.65	42	9	22
Sr	**4.790**	0.620	1.470	0.070	0.900	0.510	3.720	8.240	633	68	172
Ti	<0.010	0.010	0.040	0.040	<0.010	<0.010	0.020	<0.010	268	46	33
V	0.050	0.050	0.150	0.030	<0.02	0.04	0.130	0.170	39	5	13
Zn	3.100	0.18	0.170	0.260	0.04	0.010	0.060	0.160	30	2504	10
Be	0.00	0.00	0.00	0.00	0.00	0.00	0.00	0.00	-	-	-

ICP Results (Solution/S and Powder/P)

Appendix 17.4

Luxor (continued)

Metals PPM	T. 192 18th Dynasty – Sample name							
	S.27/3/P	S.129/P	S.131/P	S.27/1=/P	S.27/2/P	S.32/2/P	S.33/2/P	S.127/P
Al	**17310.4**	**66467.8**	**71256.2**	**39595.2**	**19200.8**	**16317.4**	**21517.0**	**23927.7**
Ba	86.7	384.6	427.4	158.3	72.0	78.7	91.7	110.0
Co	4.7	23.1	21.8	14.6	1.1	2.3	4.4	8.6
Cr	44.1	139.9	136.7	71.0	43.3	37.4	50.3	51.0
Cu	<4	54.2	66.9	57.8	<4	<4	<4	<4
Ca	**289610.7**	**50086.9**	**36201.7**	**145494.6**	**318481.9**	**314947.7**	**29049.4**	**252109.9**
Fe	**9125.7**	**59545.0**	**60957.5**	**27758.4**	**9736.6**	**9048.0**	**12183.7**	**14044.9**
K	**4611.8**	**10590.8**	**12193.8**	**7689.3**	**4601.3**	**4648.8**	**5556.7**	**5474.0**
Mn	75.3	899.4	1059.9	221.8	81.6	100.6	103.4	147.5
Mg	**5834.6**	1568.7	**14933.6**	**8187.3**	**6383.4**	**6398.8**	**6802.6**	**9219.5**
Na	**3215.5**	**19594.3**	**17983.3**	**13417.7**	**4281.4**	**3723.3**	**5901.4**	**3780.1**
Ni	<6	43.3	36.9	26.1	<6	<6	<6	12.7
Sr	**1028.4**	440.5	306.6	**1178.7**	**1141.2**	**1066.5**	**1154.3**	**1098.7**
Ti	974.9	**8394.4**	**9222.5**	**2388.2**	**1046.7**	982.3	**1332.8**	**1693.3**
V	44.8	165.0	162.4	102.1	42.6	43.7	52.3	65.2
Zn	22.0	38.2	50.9	48.8	23.1	30.4	29.7	19.0
Be	0.2	0.9	1.0	0.8	<0.1	<0.1	0.2	0.4

ICP Results (Solution/S and Powder/P)

Appendix 17.5

Luxor (continued)

Metals PPM	T. 255 18th Dynasty					T. 13 19th Dynasty				
	Sample name									
	S.83/1/P	**S.122/S**	**S.121/P**	**S.120/P**		**S.118/P**	**S.115/S**	**S.116/S**	**S.119/3/S**	**S.119/S**
Al	**1514.0**	0.260	**18003.1**	**12399.6**		**2698.1**	0.420	1.280	0.180	0.370
Ba	2.9	0.110	89.8	37.6		5.2	0.130	0.06	0.05	0.060
Co	<1.0	0.030	6.3	11.6		2.5	0.030	0.060	0.020	0.030
Cr	47.9	<0.10	186.4	766.5		33.5	<0.10	<0.10	<0.10	<0.10
Cu	<5	0.170	11.5	39.4		<5	0.16	0230	0.80	0.140
Ca	**294309.0**	**79.410**	**196118.1**	**160456.3**		**268006.0**	**126.850**	**425.520**	**84.220**	**133.960**
Fe	638.8	<0.10	**10044.3**	**8198.4**		1496.9	<0.10	<0.10	0.250	<0.10
K	43.4	15.40	**4445.2**	2211.5		-56.7	13.910	46.480	17.420	19.860
Mn	24.5	0.560	**8899.6**	67.8		73.5	0.730	0.090	0.040	0.030
Mg	**17351.5**	22.180	122.7	**4397.0**		**34016.3**	**24.830**	**30.840**	6.550	11.56
Na	**3295.4**	**360.170**	**3966.6**	**2736.7**		399.3	**324.380**	**558.920**	**416.140**	**334.45**
Ni	<6	0.160	132.3	478.8		<6	0.120	0.260	0.060	0.060
Sr	**1707.6**	0.720	790.7	**1365.5**		**1182.2**	0.930	2.060	1.790	2.410
Ti	55.2	0.010	121	595.0		107.9	0.010	0.020	0.010	0.010
V	10.3	0.070	59.8	53.9		31.1	0.080	0.130	0.100	0.090
Zn	10.3	4.440	42.2	53.9		29.1	0.050	0.050	0.040	0.050
Be	-0.4	0.00	0.3	0.3		-0.2	0.00	0.00	0.00	0.00

ICP Results (Solution/S and Powder/P)

Appendix 17.6

Luxor (continued)

Metals PPM	T. 14 19th Dynasty						T. 15 19th Dynasty		T. 11 20th Dynasty	
	Sample name									
	S.RT/S	**S.61/S**	**S.3/S**	**S.86/S**	**S.88/S**		**S.4/S**		**S.59/S**	**S.64/S**
Al	0.340	0.300	2.260	2.010	0.950		0.570		1.330	1.710
Ba	0.030	0.040	0.060	0.060	0.130		0.020		0.170	0.030
Co	0.010	0.010	0.100	0.080	0.010		<0.010		0.040	0.040
Cr	<0.10	<0.10	<0.10	<0.10	<0.10		<0.10		<0.10	<0.10
Cu	0.090	0.080	0.330	0.310	0.220		0.050		0.040	0.370
Ca	**52.150**	**92.000**	**768.440**	**683.100**	**344.590**		**103.700**		**189.950**	**319.410**
Fe	<0.10	<0.10	<0.10	<0.10	<0.10		<0.10		<0.10	<0.10
K	<120	<120	**23.300**	**142.10**	**71.680**		**14.080**		**20.000**	**29.760**
Mn	0.00	0.00	0.020	0.010	0.020		0.010		0.010	0.050
Mg	**15.170**	**20.01**	**22.410**	**34.620**	**12.650**		3.930		43.580	7.760
Na	**179.250**	**319.640**	**127.350**	**323.730**	**17670.43**		**261.560**		**74.140**	**254.120**
Ni	<0.050	<0.050	0.370	0.350	0.210		0.100		0.100	0.370
Sr	1.400	1.340	6.030	3.070	2.720		0.950		3.430	1.180
Ti	0.010	<0.01	0.050	0.030	<0.01		<0.010		<0.010	0.020
V	0.210	0.590	0.220	0.200	0.080		0.050		0.150	0.120
Zn	0.010	0.010	0.130	0.090	0.180		0.060		0.140	0.190
Be	0.00	0.00	0.00	0.00	0.00		0.00		0.00	0.00

ICP Results (Solution/S and Powder/P)

Appendix 17.7

Luxor (continued)

Metals PPM	T. 359 20th Dynasty					T. 414 26th Dynasty					
	Sample name										
	S.1/P	S.8/S	S.6/S	S.40/S		S.14/P	S.15/S	S.16/P	S.18/S	S.19/S	S.20/S
Al	**69253.0**	0.720	2.700	1.190		**26389.4**	2.000	**18684.0**	3.400	2.070	0.490
Ba	99.1	0.030	0.090	0.070		78.9	0.090	103.7	0.080	0.080	0.100
Co	10.4	0.010	0.030	0.070		1.5	0.020	2.8	0.030	0.030	0.030
Cr	77.4	<0.10	<0.10	<0.10		33.6	<0.10	35.1	<0.10	<0.10	<0.10
Cu	48.3	0.050	0.270	0.270		<5	0.10	<5	0.14	0.09	0.140
Ca	**38986.7**	**61.210**	**57.280**	**386.490**		**241842.9**	**39.140**	**249301.6**	**23.380**	**32.890**	**127.160**
Fe	**43186.3**	0.170	3.100	2.600		**11965.2**	3.270	9813.2	4.680	3.640	0.680
K	**8662.2**	<120	18.13	**67.340**		4557.2	<120	3409.3	<120	<120	**14.960**
Mn	381.0	0.020	0.140	1.750		60.9	0.220	79.9	0.170	0.240	0.210
Mg	**12959.9**	2.080	5.750	**39.660**		**5844.0**	5.420	5914.8	5.880	6.950	11.620
Na	**13846.5**	**624.890**	**455.280**	**759.590**		**3929.5**	**315.890**	**4683.5**	**272.590**	**280.830**	**461.760**
Ni	127.4	<0.05	<0.050	0.270		20.4	<0.050	<6	<0.050	<0.050	0.100
Sr	155.8	0.170	0.280	2.210		673.7	0.270	1000.8	0.190	0.250	0.690
Ti	**3802.1**	0.030	0.240	0.030		1169.0	0.300	877.8	0.540	0.310	0.020
V	197.4	0.030	0.070	0.150		51.2	0.140	37.2	0.180	0.130	0.090
Zn	246.1	0.020	0.080	0.070		62.9	0.050	41.7	0.030	0.030	0.030
Be	2.1	0.00	0.00	0.00		0.3	0.00	0.1	0.00	0.00	0.00

ICP Results (Solution/S and Powder/P)

Appendix 17.8

Alexandrian Sites

Metals PPM	Mustafa Pasha T. 1		
	Sample name		
	S.4/S	**S.5/S**	**S.14/S**
Al	0.370	9.030	0.260
Ba	0.070	9.920	0.030
Co	0.060	1.410	0.030
Cr	0.010	1.850	0.040
Cu	0.130	2.190	0.070
Ca	**183.840**	**132.340**	**61.390**
Fe	0.000	1.260	0.010
K	3.180	**329.340**	10.140
Mn	0.010	0.120	0.000
Mg	24.160	41.060	7.290
Na	**111.820**	**303.320**	**43.490**
Ni	0.330	3.330	-
Sr	20.550	16.150	8.900
Ti	0.510	2.050	0.000
V	0.020	2.240	0.070
Zn	0.080	0.680	0.040
Be	0.000	0.080	0.000

ICP Results (Solution/S and Powder/P)

Appendix17.9

Alexandrian Sites (continued)

Metals PPM	Anfushi T. 1										
	Sample name										
	S.1/S	**S.2/1/P**	**S.4/A1/S**	**S.4/A2/S**	**S.4/A3/S**	**S.4/A4/S**	**S.4/A5/S**	**S.5/1/S**	**S.6/S**	**S.7/S**	**S.8/S**
Al	1.840	**1627.39**	0.480	0.650	1.420	1.810	1.900	3.50	7.180	0.080	0.260
Ba	0.120	13.66	0.050	0.090	0.120	0.210	0.110	0.25	0.270	0.010	0.040
Co	0.170	9.92	0.040	0.080	0.170	0.150	0.160	0.230	0.130	0.010	0.050
Cr	0.240	16.01	0.050	0.060	0.130	0.220	0.110	0.320	1.160	0.020	0.020
Cu	0.840	42.72	0.150	0.300	0.710	0.720	0.730	1.760	2.740	0.240	0.100
Ca	**749.860**	**69698.05**	**110.680**	**233.370**	**732.180**	**433.850**	**531.000**	**1239.070**	**789.560**	**447.720**	**529.630**
Fe	0.090	**1072.51**	0.000	0.030	0.050	0.660	0.050	0.45	0.270	0.070	0.020
K	**488.740**	**472.85**	**235.700**	**327.060**	215.330	**2311.360**	**304.190**	**1599.610**	4.070	**145.830**	**78.980**
Mn	0.020	37.83	0.000	0.010	0.010	0.010	0.570	0.030	0.030	0.000	0.000
Mg	**455**.730	**1833.36**	53.520	51.470	67.910	**499.430**	55.050	**744.640**	**3772.650**	13.590	19.130
Na	**1130.170**	**170335.4**	**261.400**	**582.400**	**1377.540**	**559.240**	**1541.480**	**4012.29**	**328.410**	**20527.680**	**19996.160**
Ni	1.920	43.06	-	-	-	-	-	3.000	4.210	-	-
Sr	47.050	**216.33**	26.410	23.360	16.490	43.990	12.530	47.69	26.340	9.500	13.420
Ti	0.470	291.36	0.000	0.000	0.010	0.010	0.000	0.440	0.270	0.030	0.020
V	0.390	21.18	0.100	0.150	0.310	0.400	0.240	0.610	1.580	0.090	0.050
Zn	0.400	10.81	0.080	0.110	0.260	0.270	2.950	0.890	1.070	0.010	0.010
Be	0.010	0.35	0.000	0.000	0.000	0.000	0.000	0.010	0.020	-0.010	0.010

ICP Results (Solution/S and Powder/P)

Appendix 17.10

Alexandrian Sites (continued)

Metals PPM	**Anfushi T. 2, 1997**							
	Sample name							
	S.10/1/S	**S.10/2/S**	**S.10/3/S**	**S.10/4/S**	**S.10/5/S**	**S.9/S**	**S.12/2/S**	**S.14/1/S**
Al	0.130	0.320	0.950	0.760	0.370	5.160	2.100	2.290
Ba	0.110	0.100	0.480	0.030	0.150	0.270	0.140	0.220
Co	0.530	0.550	0.540	0.410	0.140	0.060	0.230	0.260
Cr	0.610	0.660	0.510	0.550	0.120	0.810	0.260	0.190
Cu	0.450	0.030	0.450	0.150	0.360	2.080	0.940	1.150
Ca	**267.720**	395.05	**166.940**	**417.320**	**351.500**	**837.310**	**802.660**	**1054.960**
Fe	0.180	0.360	0.160	0.360	0.090	0.270	0.130	0.160
K	28.880	9.410	**288.330**	49.240	**135.140**	**9230.370**	150.450	**3009.440**
Mn	-0.040	-0.040	0.000	-0.030	0.000	0.030	0.020	0.020
Mg	16.780	2.220	1.110	10.320	12.390	**2634.660**	**306.590**	34.640
Na	**25616.770**	**25556.860**	**25088.940**	**25460.130**	**22306.400**	**11713.130**	88.160	**7160.230**
Ni	0.670	0.490	0.070	0.100	0.390	3.590	-	-
Sr	10.530	5.540	6.580	6.680	5.500	13.740	20.280	60.210
Ti	0.050	-0.030	0.090	0.130	0.450	0.280	0.020	0.180
V	0.760	0.740	0.700	0.580	0.110	1.080	0.470	0.590
Zn	0.230	0.250	0.010	0.120	-0.040	0.940	0.370	0.560
Be	-0.030	-0.030	-0.030	-0.020	-0.010	0.010	0.010	0.010

ICP Results (Solution/S and Powder/P)

Appendix 17.11

Alexandrian Sites (continued)

Metals PPM	Anfushi T. 2, 1998			
	Sample name			
	S.4/S	**S.5/S**	**S.6/S**	**S.7/S**
Al	2.90	12.30	13.50	2.90
Ba	0.10	0.40	0.40	0.10
Co	0.20	0.10	0.10	0.10
Cr	<0.10	<0.10	<0.10	<0.10
Cu	0.60	0.60	0.50	0.50
Ca	**776.20**	**764.10**	**640.80**	**577.20**
Fe	0.40	0.20	0.40	0.00
K	23.60	**9398.70**	**9803.00**	**2028.10**
Mn	0.10	0.10	0.10	<0.10
Mg	43.10	**2853.40**	**3385.10**	**123.90**
Na	**106.40**	**13964.20**	**13620.80**	**5249.30**
Ni	0.90	0.40	0.40	0.60
Sn	9.90	21.40	22.00	8.50
Sr	9.40	6.00	9.80	17.70
Ti	<0.050	<0.050	<0.050	<0.050
V	0.30	0.40	0.50	0.50
Zn	0.10	0.20	0.20	0.10
Be	<0.010	<0.010	<0.010	<0.010

ICP Results (Solution/S and Powder/P)

Appendix 17.12

Alexandrian Sites (continued)

Metals PPM	**Anfushi T. 3**		
	Sample name		
	S.19/S	**S.22/S**	**S.23/S**
Al	0.170	0.370	1.420
Ba	0.020	0.150	0.070
Co	0.020	0.030	0.170
Cr	0.030	0.030	0.130
Cu	0.040	0.190	0.690
Ca	**47.100**	**128.980**	**666.940**
Fe	0.010	0.010	0.030
K	**42.810**	**930.650**	**923.280**
Mn	0.000	0.000	0.010
Mg	11.760	79.000	126.490
Na	**126.650**	**1922.790**	**1736.42**
Ni	-	-	-
Sr	2.020	55.880	34.000
Ti	0.000	-0.030	-0.010
V	0.050	0.080	0.310
Zn	0.030	0.100	0.260
Be	0.000	0.000	0.010

ICP Results (Solution/S and Powder/P)

Appendix 17.13

Piddington

Metals PPM	Sample name		
	S.1A/P	**S.1 B 2/P**	**S.10/P**
Al	**5693.878**	**6221.497**	**29990.69**
Ba	19.078	18.261	70.796
Ca	**277349.4**	**293296.8**	**207513.6**
Fe	**4516.258**	**5816.345**	**18694.49**
Mg	**1242.926**	**1500.602**	**2357.141**
Mn	123.162	104.57	216.123
Sr	154.228	115.374	142.337
Ti	518.238	572.386	**3488.572**
Zn	7.896	3.187	22.388

The Results of Analysis using Energy Dispersive Spectrometer (EDS)

Appendix 18.1

Saqqara Samples

Sample name	File no	Element													
		Au	Al	Si	K	Ca	S	Cl	Fe	Sr	Mg	P	Na	Zn	Cu
First Dynasty (E E)															
S.101	5654A	-	322	**3147**	-	211	315	184	94	-	-	-	-	-	-
	5655A	-	1039	3353	-	**8950**	**3416**	906	697	-	-	-	-	-	-
	B	-	661	**6062**	-	380	845	466	205	-	-	-	-	-	-
	C	-	173	302	-	631	132	60	438	-	-	-	-	-	-
S.102	5578A	-	156	545	-	1305	-	117	82	-	-	-	-	-	-
	B	-	782	**2385**	-	**2312**	-	323	345	-	-	-	-	-	-
	C	-	859	2777	-	517	-	306	343	-	-	-	-	-	-
First Dynasty (N E)															
S.142	5639	-	640	1972	-	**6942**	631	457	549	-	-	-	-	-	-
	5640A	-	585	1071	-	**15557**	972	668	387	-	-	-	-	-	-
	B	-	**2329**	**8079**	-	1425	698	815	1134	-	-	-	-	-	-
	5641	-	3172	**10536**	-	1430	1324	1151	1349	-	-	-	-	-	-
	5642	-	174	729	-	84	61	51	138	-	-	-	-	-	-

The Results of Analysis using Energy Dispersive Spectrometer (EDS)

Appendix 18.2

Saqqara Samples (continued)

Sample name	File no	Element													
		Au	Al	Si	K	Ca	S	Cl	Fe	Sr	Mg	P	Na	Zn	Cu
						Ka-Nefr Tomb 5th Dynasty (N E)									
S.9	5583A	-	1755	4233	-	**11743**	-	1048	472	-	-	-	-	-	-
	5594A	-	422	**1140**	-	500	-	140	210	-	-	-	-	-	-
	B	-	25	54	-	**1611**	-	53	38	-	-	-	-	-	-
	C	-	101	136	-	**2406**	-	109	60	-	-	-	-	-	-
	5636	-	1699	3713	-	**15802**	2173	1361	847	-	-	-	-	-	-
	B	-	748	1383	-	**23632**	1791	1190	595	-	-	-	-	-	-
	C	-	68	123	-	**8799**	166	164	174	-	-	-	-	-	-
	D	-	52	88	-	**1710**	164	88	76	-	-	-	-	-	-
	5637A	-	3202	**8226**	-	**12534**	2800	1863	1208	-	-	-	-	-	-
S.145	5579A	-	**5275**	1666	-	**7493**	-	1929	1479	-	-	-	-	-	-
	B	-	1251	3423	-	**21422**	-	1114	910	-	-	-	-	-	-
	C	-	925	2053	-	**12022**	-	830	393	-	-	-	-	-	-
	5588	-	1220	3074	-	**12212**	-	723	812	-	-	-	-	-	-
	B	-	1466	**3686**	-	**3769**		574	236	-	-	-	-	-	-
	C	-	212	178	-	**13070**	-	219	278	-	-	-	-	-	-
	D	-	610	680	-	**25880**	-	960	388	-	-	-	-	-	-

The Results of Analysis using Energy Dispersive Spectrometer (EDS)

Appendix 18.3

Saqqara Samples (continued)

Sample name	File no	Element													
		Au	**Al**	**Si**	**K**	**Ca**	**S**	**Cl**	**Fe**	**Sr**	**Mg**	**P**	**Na**	**Zn**	**Cu**
		Ka-Nefr Tomb 5th Dynasty (N E)													
S.134/2	5616A	-	3370	**8534**	-	**7975**	**9082**	1481	1306	-	-	-	-	-	-
	5618A	-	382	5504	-	180	308	207	137	-	-	-	-	-	-
	B	-	1753	**11410**	-	918	1028	679	1033	-	-	-	-	-	-

Idy Tomb Late 6th Dynasty

Sample name	File no	Au	Al	Si	K	Ca	S	Cl	Fe	Sr	Mg	P	Na	Zn	Cu
S.140	5646A	-	325	1247	-	**2426**	**2803**	245	172	-	-	-	-	-	-
	5648	-	539	756	-	**4197**	**4741**	375	281	-	-	-	-	-	-
	5649	-	1158	**28035**	-	1208	-	599	305	-	-	-	-	-	-

The Results of Analysis using Energy Dispersive Spectrometer (EDS)

Appendix 18.4

Alexandrian Sites

Sample name	File no	Element													
		Au	Al	Si	K	Ca	S	Cl	Fe	Sr	Mg	P	Na	Zn	Cu
							Mustafa Pasha T. 1								
S.2	5012A	-	1030	**2014**	1263	**27652**	**1909**	**1566**	482	-	579	-	349	-	-
	B	-	792	**1316**	777	**12279**	**13387**	906	321	-	448	-	271	-	-
	5013A	-	1372	**4979**	1236	**7016**	**15064**	**3582**	713	-	830	-	521	-	-
	5025	-	50	34	38	**1197**	49	36	51	-	-7-	-	-8-	-	-
	5026	-	463	**1006**	777	**11404**	**2017**	749	542	-	254	-	138	-	-
	5028	-	792	1134	844	**15549**	**15109**	883	415	-	393	-	235	-	-
S.3	1885	-	179	283	231	**4647**	4525	-	-	-	-	-	-	-	-
	1888	-	971	1503	1302	**28731**	**30950**	-	-	-	-	-	-	-	-
	1891	-	**3705**	**15343**	**2703**	1447	**2752**	-	-	-	-	-	-	-	-
	1896	-	197	239	136	**1440**	1612	-	-	-	-	-	-	-	-
S.5	0268A	4850	2472	5112	3892	**99673**	**6794**	**8245**	2657	800	1409	7549	958	-	-
	B	**4756**	2353	**4690**	4104	**107107**	**6948**	3888	2515	729	1304	**7780**	765	-	-
	C	94	119	422	-212	**-7435**	-154	**4357**	142	70	105	-232	193	-	-
	0270	**3646**	**3512**	**19192**	2619	**55922**	**4639**	2801	2108	576	893	4722	514	-	-
	A	2845	1189	2241	2394	**70888**	**3932**	2163	1778	520	786	**4047**	393	-	-
	B	2719	1279	2252	2577	**81459**	**4310**	2167	1363	470	780	**4500**	435	-	
	0274A	2135	1389	1894	2021	**46977**	**3057**	2173	**6196**	352	660	**3525**	406	-	-
	B	1254	445	689	943	**6390**	1252	856	**25636**	264	323	1306	255	-	-
	0276C	2889	1171	1732	2158	**54910**	**3171**	2079	**8689**	486	631	**3448**	584	-	-

The Results of Analysis using Energy Dispersive Spectrometer (EDS)

Appendix 18.5

Alexandrian Sites (continued)

Sample name	File no	Element													
		Au	Al	Si	K	Ca	S	Cl	Fe	Sr	Mg	P	Na	Zn	Cu
Mustafa Pasha T. 1															
S.9	5017A	-	759	**1327**	723	**1722**	**2234**	**49508**	539	-	460	-	460	-	-
	B	-	60	70	16	123	51	**331**	**185**	-	11	-	25	-	-
	C	-	630	**1247**	814	**14608**	**7928**	1196	396	-	384	-	166	-	-
	5018	-	615	**1256**	801	**14492**	**13596**	834	436	-	384	-	178	-	-
	A	-	399	840	779	**13928**	**9450**	822	338	-	252	-	125	-	-
	5019	-	193	372	229	**515**	113	**25501**	**411**	-	140	-	526	-	-
	5020	-	183	564	864	2120	622	791	**10912**	-	93	-	58	-	-
	5031A	-	874	**3384**	**2305**	1550	**2203**	411	660	-	204	-	367	-	-
	B	-	39	56	54	**546**	**370**	41	78	-	14	-	31	-	-
	C	-	1564	**2564**	1437	1557	**5998**	**38340**	1026	-	926	-	**4568**	-	-
	5032B	-	1199	2166	1830	**19216**	**21274**	**2589**	713	-	1158	-	532	-	-
Mustafa Pasha T. 2															
S.13	8874A	-	616	1199	645	**2986**	**4167**	851	484	-	357	-	-	-	-
	B	-	1010	**1946**	1041	**5086**	**7907**	1618	606	-	697	-	-	-	-
	8876A	-	1700	**2646**	1492	**6309**	**9750**	2321	1145	-	970	-	-	-	-
	B	-	909	1341	1029	**5081**	**5983**	1329	660	-	447	-	-	-	-
	C	-	1828	**3180**	**3031**	**8873**	**14407**	2534	1100	-	1083	-	-	-	-
	D														

The Results of Analysis using Energy Dispersive Spectrometer (EDS)

Appendix 18.6

Alexandrian Sites (continued)

Sample name	File no	Element													
		Au	Al	Si	K	Ca	S	Cl	Fe	Sr	Mg	P	Na	Zn	Cu
							Anfushi T. 1								
S.2/1	0371	70	**1723**	469	138	**915**	552	1343	164	**696**	181	164	269	50	101
	0371A	70	**1933**	304	104	**623**	417	**876**	130	**653**	146	146	192	64	94
	0372B	32	**1862**	201	77	270	201	442	96	368	111	111	142	35	59
	0374A	70	720	1086	**233**	**2201**	1229	**3547**	223	234	134	269	474	60	102
	0375	153	1661	2259	467	**4477**	**2536**	**6394**	436	524	295	473	938	142	189
	0376A	701	906	962	311	**4142**	**3996**	592	656	916	327	385	246	515	520
	0377A	244	412	876	362	**2751**	**3097**	**16435**	553	112	144	415	476	273	309
	B	144	507	1582	557	**5997**	**4422**	966	525	50	196	449	134	153	165
	0378A	213	665	1406	632	**16168**	1244	**12922**	484	87	348	777	970	268	260
	B	89	193	359	211	1324	1015	**9035**	240	60	104	222	278	95	119
	0474	172	1843	**24907**	605	1602	1320	742	475	84	351	509	365	226	230
	B	144	273	589	308	**5004**	1306	**124277**	325	93	133	317	675	182	201
	C	143	81	123	50	489	151	1147	184	102	28	42	49	129	146
	D	192	87	170	45	**674**	110	203	203	204	26	26	28	163	143
	0478	104	263	330	109	723	411	**1526**	138	160	63	95	94	91	73
	0480A	427	**7993**	**17633**	1544	**12762**	2438	**11970**	1322	265	703	1042	898	501	606
	B	80	194	206	59	**487**	248	**288**	149	115	32	65	38	56	69
	0481A	651	1287	2977	1246	**12398**	**8907**	**40473**	1616	334	537	1340	2752	700	806
	0482A	651	3456	**8544**	2359	**20477**	**32557**	**27501**	2562	338	1265	2856	4191	733	874
	B	587	536	471	283	643	1775	**48386**	790	290	252	593	1147	685	695
	C	588	2765	**5620**	771	2348	2875	**34122**	1545	272	543	633	767	631	754

The Results of Analysis using Energy Dispersive Spectrometer (EDS)

Appendix 18.7

Alexandrian Sites (continued)

Sample name	File no	Element													
		Au	Al	Si	K	Ca	S	Cl	Fe	Sr	Mg	P	Na	Zn	Cu
							Anfushi T. 1								
S.2	0997	138	144	428	178	2218	527	182	130	-2-	168	587	53	78	72
	1000A	1910	2304	**3678**	2090	2415	**8586**	**13283**	1529	17	2344	**9540**	4821	942	1004
	B	163	207	597	195	**2236**	**636**	377	117	-0-	161	647	80	70	80
	D	115	45	242	117	**1962**	263	279	82	-1-	85	278	37	59	62
	1004B	197	234	405	218	490	970	**3761**	151	-0-	238	934	545	99	106
	1006A	406	489	1292	443	**4749**	**1742**	978	317	-2-	433	**1886**	179	193	210
	B	633	667	1243	554	1238	**2676**	**6715**	439	-0-	665	**2907**	990	302	332
	C	560	485	1003	476	1252	**2270**	735	342	-1-	466	**2466**	176	220	254
	E	614	591	1493	557	**4869**	2139	621	411	-1-	531	**2554**	168	256	285
	1008	606	673	1063	546	751	**2359**	**10307**	418	-2-	496	**2797**	1152	284	301
	1010A	39	809	179	76	311	295	**1207**	67	-0-	94	308	163	24	38
	B	115	830	319	152	396	474	**1706**	97	-0-	144	458	248	39	58
	C	568	1149	659	281	984	**1163**	**4642**	266	-1-	299	**1075**	698	222	209
	D	349	417	637	332	399	1416	**6885**	262	-3-	397	1910	845	178	168

The Results of Analysis using Energy Dispersive Spectrometer (EDS)

Appendix 18.8

Alexandrian Sites (continued)

Sample name	File no	Element													
		Au	Al	Si	K	Ca	S	Cl	Fe	Sr	Mg	P	Na	Zn	Cu
Anfushi T. 1															
S.2A	0529	516	270	405	367	**2428**	**1231**	394	-1-	569	58	310	354	-	-
	0531	-8-	**605**	29	22	16	26	-9-	-0-	18	12	-4-	-6-	-	-
	0533A	**797**	**640**	42	32	34	52	67	-0-	22	33	199	158	-	-
	0536	**3051**	**2552**	149	70	134	172	399	-6-	81	90	878	668	-	-
	0539A	1122	624	956	581	**4678**	**7393**	702	-7-	226	1665	566	585	-	-
	B	119	177	391	143	**848**	**1871**	108	-0-	73	453	79	74	-	-
	C	2469	641	277	222	**1007**	1036	**2359**	13	56	298	1051	1087	-	-
	0546A	1099	1383	2193	1132	1910	**5118**	**34615**	893	16	794	**3951**	**3066**	622	656
	B	936	1357	2250	1465	1875	**5800**	**2390**	760	12	815	**3775**	571	484	568
	C	1015	2205	**4258**	1391	**7366**	**11830**	2643	1021	-7-	1941	**4129**	1043	506	587
	D	1134	1037	1897	1246	**8429**	**7298**	3143	860	1-	2750	**3823**	414	574	587
	E	221	333	1470	309	1989	1057	350	282	-1-	109	489	54	127	156
	0550A	846	**5186**	**7900**	1233	**6020**	**13451**	1536	1328	-2-	877	3525	1008	497	508
	B	1066	846	1584	851	**4671**	**6618**	1046	771	-6-	627	2927	247	512	579
	C	1050	**4536**	**9916**	1594	1155	**3063**	1443	2400	-7-	655	**2619**	219	503	583
	D	1020	2046	**3700**	1248	**8525**	**14437**	1977	825	-8-	1015	4058	739	533	551
	E	887	1229	2167	1444	1863	**4992**	2264	690	-3-	636	**3181**	462	496	520
	0552A	357	499	**5534**	762	**5378**	4901	1070	568	-4-	201	1108	231	327	387

The Results of Analysis using Energy Dispersive Spectrometer (EDS)

Appendix 18.9

Alexandrian Sites (continued)

Sample name	File no	Element													
		Au	Al	Si	K	Ca	S	Cl	Fe	Sr	Mg	P	Na	Zn	Cu
							Anfushi T. 1								
S.4/A2	8883A	-	959	1709	**5424**	**13007**	1306	744	-	1123	-	-	-	-	-
	B	-	1148	1961	**5945**	**14829**	1550	730	-	1318	-	-	-	-	-
	C	-	112	1946	**5120**	**12816**	1409	850	-	1139	-	-	-	-	-
	8884A	-	1160	**2214**	**4220**	**11573**	1525	762	-	1529	-	-	-	-	-
	B	-	666	**1327**	**5262**	**11008**	961	500	-	1205	-	-	-	-	-
	C	-	474	**1038**	**2249**	**4888**	494	500	-	382	-	-	-	-	-
	D	-	408	639	**2068**	**4960**	433	467	-	359	-	-	-	-	-
	8901	-	303	425	**1264**	**2777**	241	462	-	130	-	-	-	-	-
	8919A	-	930	**1606**	**4719**	**11686**	1145	770	-	883	-	-	-	-	-
	B	-	439	551	**1354**	**3173**	395	197	-	317	-	-	-	-	-
	C	-	127	80	**216**	**547**	58	19	-	55	-	-	-	-	-
	D	-	453	495	**997**	**2503**	277	331	-	210	-	-	-	-	-

The Results of Analysis using Energy Dispersive Spectrometer (EDS)

Appendix 18.10

Alexandrian Sites (continued)

Sample name	File no	Element													
		Au	Al	Si	K	Ca	S	Cl	Fe	Sr	Mg	P	Na	Zn	Cu
Anfushi T. 2, 1997															
S.13	4914	-	1314	2094	3750	**15632**	**5871**	**22868**	817	-	971	-	2446	-	-
	A	-	1819	1805	1422	1829	**4072**	**71021**	668	-	918	-	**9020**	-	-
	4915A	-	762	1032	726	**10726**	**10287**	724	318	-	457	-	248	-	-
	4941A	-	1221	1565	1084	1366	2938	**64461**	570	-	1091	-	**9324**	-	-
	B	-	402	592	607	**1530**	**1775**	**40524**	512	-	439	-	1334	-	-
	C	-	272	563	1036	**21573**	854	**1577**	1438	-	390	-	156	-	-
	4947	-	368	512	506	**4263**	**8398**	1605	186	-	358	-	1119	-	-
	4951	-	563	1127	660	**8430**	**2002**	712	689	-	404	-	400	-	-
	4954	-	1261	2256	1470	**7055**	**16497**	1788	941	-	800	-	2011	-	-
	4954B	-	244	592	553	**6900**	**1716**	569	699	-	127	-	334	-	-
	C	-	1669	2686	**11012**	**9344**	**16369**	2959	1007	-	1089	-	1139	-	-
	4957	-	1111	2028	1522	**4174**	**4238**	**2663**	1190	-	705	-	989	-	-
	4960	-	222	342	231	**1226**	**1450**	614	120	-	127	-	246	-	-
	4963A	-	295	620	1185	**12400**	1685	952	851	-	444	-	97	-	-
	B	-	56	111	324	**5844**	367	157	543	-	229	-	34	-	-
	C	-	53	52	42	**722**	71	28	49	-	89	-	13	-	-

The Results of Analysis using Energy Dispersive Spectrometer (EDS)

Appendix 18.11

Alexandrian Sites (continued)

Sample name	File no	Element													
		Au	Al	Si	K	Ca	S	Cl	Fe	Sr	Mg	P	Na	Zn	Cu
							Anfushi T. 2								
S.15/2	0555	345	1127	1469	1699	**33170**	**19305**	1696	823	-7-	680	1990	326	464	563
	0560	332	363	845	491	**5377**	**2568**	536	236	-2-	337	1166	107	173	192
	0563	737	606	1202	1098	**16248**	**5256**	986	633	-6-	445	2373	164	458	494
	0564A	607	1179	**2279**	1273	1273	**6035**	1184	649	-6-	831	**2423**	436	699	1150
	0639A	70	91	125	114	**2079**	267	100	64	-0-	30	253	20	40	55
	C	92	842	808	820	**1689**	**2005**	1376	248	-0-	318	1097	212	156	227
	0640A	148	870	**3004**	1147	**19496**	1871	1287	425	-2-	673	1121	326	220	263
	D	**2684**	790	745	1238	**31078**	**2364**	915	783	471	255	1646	172	786	779

Sample name	File no	Au	Al	Si	K	Ca	S	Cl	Fe	Sr	Mg	P	Na	Zn	Cu
S.15/2	0645A	808	541	1055	689	11416	3481	937	388	108	589	1317	203	234	281
	B	1260	652	1298	726	11563	3378	897	399	149	581	1627	230	298	312
	C	2287	2895	**6283**	4131	**87944**	**27307**	**15489**	2069	624	2589	5849	2423	1371	1560
	D	2424	1615	2711	2005	**43935**	**39542**	2622	1070	401	1141	**4876**	619	882	863
	E	543	344	568	463	**9941**	**8709**	507	298	114	196	1055	106	187	195
	0647A	187	31	28	41	**1748**	56	24	48	40	10	31	-4-	65	56
	B	1637	468	740	995	**25747**	**16431**	77	822	328	256	1721	161	716	789
	C	1525	367	463	659	**20054**	**8224**	504	651	398	176	1097	126	618	641
	D	**1876**	384	517	917	**29289**	**1384**	632	726	365	394	1148	143	714	685
	0653A	398	212	279	407	**9577**	**3524**	272	430	126	74	456	45	221	240
	B	1164	1273	1450	1752	**26975**	**7734**	1353	1272	275	417	1914	226	610	720

The Results of Analysis using Energy Dispersive Spectrometer (EDS)

Appendix 18.12

Alexandrian Sites (continued)

Sample name	File no	Element													
		Au	Al	Si	K	Ca	S	Cl	Fe	Sr	Mg	P	Na	Zn	Cu
Anfushi T. 2															
S.15/2	0654A	778	171	166	471	**13683**	**2552**	225	456	215	70	355	67	436	446
	B	616	394	815	966	**32526**	978	755	504	197	187	999	117	327	364
	0655A	381	378	647	569	**14762**	**11511**	587	362	116	196	1085	134	247	273
S18/2 S2	9663	77	95	224	214	**558**	**834**	114	62	-0-	91	239	37	-	-
	A	85	117	**227**	115	147	**484**	143	61	-0-	84	347	47	-	-
	9665A	96	114	224	121	**621**	**1023**	162	67	-0-	66	332	32	-	-
	B	95	143	295	**1423**	1059	**2093**	215	85	-0-	111	371	57	-	-
	9666	241	196	112	175	**265**	**532**	63	45	-0-	48	136	45	-	-
S.18/2 SS	9669	61	1551	2604	**4896**	**4361**	0771	1594	474	-1-	1184	1151	499	-	-
	9670	27	2814	**4049**	3368	3167	**7960**	**5675**	632	-0-	2737	**3718**	1935	-	-
	9675	308	1583	1988	2243	2324	**14602**	**58907**	793	-6-	2288	2320	8871	-	-
	9678A	379	1302	1607	1317	1910	**4354**	**79309**	774	174	948	1773	9363	-	-
	C	274	776	1047	2203	**4768**	**25974**	1313	539	133	2491	1658	2267	-	-

Sample name	File no	Au	Al	Si	K	Ca	S	Cl	Fe	Sr	Mg	P	Na	Zn	Cu
S.18/2 SS	9679	-4-	2493	119	61	72	196	103	10	-0-	94	90	51	-	-
	D	72	1631	2557	2036	**5576**	**12533**	1905	704	-1-	1626	1462	1316	-	-
	9680A	-3-	**2991**	129	51	62	111	102	-9-	-0-	82	72	47	-	-
	B	12	**1587**	159	139	168	365	102	22	-0-	119	115	70	-	-
	C	-1-	857	61	28	36	87	40	-8-	-0-	33	33	17	-	-

The Results of Analysis using Energy Dispersive Spectrometer (EDS)

Appendix 18.13

Alexandrian Sites (continued)

Sample name	File no	Element													
		Au	Al	Si	K	Ca	S	Cl	Fe	Sr	Mg	P	Na	Zn	Cu
Anfushi T. 2															
S.18/2 S3	9686A	1179	1294	1874	1349	**4653**	**9444**	1774	747	-3-	702	4349	873	-	-
	9690A	120	258	485	242	**935**	**1538**	270	114	-1-	140	553	79	-	-
	B	91	219	280	201	**2368**	**2872**	224	84	-1-	112	522	70	-	-
	C	1574	2318	**5098**	3189	3583	**9149**	3436	1117	16	1450	**6426**	779	-	-
	D	1517	2510	3565	2227	**26473**	**31823**	2952	1153	21	1513	**6995**	905	-	-
	E	887	572	584	785	**12651**	**11819**	886	768	-6-	217	1631	161	-	-
	9695A	1102	3232	**23142**	1415	1494	**4335**	1981	1542	16	**5747**	3798	607	-	-
S.13	4914	-	1314	2094	3750	**15632**	5871	**22868**	817	-	971	-	2446	-	-
	A	-	1819	1805	1422	1829	**4072**	**71021**	668	-	918	-	9020	-	-
	4915A	-	762	1032	726	**10726**	**10287**	724	318	-	457	-	248	-	-
	4941A	-	1221	1565	1084	1366	2938	**64461**	570	-	1091	-	**9324**	-	-
	B	-	402	592	607	1530	1775	**40524**	512	-	439	-	1335	-	-
	C	-	272	563	1036	**21573**	854	1577	1438	-	390	-	156	-	-
	4947	-	368	512	506	**4263**	**8398**	1605	186	-	358	-	119	-	-
	4951	-	563	1127	660	**8430**	**2002**	712	689	-	404	-	400	-	-
	4954	-	1261	2256	1470	**7055**	**16497**	1788	941	-	800	-	2011	-	-
	B	-	244	592	553	**6900**	1716	569	699	-	127	-	334	-	-
	C	-	1669	2686	**11012**	**9344**	**16369**	2959	1007	-	1089	-	1139	-	-
	4957	-	1111	2028	1522	**4174**	**4238**	2663	1190	-	705	-	989	-	-

The Results of Analysis using Energy Dispersive Spectrometer (EDS)

Appendix 18.14

Alexandrian Sites (continued)

Sample name	File no	Element													
		Au	Al	Si	K	Ca	S	Cl	Fe	Sr	Mg	P	Na	Zn	Cu
Anfushi T. 2, 1998															
S.1	4765A	-	849	695	**1289**	**6939**	**2928**	**1534**	690	-	370	-	200	-	-
	B	-	744	864	**928**	**24895**	**952**	**926**	401	-	334	-	164	-	-
	C	-	490	**3666**	538	**13732**	**8832**	338	386	-	148	-	80	-	-
	4767	-	349	357	419	**6323**	**5578**	281	223	-	149	-	64	-	-
	4771A	-	1894	2287	2831	**50916**	**4255**	**2469**	853	-	1281	-	623	-	-
	B	-	1566	1924	**11431**	**23882**	**19403**	1719	891	-	1118	-	396	-	-
	4772A	-	340	93	1122	**15975**	695	286	393	-	35	-	50	-	-
	B	-	236	116	509	**2796**	**1440**	118	215	-	84	-	91	-	-
	4773A	-	2654	5641	**16946**	**20754**	**34070**	2430	893	-	3763	-	875	-	-
	B	-	2981	3027	**18946**	**21851**	**40288**	2911	891	-	**4835**	-	1194	-	-
	4764	-	597	644	649	**4988**	**12957**	659	228	-	318	-	1620	-	-
	4855	-	588	913	**2508**	**2175**	**4317**	648	495	-	354	-	129	-	-
	4856A	-	522	**9599**	772	612	1479	410	203	-	329	-	159	-	-
	B	-	861	1341	852	745	**2190**	1266	409	-	669	-	302	-	-
	4558	-	960	1445	**4779**	3991	**9374**	**3045**	477	-	1211	-	691	-	-
	4859	-	915	1393	966	**9077**	**9694**	1200	464	-	539	-	332	-	-
	4868A	-	827	1625	1225	**19142**	1695	1309	489	-	580	-	382	-	-
	B	-	302	672	802	**15917**	969	572	415	-	213	-	106	-	-
	C	-	380	615	581	**2946**	1025	664	322	-	235	-	127	-	-

The Results of Analysis using Energy Dispersive Spectrometer (EDS)

Appendix 18.15

Alexandrian Sites (continued)

Sample name	File no	Element													
		Au	Al	Si	K	Ca	S	Cl	Fe	Sr	Mg	P	Na	Zn	Cu
							Anfushi T. 2, 1998								
S.1	4870A	-	433	791	**3127**	**2460**	**4186**	614	278	-	661	-	135	-	-
	B	-	412	757	**2221**	**2897**	**3443**	507	224	-	269	-	149	-	-
	4875A	-	467	780	**4452**	**3227**	**6743**	634	276	-	368	-	180	-	-
	B	-	374	604	**3288**	**1837**	**4001**	531	249	-	218	-	137	-	-
	C	-	290	482	**2940**	**4255**	**4335**	430	316	-	279	-	140	-	-
	D	-	512	829	**4002**	**6369**	**5827**	670	411	-	508	-	208	-	-
	E	-	312	510	**1086**	**4963**	**6509**	473	357		203		507		
	4876A	-	504	926	**3028**	**7695**	**5950**	750	372	-	545	-	219	-	-
	B	-	276	467	**1632**	**3774**	**3681**	515	285	-	364	-	179	-	-
	4910A	-	1033	2357	1900	**47253**	1968	1957	903	-	1186	-	381	-	-
S.2	4727A	-	3288	**11826**	2887	**24228**	**21829**	2912	1798	-	2427	-	692	-	-
	B	-	1825	**3205**	2664	**56402**	**5061**	3092	988	-	1706	-	564	-	-
	4729A	-	18	32	40	**1219**	168	-7-	87	-	-9-	-	-5-	-	-
	B	-	848	**2681**	832	869	601	421	**1139**	-	223	-	43	-	-
	4731	-	526	**9539**	518	**4816**	517	488	1043	-	1722	-	163	-	-
	A	-	678	**1301**	740	**16810**	929	836	271	-	467	-	183	-	-
	B	-	619	**11787**	543	**4186**	665	660	581	-	**3600**	-	284	-	-
	4735	-	1338	3094	1054	**11536**	**11742**	1031	453	-	658	-	255	-	-
	A	-	280	345	248	**4862**	416	253	77	-	127	-	67	-	-

The Results of Analysis using Energy Dispersive Spectrometer (EDS)

Appendix 18.16

Alexandrian Sites (continued)

Sample name	File no	Element													
		Au	Al	Si	K	Ca	S	Cl	Fe	Sr	Mg	P	Na	Zn	Cu
							Anfushi T. 2, 1998								
S.2	4754	-	372	372	566	**15872**	495	444	262	-	121	-	61	-	-
	4756A	-	100	88	106	**2227**	**1626**	55	63	-	29	-	16	-	-
	B	-	**1402**	283	151	209	518	182	127	-	159	-	68	-	-
	4760	-	406	726	468	**8073**	675	450	332	-	162	-	60	-	-
	4762A	-	3962	**4929**	**6804**	**48415**	**40954**	2922	1985	-	1335	-	747	-	-
	B	-	1878	2454	1499	**50052**	**16582**	1656	772	-	972	-	430	-	-
	C	-	1338	1716	1044	**21029**	**17930**	1045	631	-	432	-	249	-	-
	D	-	225	52	89	**3682**	74	39	84	-	27	-	16	-	-
	E	-	207	121	116	**2731**	**1259**	87	59	-	45	-	29	-	-

The Results of Analysis using Energy Dispersive Spectrometer (EDS)

Appendix 18.17

Alexandrian Sites (continued)

Sample name	File no	Element													
		Au	Al	Si	K	Ca	S	Cl	Fe	Sr	Mg	P	Na	Zn	Cu
Anfushi T. 2, 1998															
S.4	8830A	-	510	1271	1316	**1847**	**3530**	821	453	-	329	-	154	-	-
	B	-	347	660	728	**1049**	**1988**	545	231	-	220	-	117	-	-
	8834A	-	2196	4034	3540	**12013**	**20042**	2167	2088	-	1110	-	503	-	-
	B	-	644	565	312	671	**1338**	221	723	-	138	-	47	-	-
	8835A	-	1418	2760	**4090**	**4317**	**12509**	1567	893	-	1267	-	399	-	-
	8843A		569	174	243	**606**	**725**	124	**786**	-	60	-	24		
	B	-	3911	**11135**	**10161**	6937	**19228**	2722	2538	-	2583	-	782	-	-
	8844A	-	2413	**5649**	2706	**8648**	**12964**	2102	1913	-	1289	-	563	-	-
	B	-	2596	**7612**	2319	**9246**	**10939**	1845	2132	-	1235	-	446	-	-
	C	-	2379	**4656**	2730	**14349**	**19032**	2532	1787	-	1390	-	667	-	-
	8849A	-	3934	**8782**	2326	**9041**	**10925**	1846	2160	-	1166	-	494	-	-
	B	-	640	1436	621	**1858**	**3320**	706	595	-	360	-	153	-	-
	C	-	2706	**4714**	3375	**22963**	**19558**	1921	**4337**	-	893	-	409	-	-
	8856A	-	794	1965	**2198**	**4227**	**6254**	919	796	-	419	-	191	-	-
	B	-	257	325	172	**952**	**861**	207	146	-	119	-	75	-	-
	8857A	-	1085	**2048**	1872	**2487**	**6274**	1155	892	-	425	-	183	-	-
	B	-	1117	**2837**	1177	**2820**	**5406**	1148	791	-	660	-	280	-	-

The Results of Analysis using Energy Dispersive Spectrometer (EDS)

Appendix 18.18

Piddington Samples

Sample name	File no	Element							
		Au	Al	Si	K	Ca	Fe	Mg	P
S.1A	6697 A	2272	**4609**	**8637**	1694	2141	2356	735	**4300**
	6701 A	4953	**22966**	**49774**	**7610**	6762	6982	5228	**26342**
	B	3189	**13567**	**35304**	6281	**8215**	**7940**	2300	**11733**
	6703 A	1409	2950	**7267**	4268	**66130**	3526	778	**4905**
	6708 A	3979	**7884**	**35188**	3576	3870	4295	1846	**14026**
	B	1396	2174	**3215**	680	1554	1215	323	**1859**
	C	3468	**11986**	**24542**	**4444**	**18738**	4539	2696	**14303**
	6709 A	-	**8820**	**35326**	3917	5009	4854	2199	**18269**
	6710 A	-	**8872**	**20137**	4600	6149	**6761**	1486	**10365**
	B	-	**17997**	**36002**	**6603**	**10446**	**6572**	3463	**18306**
	C	-	4195	**7362**	2347	**13237**	1701	1633	**10175**
S.2	6710 D	-	1383	**6384**	584	753	508	397	**2526**
	6714 A	-	**4214**	**11279**	1349	**8520**	745	965	2381
	B	-	**15736**	**31196**	**4726**	**31780**	3816	3136	3464
	C	-	402	1098	**2170**	**33803**	**1494**	498	533
	6715 A	2059	3700	**9120**	3806	**5919**	**7416**	1044	2467
	B	3377	**23989**	**52834**	**7766**	**8822**	**8083**	5515	**16522**
S.3	6723	2222	3496	**8737**	**8483**	**80383**	19272	2209	1504
	6730	1519	2052	**5942**	4819	**75712**	3697	1972	3060

The Results of Analysis using Energy Dispersive Spectrometer (EDS)

Appendix 18.19

Piddington Samples (continued)

Sample name	File no	Element								
		Al	S	Si	K	Ca	Cl	Fe	Mg	Mn
S.1B	6925 A	1289	**2484**	**2232**	1171	**17712**	1353	682	501	569
	B	**3028**	2693	**5727**	1162	**11448**	1259	1037	733	580
	C	**2909**	2254	**6325**	973	**5893**	993	886	635	449
	6926 A	1108	**1708**	**2768**	1100	**14236**	922	701	636	461
	B	4867	3854	**18111**	23377	**23481**	1899	2006	1021	1215
	C	**5624**	4151	**11663**	2436	**18827**	2093	1948	1163	1376
S.3B	6929	**2120**	2317	**4286**	1363	**9616**	1038	1509	390	1341
	A	**3480**	3818	**9459**	2742	**22315**	2000	1938	1306	1069
	D	**4876**	3584	**9068**	2109	**16720**	1791	1581	1012	956
	C	1336	**1979**	**2605**	1434	**19281**	1010	1845	252	1482
S.5B	6931 A	**10243**	4418	**20481**	3281	**16408**	2515	2779	1856	1154
	B	**3786**	5171	**8882**	2938	**36307**	2535	1203	1595	1203
	C	**3860**	5129	**30972**	2882	**17051**	2435	2181	1519	1311
	6932 A	3602	**4689**	**10374**	3494	**39752**	2829	2396	2083	1251
	B	**9235**	5126	**16597**	2744	**22699**	2539	2600	1686	1210
	C	2860	**5129**	**30972**	2882	**17015**	2435	2181	1519	1311

Appendix 19

Current Conservation Methods Used on Some Sites in Italy

Methodologies of conservation were developed at the Central Institute for Restoration (I.C.R) in Rome in the 1950s, based on the work of C. Brondi. His theory draws a synthesis between the archaeological restoration, which does not allow any pictorial intervention and the 19th century restorer's tradition of re-doing and completing every thing. There should be a balance between the historical and aesthetic requirement of improving the legibility of the conservation.

In the present study I took notes on the aesthetic presentation of the objects, depending on the specific cultural environment, case by case, to classify the different losses; e.g., the patina layer, losses in the paint layer, deep lacunae, cracks and their reintegration, according to the type of lacuna. For losses of the paint layer, two techniques of reintegration are used; the first technique is known as "*acqua sporca*" using only light grey (greyish) water colour to decrease the tonal contrast between the paint layer and the colour of the render in the lacuna. The second technique, toning back, is reintegration with transparent colour glazes, matching the hue of the original colour surrounding the lacuna.

There are two types of reconstruction and completion of the missing parts of deep lacunae: The first is known as "*tratteggio*"; vertical or horizontal painted lines are applied, in some cases, following the design of the painting. The second type is non-reconstruction of lacunae: one should simply use the same colour plaster, below the original surface level (Schmid 1998b).

These techniques are used at I.C.R. where they were explained when I visited the laboratory. One aspect of the programme of the conservation methodology in Italy is to teach the conservators, and I have seen most works being conserved not only in Italy, but also in Egypt and England, using these methods.

From the conservation sites in Italy, I noticed that the methods of conservation applied varied. **In the Sistine Chapel in the Vatican Museum,** the process is concentrated on cleaning the surfaces. The restoration was being carried out on the north wall on the works of two different artists, Botticelli and Perugino. The technique used was *fresco*, but they also found *tempera* techniques in some areas. The problems were of the humidity, dirt and human factors.

Cleaning was the main operation in the Sistine chapel, divided into two methods: First, was cleaning of *tempera* which was not easy, because the painted layer was very sensitive to chemicals. They could not use water, as it would dissolve the colours. Ammonium bicarbonate was used, mixed with water as the first stage. Japanese paper was first put on the surface, then ammonium bicarbonate solution applied with a brush. 3% Paraliod B72 in toluene was used for the consolidation of the pigments and left for one week. The same process was applied a second time, as the same pigments were used for *fresco*, the conservators had to distinguish between *tempera* and *fresco* techniques because some colours, e.g. Lapis lazuli and green (copper based) were mostly applied in the *tempera* technique. The second process for cleaning *frescoes*, used ammonium carbonate and carboxyl methylcellulose, mixed and applied to the wall through Japanese paper, then removed with water. The same techniques for lacuna and the reintegration of watercolours were used, as mentioned above. The mixture of lacunae was different from Michelangelo's painting "final judgement" where they used a mortar mixture of pozzolana and chalk 1-2 for lacuna, and applied the technique of "*acqua sporca*" without *tratteggio*.

On Botticelli and Perugino paintings they used marble dust and chalk, as that was used in the original mixture of the mortar used. It was interesting to note that the lower part was restored by re-painting with the same colours, but I noticed differences between the old paint and a new, as the colours were so strong compared with the general view. In the case of Botticelli and Perugino paintings on the north wall restoration, a restorer has shown some areas executed in *fresco* and some in *tempera* technique. It might be possible to distinguish the techniques used in the case of *giornata* and direct incision, indicating *fresco* technique.

Some studies have been done on the Sistine Chapel by Mancinelli (1994). He indicated the presence of the *secco* technique on Michelangelo's paintings through the systems used to transfer the drawing from the cartoon, indirect incision upon plaster, the presence of wire or roller marks, and the holes produced by nails to keep the cartoons in place.

The Tarquina tombs painted Etruscan tombs in Tarquina faced two kinds of decay. Firstly, natural alteration caused by the atmosphere, evaporation of water from the wall and the crystallization of salt, growth of micro-organisms, infiltration of rainwater and instability of the rock support. Secondly the impact of human factors; visitors, illegal detachment of fragments and, most interesting, that some tombs are still on private agricultural land, which caused most damage by salt crystallization. Agriculture affected the porosity of soil and contamination by animals and fertilizer. Not only these factors, but also one of the most common reasons for damage was the use of cement mortar for conservation in the 1950s, as discussed in Chapter 3 (part 1, see 4.1). Between the 1950s and 1960s, the detachment of paintings and their transfer to museums appeared to be the best solution. The *strappo* technique was used to remove the lime layer with the paint; however, that was not successful enough to keep all the paintings as they were, as some have been lost.

Paraloid B72 was again used to treat some tombs, for consolidation of the paint layer with the injection of PVA emulsion for the re-adhesion of the rendering. This method caused damage to wall paintings. The lecturer explained how the pre-Hellenistic tombs were prepared with layers of lime, crushed limestone plaster of irregular thickness and a lime wash of low porosity. Minimal penetration of acrylic resin was obtained and in many causes the lime tended to flake. The conservation of several tombs from 1985 by ICCROM, with treatment to remove acrylic resin, and a new system of climatic control, has been introduced. Another thing, which is very important, is a maintenance programme of regular inspection and periodic biocide treatments (Schmid 1998c: Pers. Comm).

The Typhoon Tomb in Tarquina

The tomb is dated to the 2nd century BC and was under conservation. It is so beautiful and unique in its decoration, especially the ceiling, with incisions on the wet plaster coloured with red ochre. The technique used was *fresco*, and I noticed the incision of the drawing, which indicated true *fresco*; the colours used for painting were red, black, yellow and blue. These colours were widely used in antiquity. The conservators explained the problems of the tomb and also the process of the conservation treatments.

The problem in the tomb was the humidity and changing climate between hot summer and cold winter, allowing the walls to dry, producing salt crystallization. The paintings were mostly covered with a layer of very soft calcium bicarbonate and some areas of hard calcium carbonate. The lamp that was fixed directly to the wall had a bad effect, producing a black layer on the painted layer. A little dry brush was used to remove salts, but the conservators were not able to do any cleaning on the hard crystals. Even mechanical cleaning, would damage the painted layer, so these areas were left. The tomb was rock cut, the stone called Mucco, is quite yellow with some red calcarenite, withgrains of calcareous stone and cemented fossils. They used the same composition for new mortar to reconstruct lacunae and fill the edges, injecting the flaking areas of the painted layers with a different proportion for each process. However, it is archaeologicaly aesthetic correct to leave every thing as it is, so no intervention is required to complete the missing areas or to use any method of conservation.

Arezzo and Florence

In 1997 there was a day conference at the Courtauld Institute of Art, on "Piero Della Francesca's Arezzo *frescos* Techniques and Conservation", so it was very useful to visit that church work site. The conference detailed the problems of the church and showed not only the conservation processes and treatments, but also the methodology of analysis, to examine and determine the original techniques used.

In the 19th century the French occupied the church. It was restored in 1858 and 1901 and re-constructed and opened to public in 1911. The last restoration was carried out in 1960-63. On a visit to the church one can see the differences between old and new restoration at the end of the right side of the entrance to the church. The mortar used for lacunae was not matched with the original, and the appearance is too heavy and strong in colour. This indicated the problem of using cement for restoration in 1950s-60s, which produced sulphation, covering the surface. In addition it was treated with Paraliod B72. Both *tempera* and *fresco* techniques were used, which are not suitable for resin treatment. The methods used for analysis were varied.

The conservation treatments applied to the church were of two different techniques using exchange resin and pre-consolidation of the painted surface, which was carried out before cleaning with barium hydroxide. It was used many times, because of its safety for working, but care must taken with copper based colours. These methods are applied, not only because of the techniques used, but also because of the kind of decay, as previously stated in Chapter 6 (see, 2). Through visits to the church I saw some documentation, indicating *fresco* technique such as *giornata*. A restorer has identified *tempera* technique inferred from to the colours used, e.g. lapis lazuli, as the artists sometimes painted on *fresco* using organic media (dry *fresco* as discussed in Chapter 2, see 4.8.2 and 4.8.3)

They used the same methods of intervention with *acqua sporca* and *tratteggio,* when required. A question was raised about the methods that should be used for cleaning the *tempera* painting, as the painted surface was very sensitive to chemicals. The answer was ammonium bicarbonate, the same as was used in the Sistine Chapel. Some case studies show that the best way is to clean the surface painting through Japanese paper. Many tests were required for the cleaning of walls painted using the *tempera* technique. The type of organic binding media used was identified in the study by Dr. Moura Mateini and others, which also shows all the methodology of analysis and the conservation processes (Conference 1997).

The use of barium hydroxide and ammonium carbonate is a common treatment for cleaning wall paintings in Florence, on both internal and external walls. These methods are stable and resistant to weathering, without any change occurring since the time the application was carried out. In Plazzo Medici Riccardi Cappella dei Magdi, which was executed in *tempera* technique and restored in 1993-94, calcium caseinate was used for the treatment. Other examples in Florence of the cleaning treatment using barium hydroxide and ammonium carbonate were the church of S. Marco, which was an important example restored since 1970, and the external wall of a house near to Piazza S. Croce St., which was completely covering with a layer of black crust of pollution decay. The external walls of this house were decorated with Roman ledges, and although the house was exposed to weathering decay, the painted surface was still resistant and in a good condition.

Appendix 20

Experimental Mixes for New Lime Mortar

(Practical experiments at the Woodchester & Orton trusts)

Various mortars were prepared to test the quality of the mixtures and application methods. The new mortar should be less strong than the old for re-plastering/re-pointing because of great strength of these materials.

Experimental Details

•-Mixtures

(Parts by Volume)					
No 1 (hot mix)	**No 2**	**No 3**	**No 4**	**No 5**	**No 6**
1 quick lime	1 lime putty	1 lime putty	1 hydraulic lime	The same as No 1, with hair	2 lime putty
1 fine sand	1 fine sand	1 fine sand	1 fine sand		3 sharp sand
2 coarse sand	1 stone dust	2 coarse sand	2 coarse sand		3 stone dust

We also carried out a trial with a casein mixture, “calcium caseinate,” for **repairing stone** (Woodchester trust) using ½ lime putty to ½ natural cottage cheese (low fat).

Mortar No 1, was prepared for pointing. In the case of re-plastering, the sand and lime mixture should be left for 6 months before use. The advantage of this process is that the mixture is rich in lime and more suitable to use and has better application properties. Quick lime = 1½ putty lime.

Mortar No 4, Lime (quick lime) is put in water at the same time as making the first mortar, great care should be taken when using hydraulic lime in the mixture.

Mortar No 5, was prepared a year earlier, with the same mix as mortar No 1. The mortar was diluted with water to obtain a thinner mixture. As it had lost some strength, hair (goat) was added to give strength to the mixture. The amount of the hair needed depends on the way it spreads out to cover the surface of the mortar, which is then mixed on the floor. The method of mixing the mortar was by means of placing layers of mortars and hair, then stirring. If there is any need to add more hair then it should be done using the same method. To test the quantity of hair, some mortar is taken by the “gauging trowel”, when there should be one hair for each 1 mm of the mortar.

However, mortar No 6 (Woodchester trust) was prepared without adding water, but if water is needed, it is possible to spray some water on the surface and then mix it up, kneading it carefully.

•-Application
Before the application of these mortars, the walls ware cleaned. Then water sprayed on to the areas that would be re-pointed to prevent the mortars drying quickly. Crushed stone was used in the case of deep holes as a support and then mortar applied. Plastic brushes were used to clean off the extra mortar and then to scrub of the surface to get the same impression as the old mixture, however, this stage should be applied on the next day to allow for setting. The wall should dry slowly, spraying lightly with water, and should be covered to keep off the wind and sun for two day. In the case of small walls, it is possible to cover it with wet rag/hessian to keep it damp.

1-Mortars 1, 2 and 3 were used for re-pointing and re-pair of old walls.
2-Mortars 2, 1 and 5, were applied on wet and dry walls.
3-Mortars 3 and 4 were used on a dry wall.
4-Some trials were carried out for re-pointing and re-plastering using different combinations, e.g., mortar No 5, was used as a first coat for re-plastering the wall and then the fine mixture of mortar No 2, was applied next day. However, before the application of the next layer mortar, the surface was scraped off, to remove the crust that had formed. This helps the air penetrate the surface, assisting the carbonation process inside the mortar.
5-Another trial used mortar 3 and then mortar 2 for re-pointing of old walls; stone, brick etc.
6-Mortar No 6, was used for re-pointing stone.

Casein Mixture " Calcium Caseinate" for Repairing Stone
The mixture used was as glue to stick the stone together. Before preparing the mixture, both the lime putty and cheese was sieved separately several times to make them smooth, then mixed together. The whole mixture then was sieved many times in order to obtain a pure and fine mixture. A little water might be used for sieving, however, care must be taken, as the mixture should be thicker than lime-wash.

The stone/wall was cleaned with a brush and then sprayed with water. Both the broken area and the wall were painted with the casein mixture and stuck together. Mortar No 6 was used for re-pointing. A plastic brush was used to remove the extra mortar. However, this mixture can promote biological growth.

•-The Results (next day)
Mortar 2, showed cracks on the dry and wet wall.
Mortar 4, showed no cracks, but the mortar was still damp and soft, needing more time to dry.
Mortar 5, had no cracks at all.
Other mortars mixed with lime putty gave good results. The various mortars used for the re-plastering and re-pointing were successful. However, more experimental work would be required in different environments.

Photographs Part I

Captions

Photographs 1 A, B & C show different biological growth. A shows the water inside the tomb and which is highly coloured. Algae have reproduced extensively in the conditions suitable for their growth.

B shows a heavy colonization of lichens and algae on the outer surface, whereas, on C, different kinds of bio-deterioration and decay are shown; efflorescence, plants, lichens as well as a blackish crust (Tombs of Cerveteri in Italy).

Photographs 3 A and B. In photograph A, the salt can seen in the top of the picture. It is also possible to see the decoration style, which is in imitation of marble, which was common in Greek times. B, a detail showing micro-organisms on the painted layer, notice also the cracks filled with gypsum plaster, identified using a spot test (Mustafa Pasha, T. 2, Alexandria).

Photographs 4 A & B showing the underground-water, as well as salt crystallisationin tomb 1. Notice also the decoration style on B (Anfushi, Alexandria).

Photographs 5 A & B show salt the same as in tomb 1 where salt has covered not only the wall, but also the door B (Tomb 2 Anfushi, Alexandria).

Photographs 6 A & B. A shows that although the wall render has started to dissolve, one can see theoriginal division of the wall by incision suggesting buon frseco (tomb 3). B shows that micro-organisms of alga growth has covered the wall due to the light and humid conditions, which are very suitable for microbiological growth (T. 5, Anfushi, Alexandria).

Photographs 7 A and B, showing samples of the mud plaster/mortar, first dynasty (E E). A, shows the mud plaster, based on a mixture of mud and straw, as well as other aggregates (see Appendix 8.1). On B, it is possible to see that the clay plaster/mortar has covered mud plaster/mortar; some traces of red and yellow were noticed on top (first dynasty, E E).

Photograph 7 C, a sample of clay plaster/mortar, first dynasty (N E) showing that the plaster/mortar is covered with another finishing layer of plaster (see X-RPD, Appendix 16.11). Note the size of the bubble found in the mixture (late first dynasty, N E).

Photographs 8 A & B, showing very thin pieces of gypsum plaster collected from the mastaba of Dahshur; note the colour of the plaster, encrusted by the reaction factors; e.g. chemical reaction or polluted atmosphere. The plaster was in the form of a glassy structure and the surface structure appeared very crumbly, as shown in Photograph 4 B, using SEM (Dahshur 4th dynasty N E). Scale 1-cm divisions.

Photograph 9, showing a few fragments of the clay plaster covered with another layer, it is possible to see two layers on the right of the photograph. This layer is painted bluish/blackish; the plaster was very soft. Top left, the original colour of the layer (a finishing layer for painting), can be seen which appeared finer than the clay plaster (Ka-Nefr tomb, N E).

Photograph 10, shows small pieces of mud plaster covered with layers of gypsum plaster; on the top right it is possible to see the traces of the gypsum plaster (Ka-Nefr tomb N E).

Photographs 11 A & B. A, showing the rock before covering with mud plaster/mortar. The yellowish/brown areas are due to oxidization. B, shows traces of the whitish layer that covered the plaster. The straw which was used as organic material in the mixture of the plaster can be seen (Tomb 56, Sheikh ABD El-Qurna, Luxor).

Photographs 12 A and B. A, shows the broken column in tomb 55, carved from rock and covered with mud plaster/mortar, then a layer of gypsum plaster (identified by spot test (S. 76), which was coated with very thin whitish layer, as shown under the microscope. B, collected from the north wall showing a small piece of the salt. The structure of the salt can be seen in forms of ancient dust. Some areas were in the form of fibres or needles (Sheikh ABD El-Qurna, Luxor).

Photograph 13 shows small fragments of the plaster painted in Egyptian blue with a red panel. The fine cracks on the surface and the blackish layer underneath can be seen. (T. 192, 18th dynasty, Khokha, Luxor).

Photographs 14 A and B, showing the two layers of the mud plaster/mortar covering the middle of ceiling, and a layer of clay plaster (better plaster) covering the mud plaster. B, is a detail of A, showing the cracks present on the surface, and the organic material, straw, used with the mixture (T. 13, Drah Abu El-Naga, Luxor).

Photographs 15 A and B, showing a small fragment from T. 13. A shows the surface before the cleaning off of the darkened layer; note the crack and the small holes. B, shows the plaster after removing the PVA film and cleaning the surface (Drah Abu El-Naga, Luxor).

Photographs 16 A, B and C. A, showing the stone covered with a layer of iron oxide as shown on the right. A layer of gypsum formed between the stone can be seen, note also the yellowish and blackish areas on the left. B shows a sample of mud plaster covered with a brownish/reddish layer. C shows small fragments of the mud plaster covered with a thin layer of clay plaster (better plaster), T. 414, 26th dynasty, Asasif, Luxor.

Photographs 17 A, B & C, showing the surface structure of (S.4) caused by dissolving the lime binder. This resulted in a very rough and un-homogenous structure as seen on 17 B, a detail of the back of the first left and C, the back of the middle of S.4, notice also the shell that was found in the mixture of the mortar, also noticed on the investigation of sand from the site (T. 1, Anfushi, Alexandria). .

Photographs 18 A & B (S.1/1 & S.1/2) and C showing different kinds of deterioration resulting from salt crystallization and some other sources, on a very soft and powdery whitish layer.
On B, a hard layer of fossil dust in a yellowish colour. B, the middle sample before cleaning both samples on left and right after cleaning which show the cream colour of the painted layer.
Whilst C shows on a polished section of two different kinds of calcite consisting of two different layers of different minerals these samples were analysed to determine the composition of these layers (see X-RD, mineral identifications, Table 8, page 152) (T. 1, Anfushi, Alexandria)

Photographs 19 A-F, show different sections of the plasters/mortars used from Luxor, A and B, show samples of the clay plaster T. 192, which was covered with gypsum plaster. In spite of the section being a bit dark, it is clear to see the difference of the finishing plaster on the two sections and also the mixture of the clay, which looked darker.
C, is a sample of T. 13, showing the clay plaster, which was coated with another layer of finishing plaster (see X-RPD, Appendix 16.12-13), and shows the ordinary plaster. D, T. 359 and E, T. 414 showed the mud plaster samples covered with a thin layer of clay plaster being different from the previous one due to the different mixture used in the preparation of the walls.

Photographs 20 A-E, show the different types of plaster/mortar used in the Mustafa Pasha T. 1 and T. 2, as well the techniques applied. The intonaco layer can be seen in yellow on A and in red on B, in different thicknesses.
C, shows a coccolith, indicated by SEM, which was found on B.
D, shows the splashed technique, as found on the sample from Mustafa Pasha T. 1.
E, shows the two different layers of plaster used in T. 2; the first layer contained black sand, and the second was finer with more lime.

Photographs 21 A-E, show different types of the plaster/mortar from the Anfushi T. 1 & T. 2. A, shows the mortar is a mixture of brick and other grains (see Appendix 12-3). The whitish layer that covered the surface, the cracks inside the mortar can also be seen.
B, is similar to photograph 20 E, as the first layer contains black sand and is nearly the same mixture, whilst C, consisted of two layers with the same mixture and a white piece of chalk/lime can be seen.
D, a different plaster is indicated (gesso plaster). Notice ironstone as well as the lime/chalk that was in the original mixture of gypsum. E shows the intonaco layer that covered the mortar, similar to the Piddington samples (photograph 22 B).

Photographs 22 A-D show different sections of the plaster/mortar which contained the paint layer, intonaco layer in different thickness (22 B & D). The plaster layers as shown on D consisting of two layers, with, the calcite film as showing on top D. Notices also the different composition of the plaster, as shown on C, Piddington.

Photographs 23 A and B, cross-polish section showing the structure of the plaster (sample S.1) using SEM. The calcite film covering the painted layer had a thickness of between 0.62 mm-0.26 mm, Piddington.

Photographs 24 A and B, show the additives to the mud plaster/mortar (coarse plaster), which are shown in the graphs. The sand used in sample S.144 was yellowish indicating the presence of more iron, in addition to a different source, as this sample belongs to the southern area, whereas the other sample (S.100) is to the north. However, analyses have shown that the same additives were mixed with the mud, as reported above (S.100, from first dynasty E E and S.144, first dynasty N E, Saqqara).

Photographs 25 A-C showing the microfossil shells found with other additives in the mixture of powdered stone, A and B from a sample S.16, and C from a sample of the limestone (S.105). D shows detail of C, the surface structure consisted of many layers, EDS analysis identified the main elements as calcium carbonate.

Photographs 26 A and B. A shows seeds found in the mud plaster and B shows a piece of bone found in the mud brick. It was impossible to determine whether the bone was human or animal. The identification of seeds is given in Table 1 (Idy tomb, dated to the late 6th dynasty).

Photograph 27 shows the form of the calcite film as a botryoidal group, which covered most of the Piddington fragments.

Photograph 28, A shows a sample of limestone from Dahshur, which had a very fine grain size and calcite. B shows a sample of mud brick from Saqqara which contained quartz, calcite, microfossil and clay.

Photograph 29, sample of very fine limestone from the site of Luxor showing microfossils, calcite and tiny rhombus. Notice the difference compared with 28 A.

Photographs 30 A & B, showing sample of clay plaster T. 255, Drah Abu El-Naga. A shows the cells of the organic materials used in the mixture with clay. B, shows calcite, quartz, micro-fossil and gypsum.

Photographs 31 A & B, show a sample of plaster from T. 192, Khokha, containing four layers, the first and third being reddish layers 31 A, as in A showing gypsum and quartz. The second and fourth layers were lumpy 31 B, containing organic materials, quartz and clay.

Photograph 32 shows samples of finishing plaster from T. 255, Drah Abu El-Naga, containing gypsum and a few calcite grains.

Photograph 33, shows gypsum plaster from T. 192, Khokha, containing calcite and quartz. Note the differences between the two sections compared with 32.

Photograph 34 A & B (S.14), showing the structure of limestone from the site of Mustafa Pasha, quartz, shell fragment, rounded sand grain and micrite around the structure. Notice in B the angular quartz grain which has not been affected by the wind.

Photographs 35 A & B (S.4) Mustafa Pasha T. 1, show the internal structure of the plaster, A shows quartz grain, and on B micrite and calcite spar.

Photographs 36 A & B, show the structure of mortar samples from Anfushi T. 1, A showing oolitic limestone (egg shape) full structure and on B calcite spar.

Photographs 37 A & B show the internal structure of samples from Anfushi T. 3, shows A, pieces of shell fragments and in B, the same shell fragments and foraminifera (tests), the particle size of the structure is quite big and porous.

Photographs 38 A & B, the Libya sample (Tolmeita beach) showing; foraminifera, shell fragments, alga, ferruginous cemented limestone. A, shows calcareous alga and shell fragment and on B different shell fragments. Notice the similarity between Anfushi and Libya indicating the youngest ridge, resulting in incomplete structural spaces around the grains.

Photographs 39 A & B the Cyprus sample (Queens bay). The structure consisted of foraminifera, Golbigerna, isolated shell, quartz and ummulites. A, shows shell fragments and foraminifera and B shows quartz and lithclast. Notice that the structure is nearly full, suggesting that this sample is from an older ridge than the Anfushi and Libyan samples.

Photographs 40 A-H show general structures of lime-based plasters/mortars from Piddington Roman villa. A, the top of the intonaco, covered with a very thin calcite film and sheared quartzite, B, showing limestone (bottom right) with microfossils, quartz and other shell fragment. C, quartz and shell fragments. D, shows a structure of a piece containing fine grains of quartz and shell fragments. E-H showing different grain sizes of, quartzite and shell fragments.

Photographs 41 A & B. A shows limestone covered with a layer of salt, NaCl (halite), confirmed by EDS analysis. High magnification of the structure of the stone showed microfossils. B shows small shell fragments, as well as cupped shapes of Ca. Analysis indicated the elements found as Ca, the highest peak, and Mg, Al and Si (T. 255, Drah Abu El-Naga, Luxor).

Photographs 42 A & B, showing a sample of mud plaster/mortar. A shows a crack, the structure of the sample at different levels, and some holes. On the top right is some quartz. Higher magnification in B shows an inhomogeneous structure. The elements present are; Ca, Si, S and Fe (first dynasty E E).

Photographs 43 A, B & C show the structure of clay plaster sample from T.414. Note the holes indicating the high porosity of the plaster. With high magnification it was possible to see that the mixture contained different shell fragments, (B) and other organism such as coccoliths (C). It was noticed in other samples as well, that the powdered limestone was full of these organisms. EDS analysis indicated S, Ca, Si, Fe and Al (T. 414, Luxor).

Photographs 44 A-B (Mustafa Pasha, T. 1) C and D (Anfushi, T. 1), showing the structure of the plaster on A, and B indicating the presence of Ca and S. On B, notice the shape of the crystal of $CaCO_3$*; analysis has indicated the presence of Cl as well. Whilst, C, and D, show the bio-deterioration effect, in terms of fungal growth as on D, and the left of C, which is covered with salt in the form of halite (NaCl), in addition to the presence of S, Ca and Cl.*

Photograph 45 shows the structure of the plasters, which contained a coccolith, showing the composition of the plaster/mortar with microfossil structures. (T. 1, Anfushi, Alexandria).

Photographs 46 A-D, showing the bio-deterioration and decay of Piddington samples due to fungal hyphae is shown on A and B as well as the shapes of the calcite film on the fragment surfaces (C-D). Most of samples were examined using SEM. These show more or less the same structures of covering with layers of calcite film in different thickness. EDS indicated the presence of Si, Al as well as P, which has shown on all samples.

Photographs 47 A- C, show the different shape of salt crystallization. On A, the top is covered with Si. B shows a different structure of salt crystallization. C shows different crystals in the form of whiskers. EDS analysis showed that the most common elements present were; Cl, Ca and Na (T. 55 Sheikh Abd El-Qurna, Luxor).

Photographs 48 A, B and C, show the structure of the salt formation from T. 14. A shows very long needles as well as tubes of salt. EDS analysis showed Cl, Ca and Na. It is possible to see on B a very smooth surface, detailed in the top of A and C, and the middle in B, indicating CI and Na (halite) (T. 14, the Valley of Kings, Luxor).

Photographs 49 A-C, showing the different character of salt crystallization. C shows a different face of the salt growth in the form of tubes. Analyses on different areas gave a full picture of the elements present such as S, Ca, P, K, Si and Cl (T. 2, Anfushi, Alex.).

Photographs 50 A-D, show the fragments of wall painting before and after cleaning, these fragments were painted in cream colour, all from the same room.

Photographs 51 A & B, shows the fragments before and after cleaning, showing the marks which appear differently from the original colour, as well as the micro-cracks on A.

Photograph 52 A-C, show the scheme of the decoration. The colour of painted layer mixed with green and Egyptian blue, and the marks left on C.

Photograph 53 shows the whitish layer of calcite film on top, another layer of calcite covering the red intonaco layer in a yellowish colour, and the cracks.

Photographs 54 A & B show the painted layers, which contained two layers as indicated. You can notice the decoration scheme, as well as the black specks on the surface, which caused damage to the painted layer.

Photographs 55 A & B showing on A, the surface of the fragment covered with different calcite film as you can see and on B, showing the deterioration and decay on the back of the fragment.

Photograph 56, showing the fragments after cleaning and lifting of fragment A as the first layer from bottom, painted in yellow colour with black and brown bands and B, the second painted in cream colour, notice the holes on the surface to prepare for another layer to key on.

Photographs Part I

P. 1 A ***Bar is 0.5 m***

P. *1 B* *X 1*

P. 1 C *X 1*

P. 3 A ***Bar is 1.5 m***

P. 3 B ***Bar is 0.5 m***

P. 4 A ***Bar is 1.40 m***

P. 4 B ***Bar is 0.5 m***

P. 5 A Bar is 0.5 m

P. 5 B Bar is 0.88 m

P. 6 A X 2

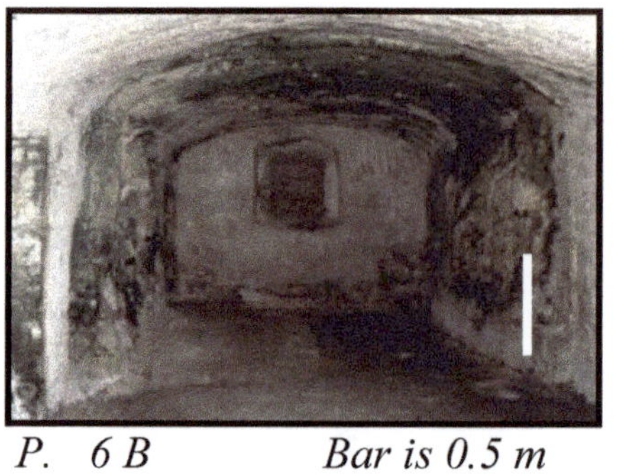

P. 6 B Bar is 0.5 m

P. 7 A (S.100) Scale 1-cm division

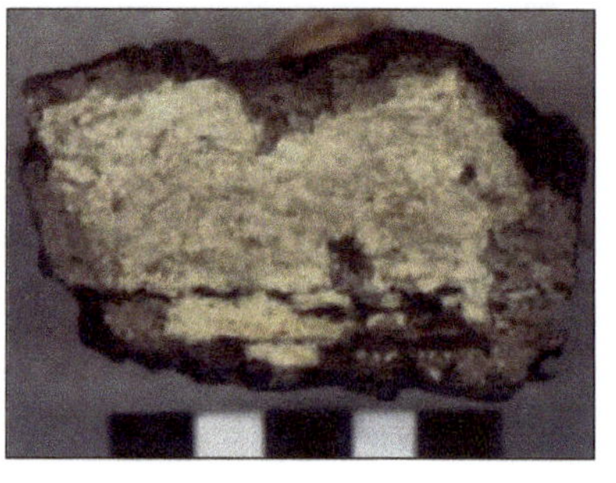

P. 7 B (S.102) Scale 1-cm division

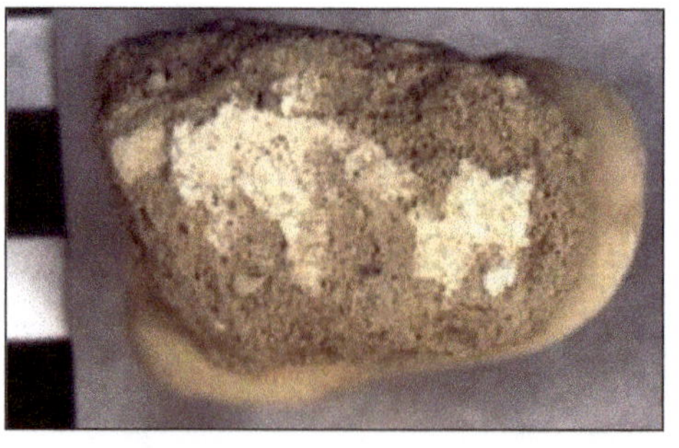

P. 7 C (S.143/1) Scale 1-cm division

P. 8 A (S.55/) Scale 1-cm divisions P. 8 B (one piece of S.55/3)

P. 9 (S.46) Scale 1-cm division P. 10 (S.5)

P. 11 A (T.56) X 2

P. 11 B X 2

P. 12 A (T.55) Bar is 1.20 m

P. 12 B X 7

P. 13 Scale 1-cm division

P. 14 A (T.13) X 1

P. 14 B (detail)

P. 15 A Scale 1-cm division P. 15 B

P. 16 A (S.84) Scale 1-cm division

P. 16 B (S.85) Scale 1-cm division

P. 16 C (S.18) Scale 1-cm division.

P. 17 A (S.4) ***P. 17 B*** ***P. 17 C***

Scale 1-cm division

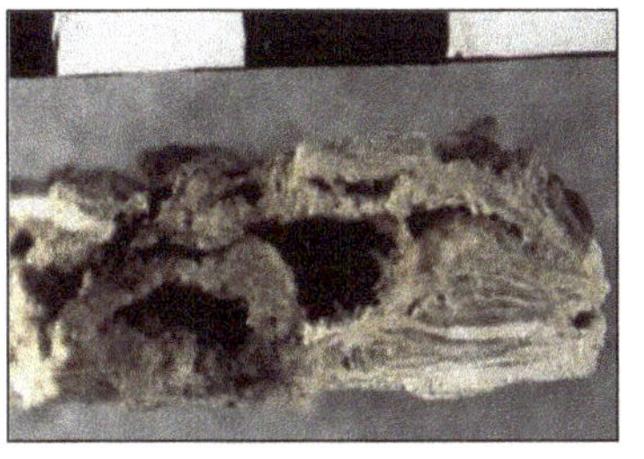

P. 18 A (S.1/1) *P. 18 B (S.1/2)* *P. 18 C (S.2)*

Scale 1-cm division

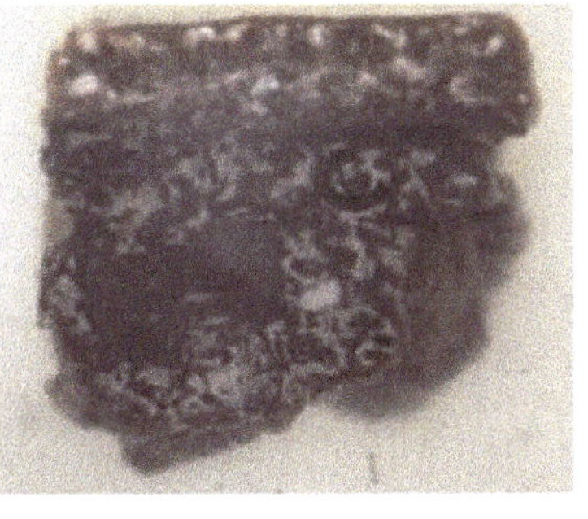

P. 19 A *(S.29) X* ***3*** ***P. 19 B*** *(S.32) X 3* ***P. 19 C*** *(S.119) X 2*

P. 19 D (S.40) X 6 *P. 19 E (S.18) X 2.5*

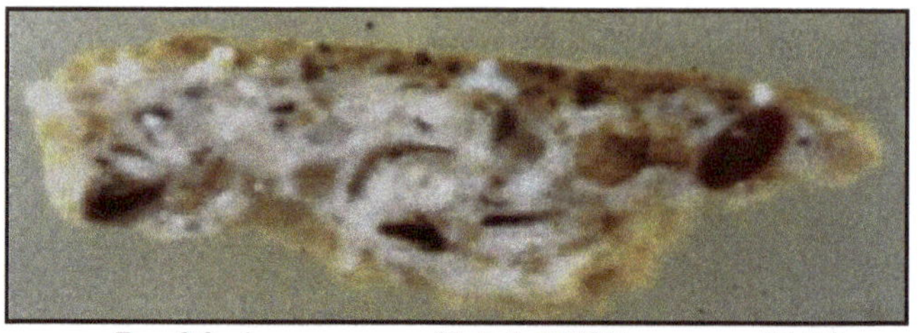

P. 20 A (S.2 MPT1) X 10

P. 20 B (S.5 MP T.1) X 4

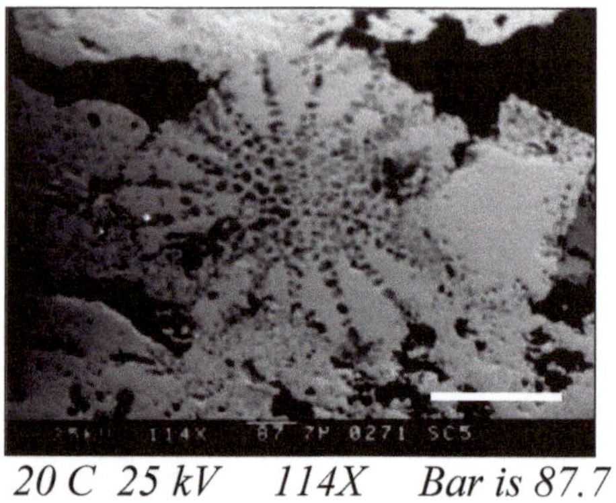

P. 20 C 25 kV 114X Bar is 87.7 µm

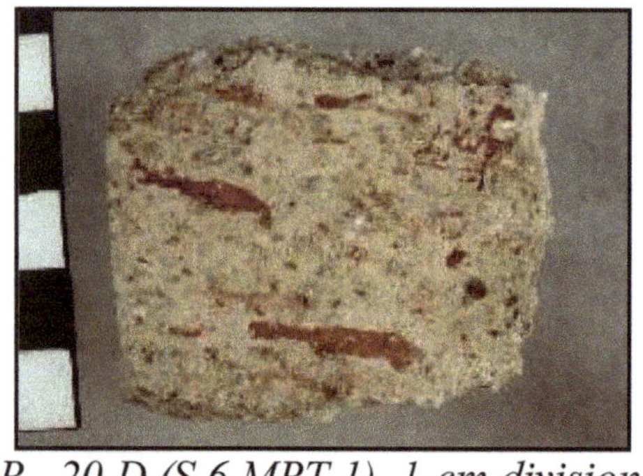

P. 20 D (S.6 MPT 1) 1-cm division

P. 20 E (S.13 MP2) X 5

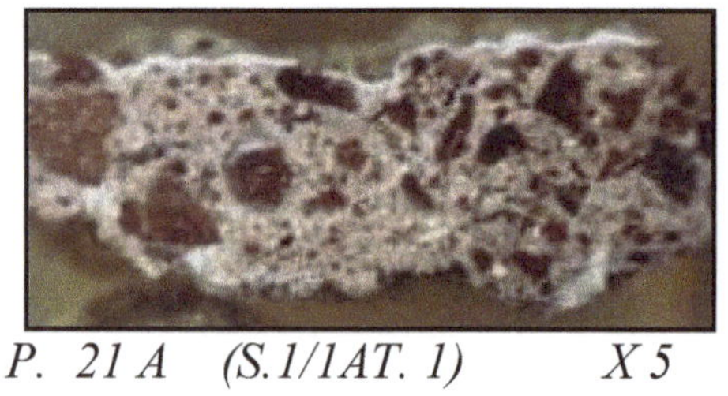

P. 21 A (S.1/1AT. 1) X 5

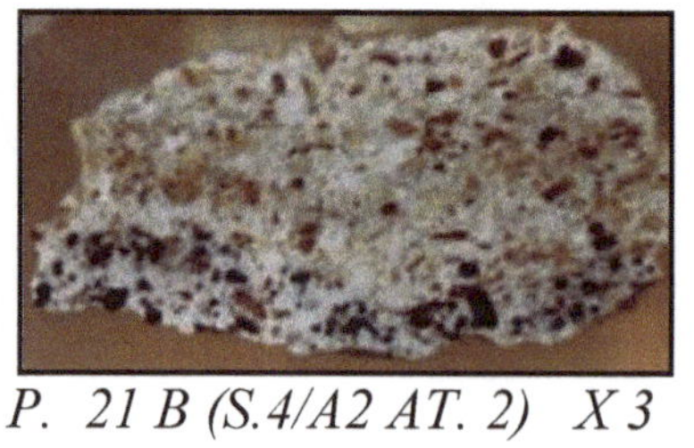

P. 21 B (S.4/A2 AT. 2) X 3

P. 21 C (S.4/A4 AT. 2) X 3

P. 21 D (15/3 AT. 2) X 3

P. 21 E (S.21 AT. 3) X 2

P. 22 A (S.1) X 3

P. 22 B (S.9) X 4

P. 22 C (S.10) X 3

P. 22 D (S.11) X 2

P. 23 A 25 KV X 35.3 Bar is 283 μm

P. 23B 25 KV X 35.3 Bar is 283 μm

P. 24 A (S.100) Scale 1-cm division

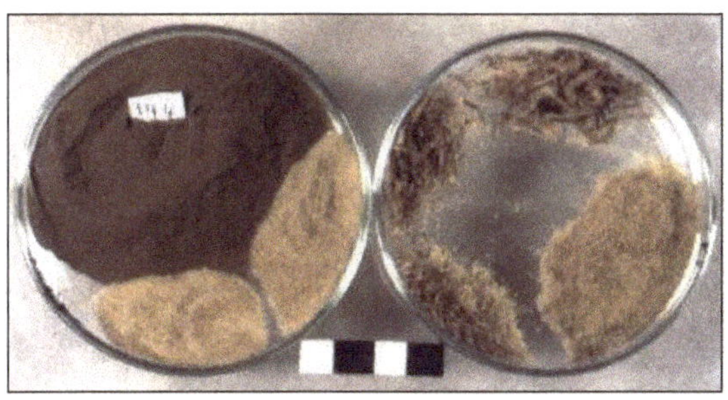

P. 24 B (S.144) Scale 1-cm division

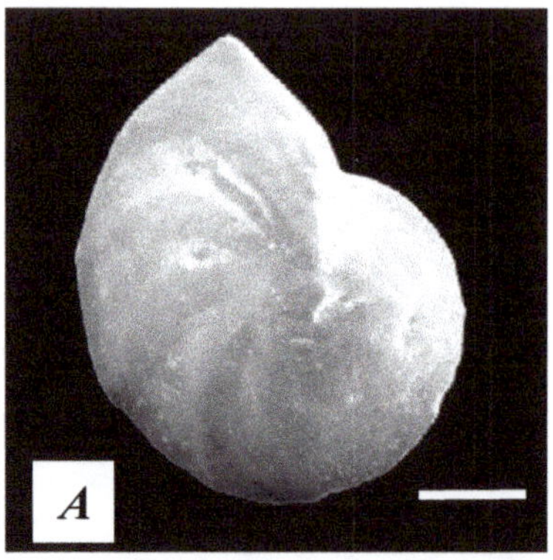

P. 25A ***20 KV X 60.0 Bar is 500 µm***

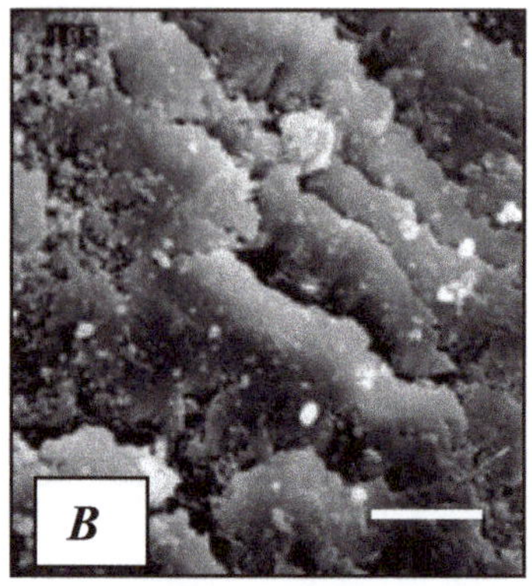

P. 25B 20 KV X 60.0 Bar is 500 µm

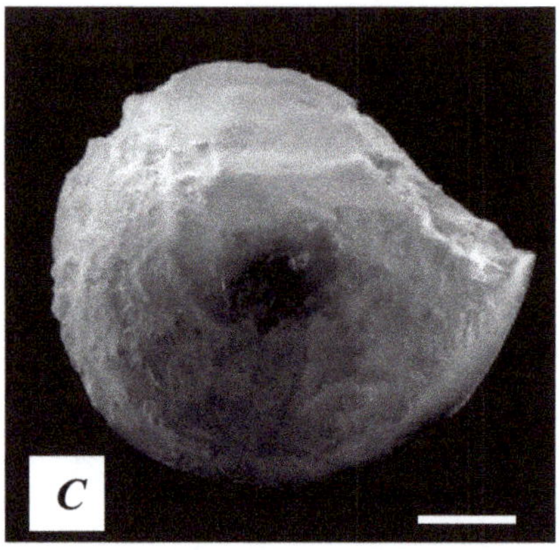

P. 25C 20 KV X 49.0 Bar is 610 µm

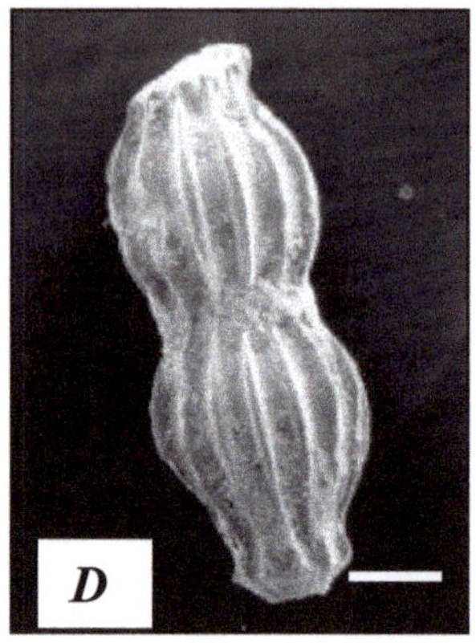

P. 25D 20 KV X 2.20 Bar is 13.6 µm

P. 26 A (S.137) Scale 1-cm division

P. 26 B (S.139) Scale 1-cm division

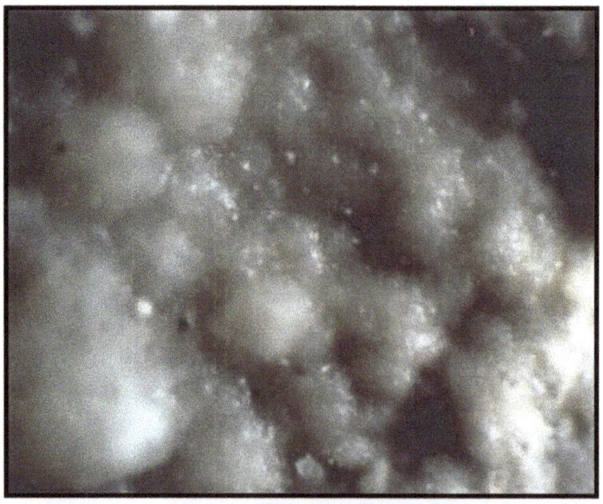

P. 27 (S.3) (Magnification 400 X)

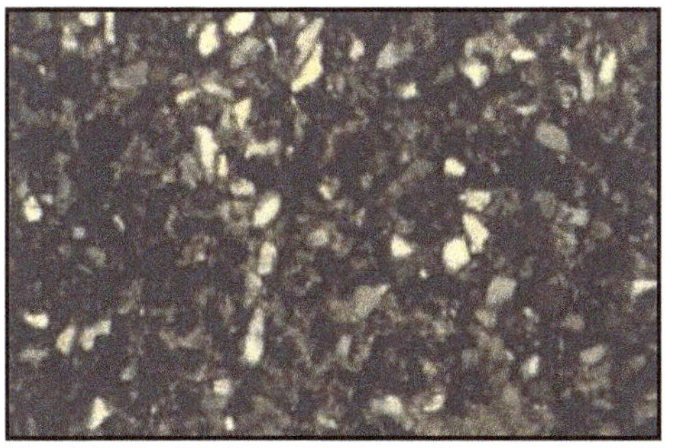

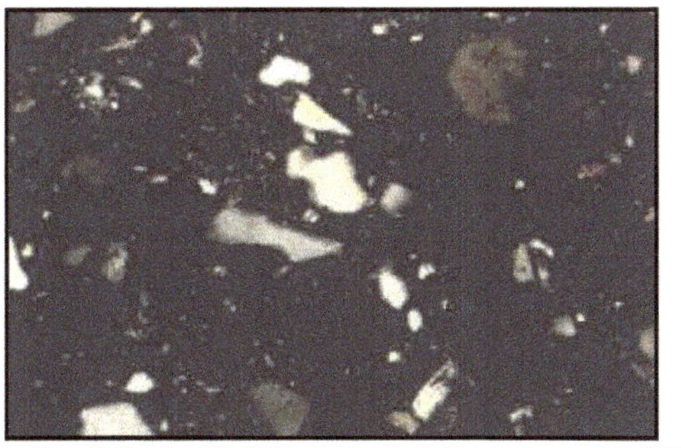

P. 28 A (S.54) Magnification 40 X ***P. 28 B (S.51) Magnification 40 X***

P. 29 (S.27/3) Magnification 40 X

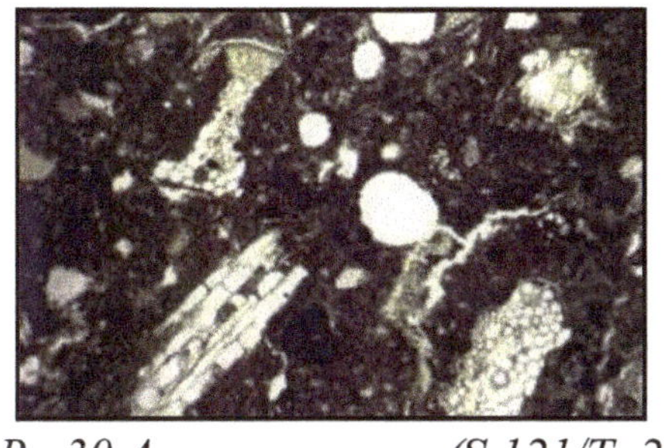

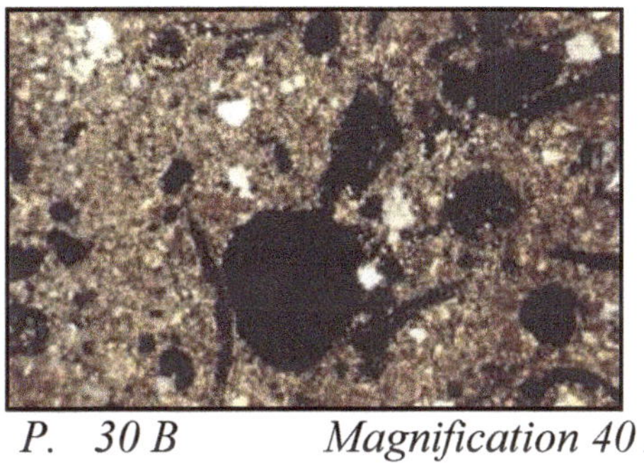

P. 30 A (S.121/T. 255) P. 30 B Magnification 40 X

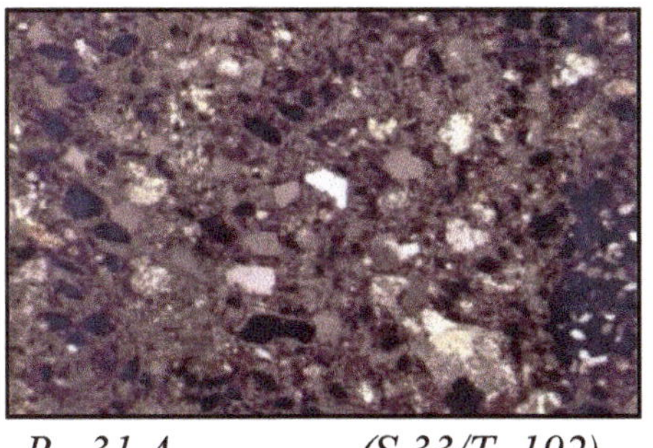

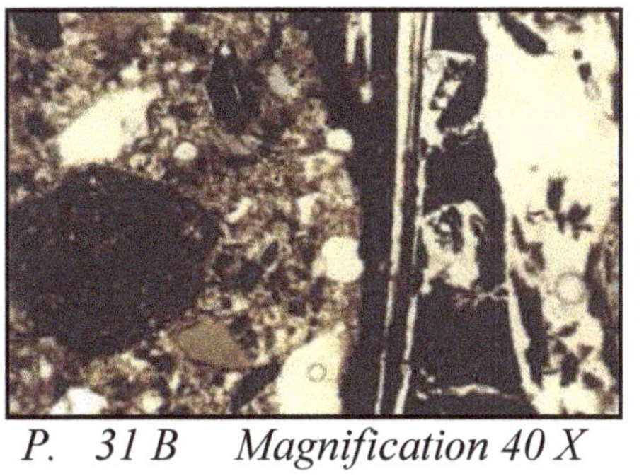

P. 31 A (S.33/T. 192) P. 31 B Magnification 40 X

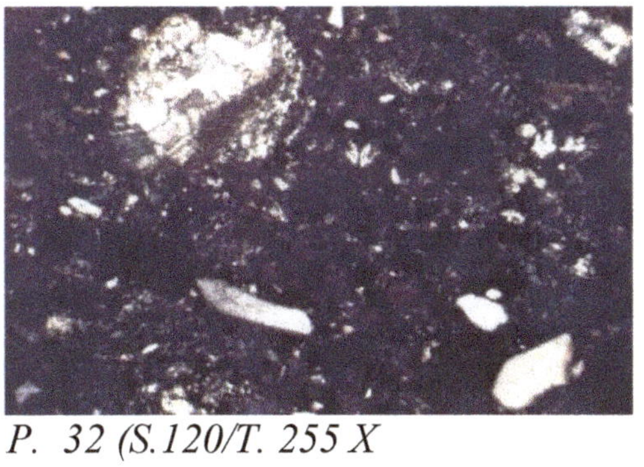

P. 32 (S.120/T. 255 X

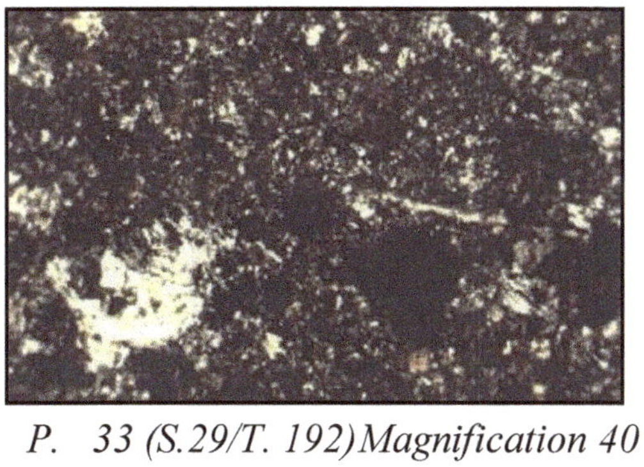

P. 33 (S.29/T. 192) Magnification 40 X

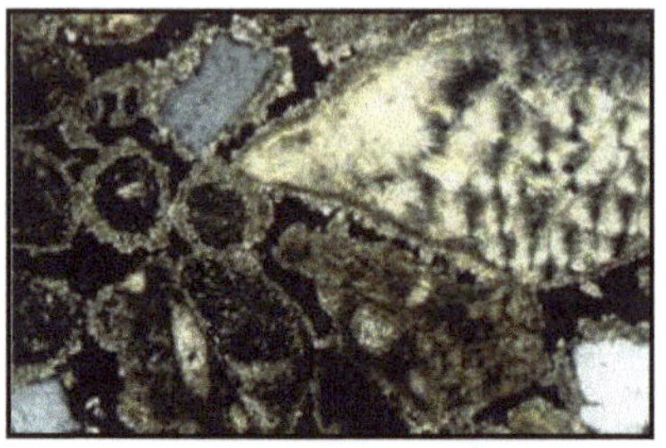
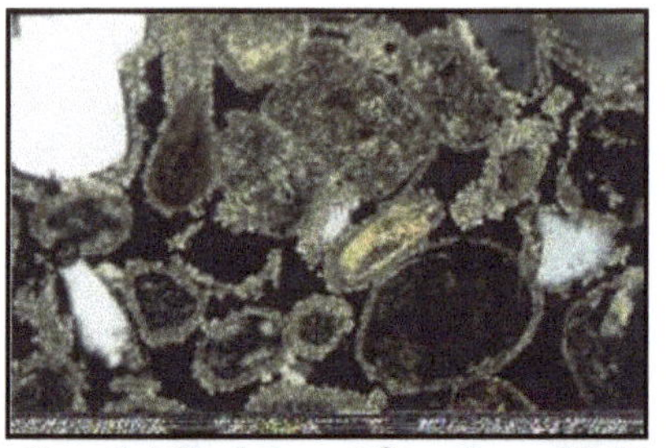

P. 34 A (S.14) *P. 34 B Magnification 40 X*

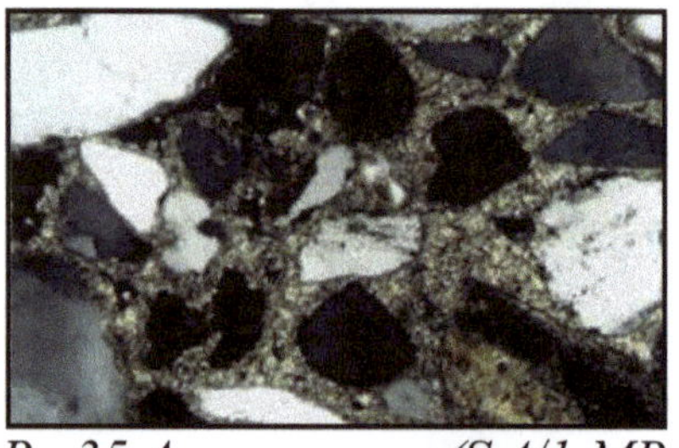

P. 35 A (S.4/1 MP T. 1) *P. 35 B Magnification 40 X*

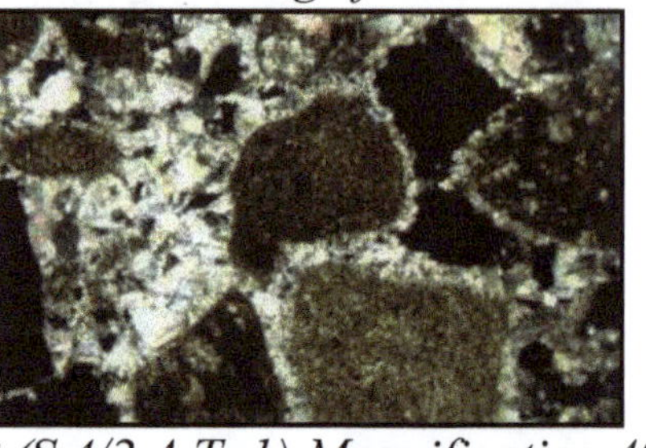

P. 36 A (S.4/1 A T .1) *P. 36 B (S.4/2 A T. 1) Magnification 40 X*

P. 37 A (S.19 A T. 3) *P. 37 B Magnification 40 X*

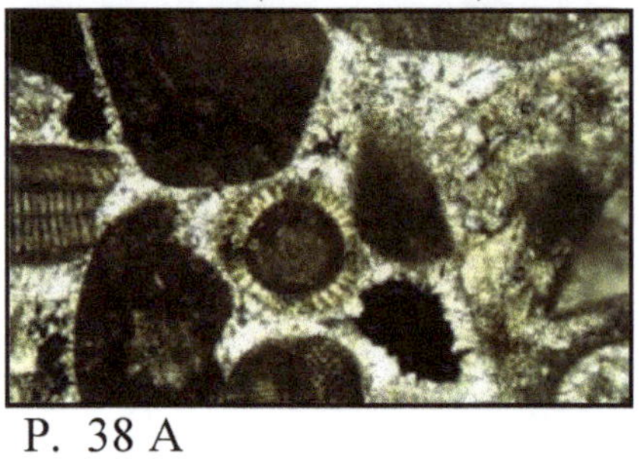
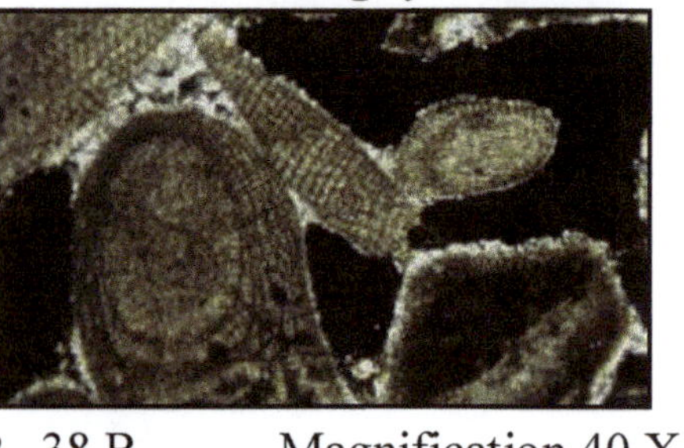

P. 38 A P. 38 B Magnification 40 X

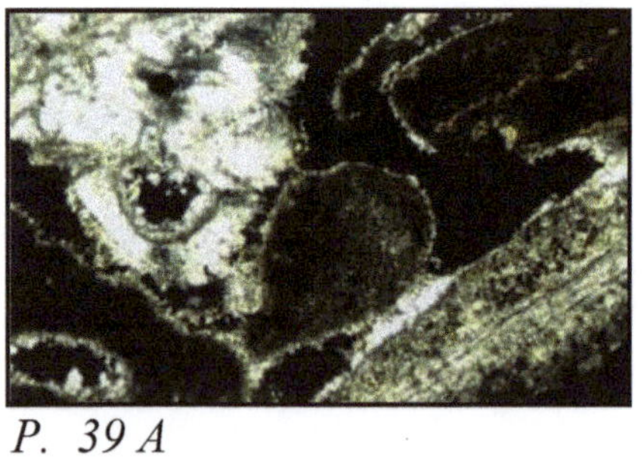
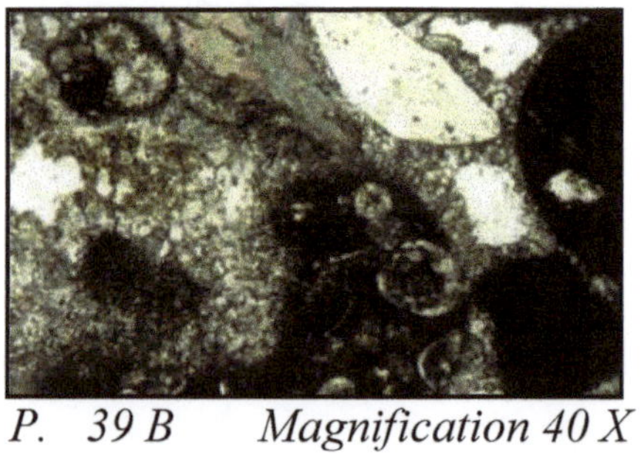

P. 39 A *P. 39 B Magnification 40 X*

P. 40 A (S.9)

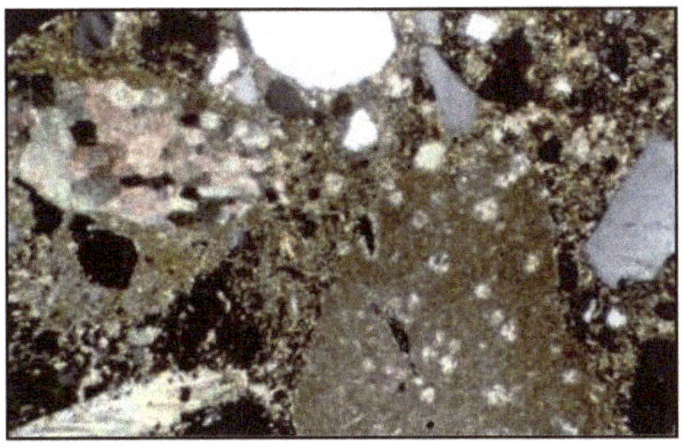

P. 40 B Magnification 40 X.

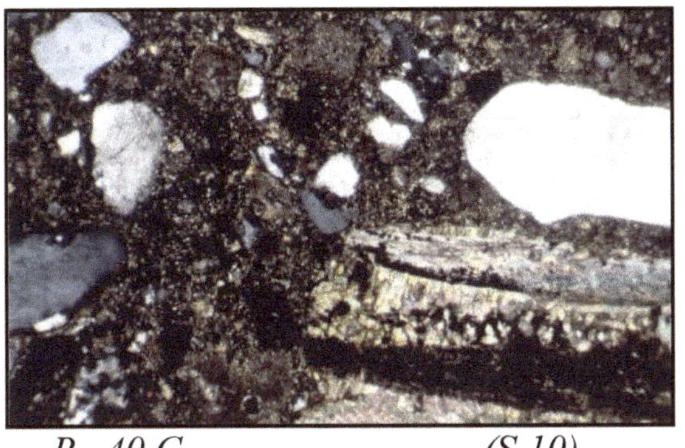

P. 40 C (S.10)

P. 40 D Magnification 40 X.

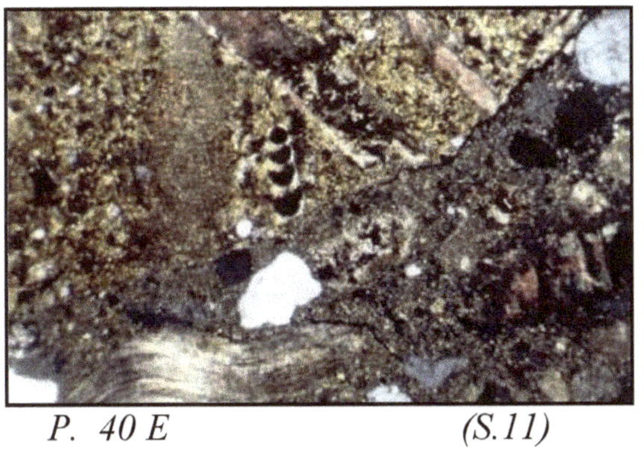

P. 40 E (S.11)

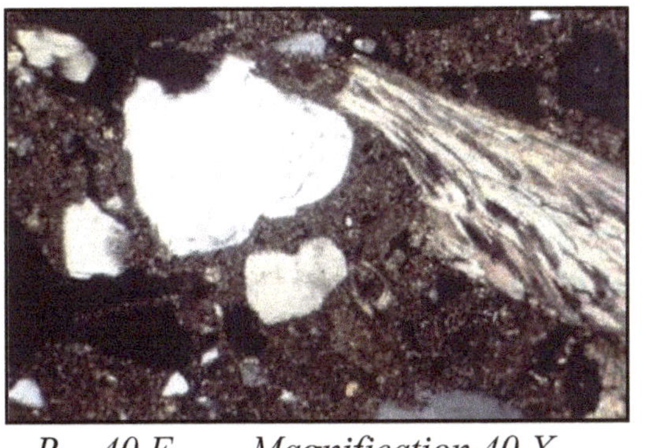

P. 40 F Magnification 40 X

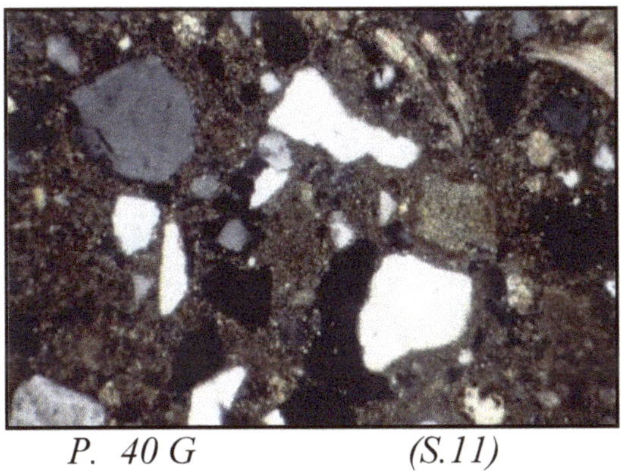

P. 40 G (S.11)

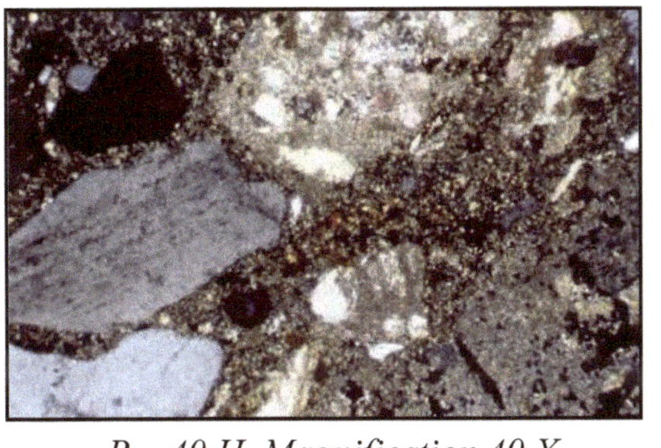

P. 40 H Magnification 40 X

15 KV X 208 Bar is 48.1 µm

15 KV 1.38 KX Bar is 7.25 µm

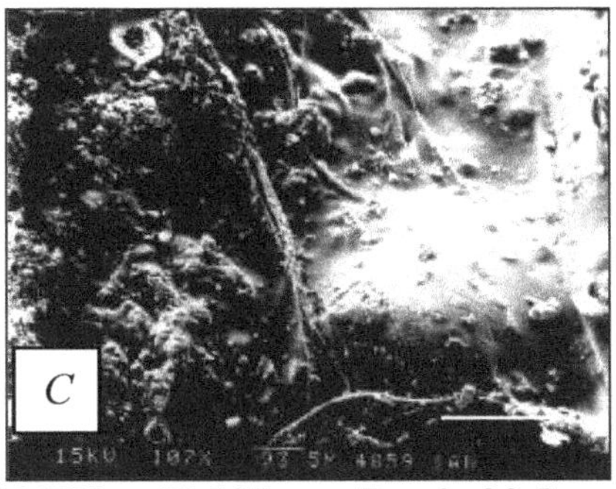

15 KV X 107 Bar is 93.5 µm

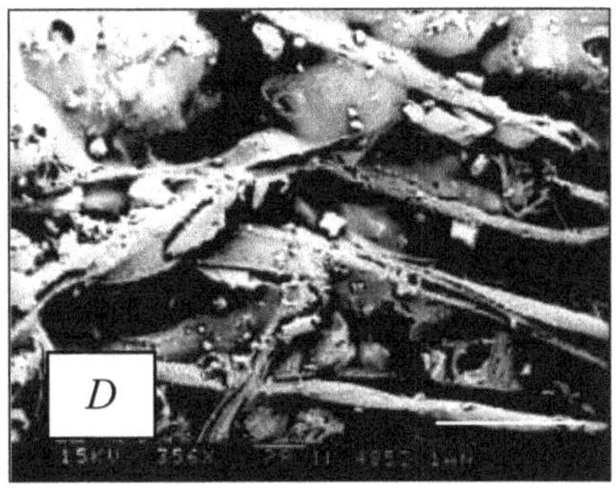

15 KV X 356 Bar is 28.1 µm

P. 44 A-D

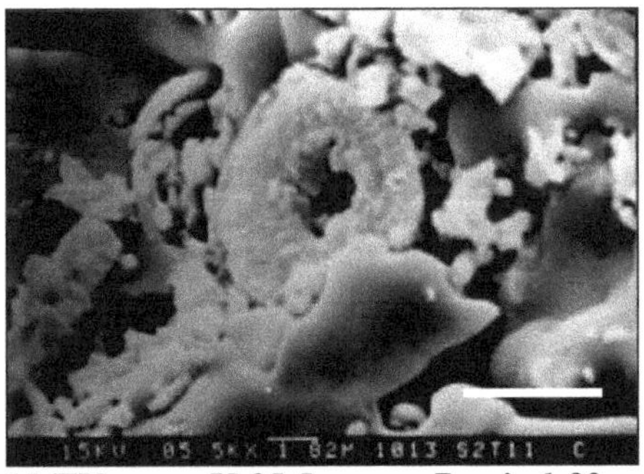

15 KV X 05.5 Bar is 1.82 µm

P. 45 (S.2/1)

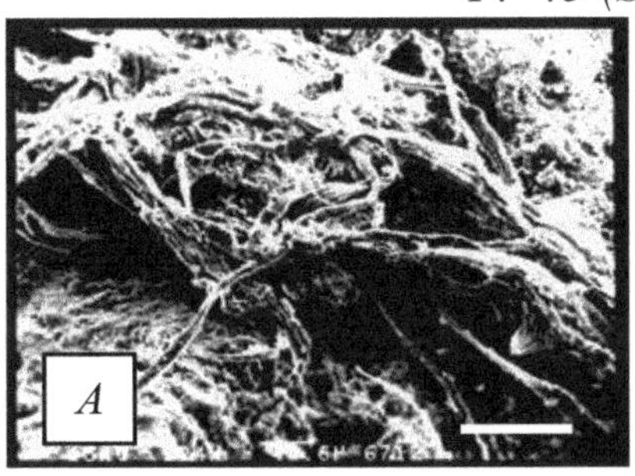

15 kV X 224 Bar is 44.6 µm

15 kV X 075 Bar is 133 µm

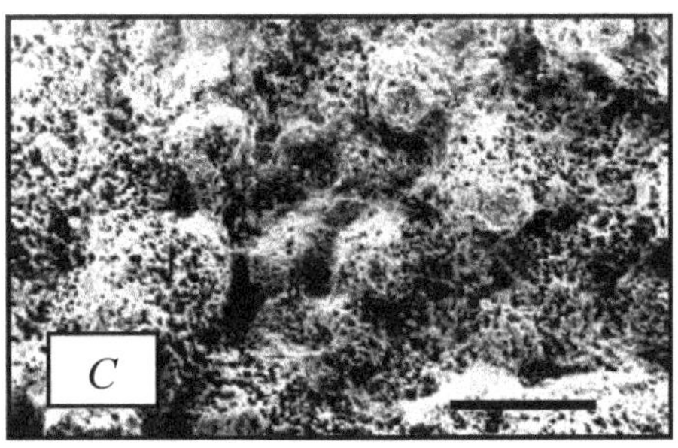

15 kV X 21.9 Bar is 457 µm

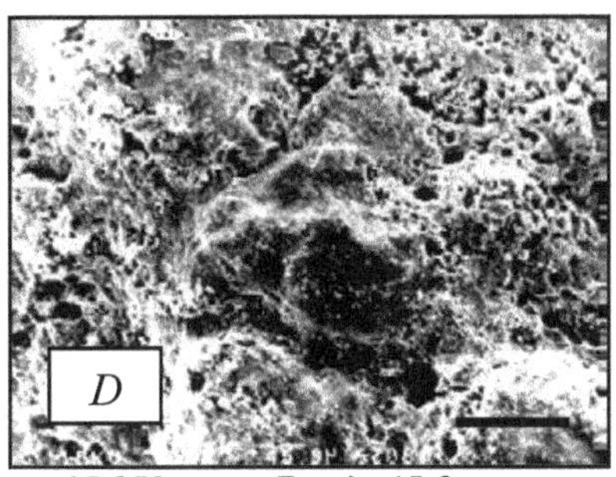

15 kV Bar is 45.8 µm

P. 46 A-D

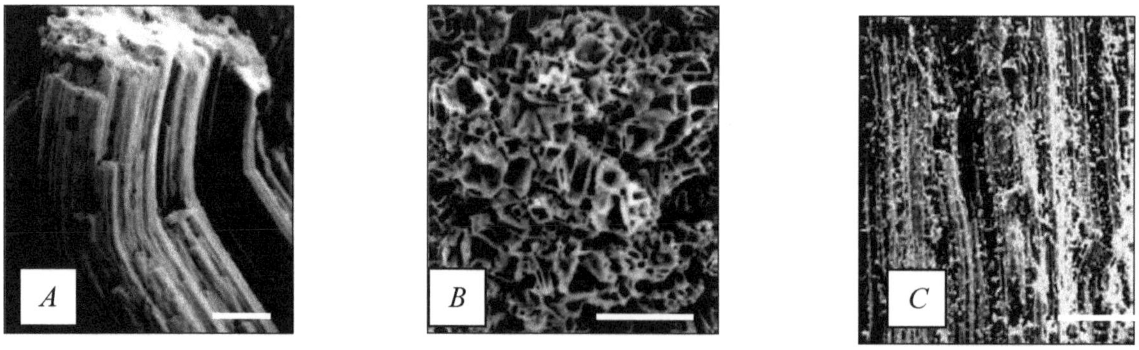

20 KV X 25.0 Bar is 1200 µm ***20 kV X 500 Bar is 60 µm*** ***20 kV X 199 Bar is 158 µm***

P. 47 A-C (S.110/1)

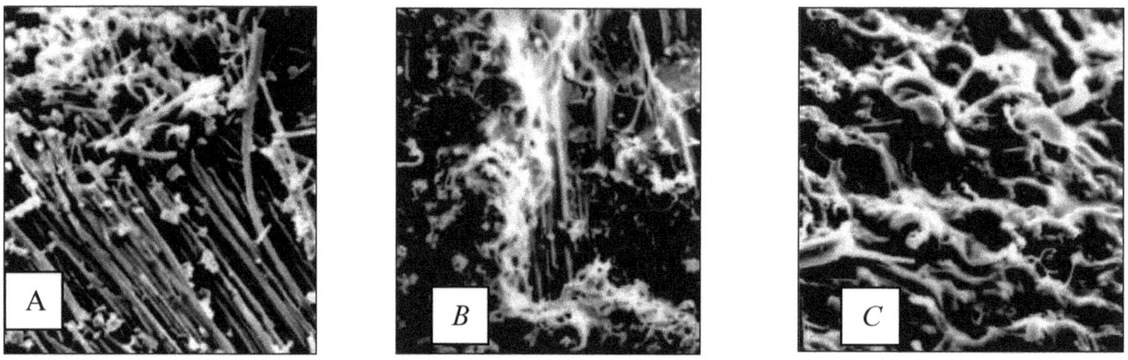

20 KV X 400 Bar is 75 µm ***20 kV X 350 Bar is 86 µm*** ***20 kV X 1.20 Bar is 25.0 µm***

P. 48 A-C (S.88)

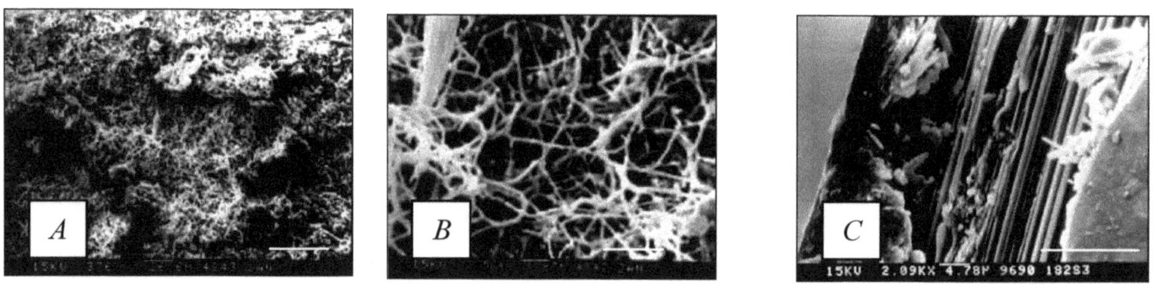

15 KV X 376 Bar is 26.6 µm ***15 KV 2.52 KX Bar is 3.97 µm*** ***15KV 2.09 KX Bar is 4.78 µm***

P. 49 A-C

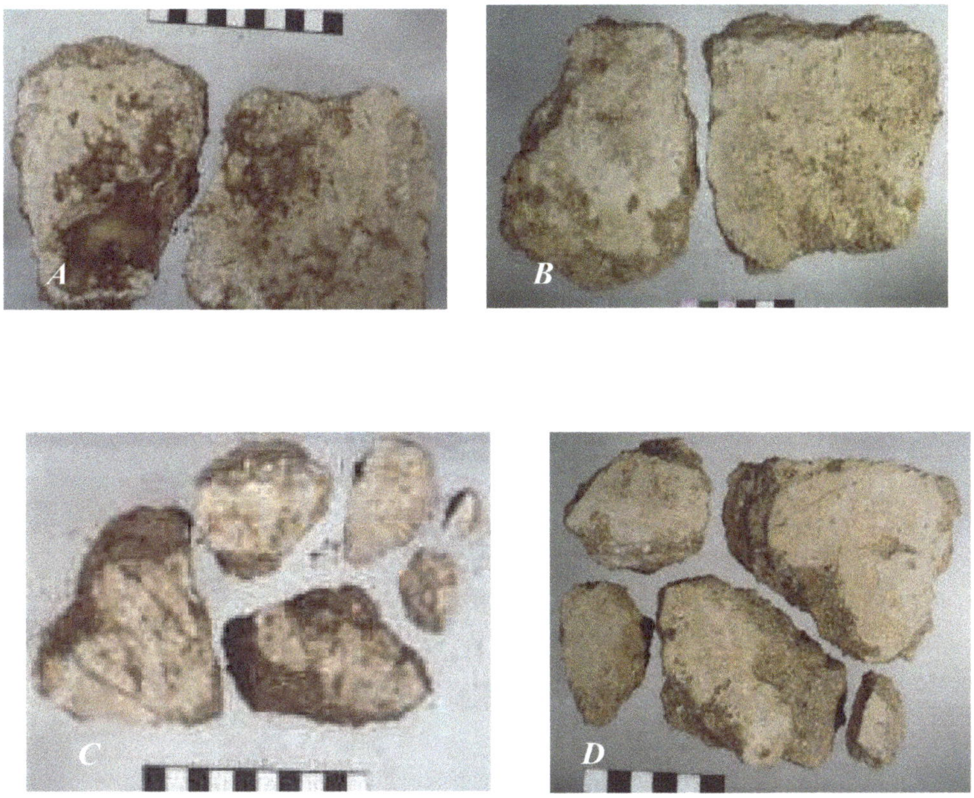

P. 50 A-D (S.1 top & S.3) Scale 1-cm division

P. 51 A *(S.4)* ***P. 51 B*** *Scale 1-cm division*

P. 52 A (S.5) P. 52 B P. 52 C (S.6) Scale 1-cm division

P. 53 (S.7) Scale 1-cm division.

P. 54 A (S.9)

P. 54 B Scale 1-cm division

P. 55 (S.10) X 0.5 Scale 1-cm division

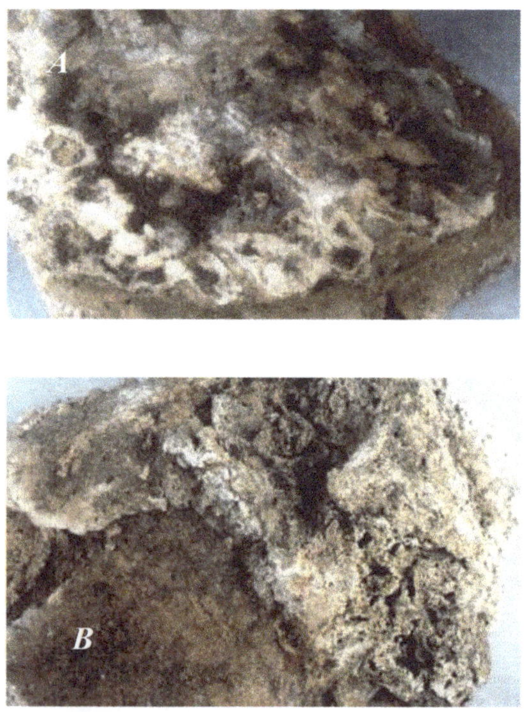

P. 56 ***Scale 1-cm division***

Photographs Part II

Captions

Photograph 57, Hieraknoplis wall painting, which was executed on a surface of mud plaster and painted in red, green, yellow, black and white. Pre-dynastic about 3300 BC. Cairo Museum.

Photograph 58, showing about 16 squares were used for the drawing of a portrait which was painted on a piece of limestone (Ostraka) 18th dynasty New Kingdom. (Abd El Salam 1992).

Photograph 59, shows the grid lines used, in red colour on a layer of the plaster, which covered the stone, in the tomb of Akhhotap I at Mair near Asyut. (Abd El Salam 1992).

Photograph 60, shows the correction of the drawing made up in red colour before painting from the tomb of Horemheb 18th dynasty Thebes. (Abd El Salam 1992).

Photographs 61 A & B, one underground below the southern building of the step pyramid (Mastaba of King Zoser). A shows one of the series of a low relief panel carvings with a combination of faience. (Abd El Salam 1992). B, shows the decoration style of one of the panels surmounted by Djed pillars supporting an arch reconstructed from elements found in subterranean chambers in the step pyramids. A room in the underground of Zoser's tomb was decorated in the same decoration, which consisted of several panels. You can see that most of the faience are greenish comparison with A, due to the manufacture of the faience. 3rd dynasty, the Old Kingdom (B in Cairo Museum).

Photograph 62 from the tomb of Hasy-Re, shows the low relief has been executed on wood panel. The King is seated at an offering table and carries his staves of office; third dynasty Old Kingdom. Cairo Museum .

Photograph 63, shows a part of the wall painting from the tomb of Atet, 4th dynasty Madium the Old Kingdom. (Abd El Salam 1992).

Photograph 64, shows a part of the wall painting from the tomb of Atet, which was executed using the inlay process; 4th dynasty of the Old Kingdom. (Abd El Salam 1992).

Photograph 65, showing the crossbeams in the large areas of stone which are then divided into small squares; tomb of Nefer-Maat, 4th dynasty Old Kingdom. Cairo Museum.

Photographs 66 A & B showing the inlay process on the remains of the decoration of the palace of Ramasses III, representing some prisoners; Medient Habo 20th dynasty New Kingdom. Cairo Museum.

Photograph 67, showing the two different decoration styles as well as the plasters. The first was in vertical stripes, and the second was the incrustation style with imitation of marble on the secolo right side of room 1. (T. 2, Anfushi, Alexandria).

Photographs 68 A-D, show a sample of gypsum plaster (S.55/3). A shows the surface. Note some spots on the surface and a yellowish crust. B shows a detail of A. D is a higher magnification of C showing the structure, which is mainly gypsum (Dahshur, 4th dynasty N E).

Photographs 69 A & B, showing a deposit of gypsum, which was found inside the plaster/mortar. EDS indicated the presence of S, Ca and Si.

Photographs 70 A, B and C, shows clay plaster/mortar (S.134/2), with a crack on the surface, which has an inhomogeneous structure as a result of mixture. EDS indicates the highest elements present; Al, Si, Fe, Ca. It was noticed that some areas contained S, Cl and K. C shows a detail of B, with the main elements; Al, Si, K, Ca and S; another area shows the presence of S, Si and Ca (Ka-Nefr tomb 5th dynasty N E).

Photographs 71 A & B, show a limestone sample with cracks inside (A) noted on different samples of stone as well as plaster. Analysis showed very high peaks of Ca in addition to the other elements of Al, Si and Fe. Different microfossils as well as shells and coccoliths were the most common fossils found in the stone (T. 56, Sheikh Abd El-Qurna, and Luxor).

Photographs 72 A, B and C, showing samples of limestone from T. 359, with a very interesting structure, as the stone was covered with a brownish/reddish layer of iron oxide, confirmed by spot tests. Low magnification area analysis of A showed; Si, Al, K and Fe and on B the highest peaks were Fe, Al and Si. C, shows a detail of one of the rounded shapes in B, indicated by EDS to be Fe. Crystals of Ca were also visible in some areas (T. 359, Deir El-Medina, Luxor).

Photographs 73 A & B, show samples of mud plaster covered with a layer of clay plaster. In A, note the cracks on the surface. EDS analysis showed the elements present were; Ca, S, Si, Al and Fe. B shows the surface of the clay covered with a dust layer. Different areas show the presence of K (T. 414, Asasif, Luxor).

Photographs 74 A & B, show the surface structure of finishing plaster which looks fine in A. High magnification shows the different mixture of the grain sizes. EDS analysis indicated S and Ca. Analysis of different areas has shown that the most common elements were; S, Ca and Si, in addition to Cl which was visible as well (T. 55, Sheikh Abd El-Qurna, Luxor).

Photographs 75 A- C, show a sample of salt crystals that has grown on the wall. A shows a very smooth structure of the salt growth, which was shown by EDS analysis to be sodium chloride. Notice the different form of the crystals found, which grew as tubes, analysis showing Al and Si. (T. 56, Sheikh Abd El-Qurna, Luxor).

Photographs 76 A-C, show that different crystals have grown underneath the plaster, as seen on right of A, which grew in the forms of needles. B shows disintegration of the surface caused by salt and C, shows a detail of B. The smooth areas of salt, identified as Cl can be clearly seen. EDS analysis showed that the most elements present were Cl and S, but the highest peak was Cl. (T. 55, Sheikh Abd El-Qurna, Luxor).

Photographs 77 A & B, show the cracks between the plaster. On A, Ca and S are the main elements are present. B, is a detail showing the effect of the salt (Cl) (Anfushi, T. 1).

Photographs 78 A- D. A shows the surface structure deterioration and decay, B, fungal growth and holes, C shows the fungi and Na Cl, which covered the surface, D is a detail of C. Analysis indicated the presence of S, Ca, P and Cl (T. 2, Anfushi, Alexandria).

Photographs Part II

P. 57

P. 58

P. 59

P. 60

P. 61 A

P. 61 B

P. 62

P. 63

P. 64

P. 65

P. 66 A

P. 66 B

P. 67

15 KV X 056 Bar is 1.79 µm

15 KV 0.270 KX Bar is 37.0 µm

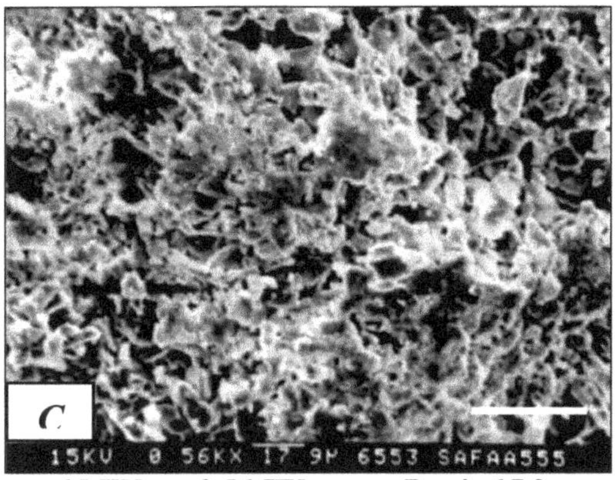

15 KV 0.56 KX Bar is 17.9 µm

15 KV 01.80 KX Bar is 5.56 µm

P. 68 A-D (S.55/5)

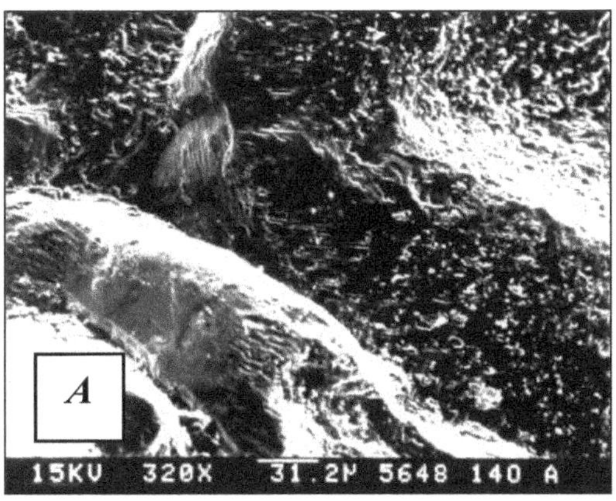

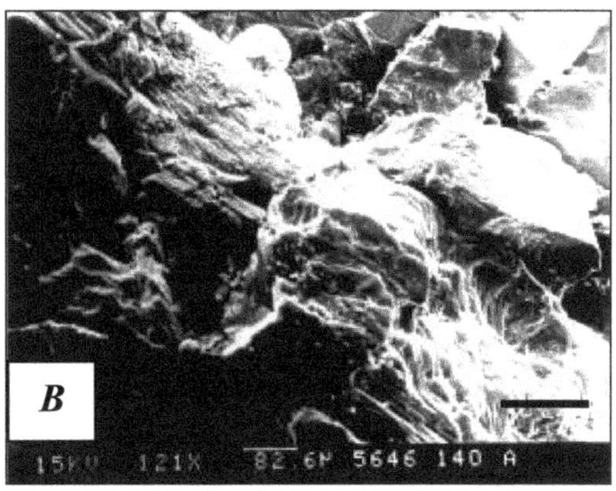

15 KV X 121 Bar is 82.6 µm

P. 69 A - B (S.140)

15 KV X 29.6 Bar is 350 µm

15 KV 1.09 KX Bar is 5.17 µm

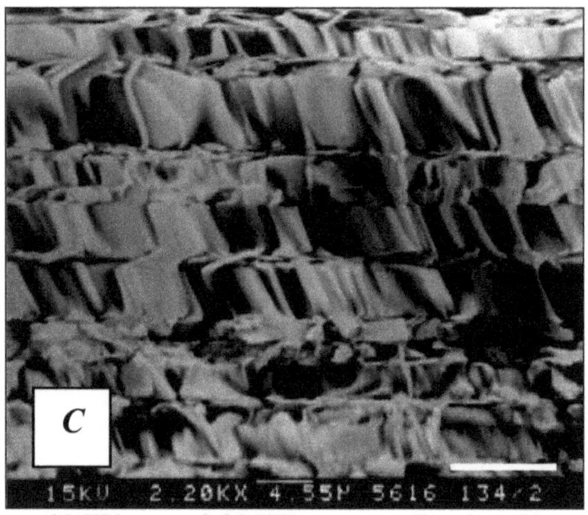

15 KV 2.20 KX Bar is 4.55 µm

P. 70 A- C (S.134/2)

20 kV X 600 Bar is 50 µm ***20 kV X 8.000 k Bar is 3.8 µm***

P. 71 A - B (S.107)

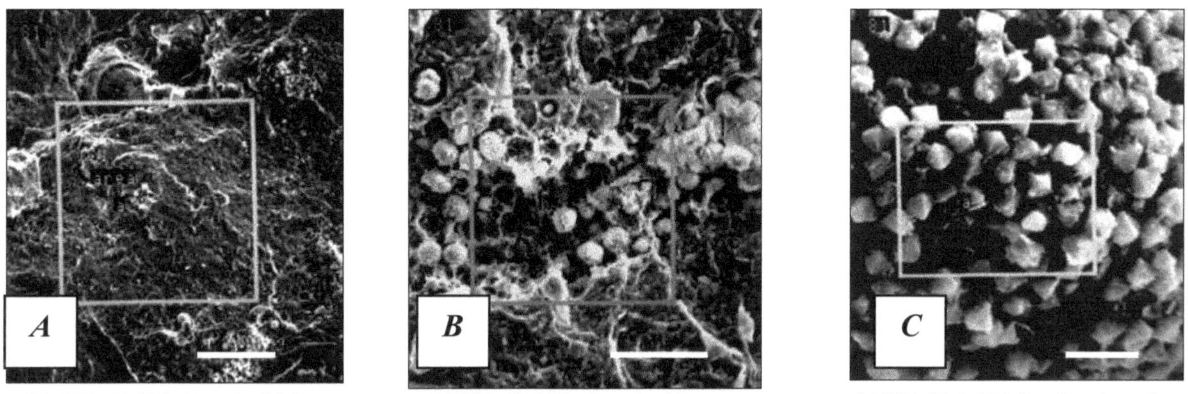

20 KV X 152 Bar is 200 µm ***20KV X 500 Bar is 60 µm*** ***20KV X 4.000 k Bar is 7.5 µm***

P. 72 A- C (S.8)

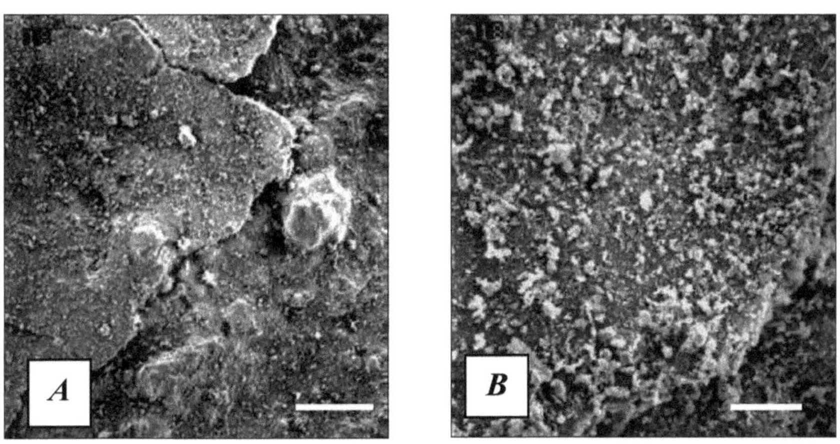

20 KV X 40.0 Bar is 750 µm ***20 KV X 300 Bar is 100 µm***

P. 73 A - B (S.18)

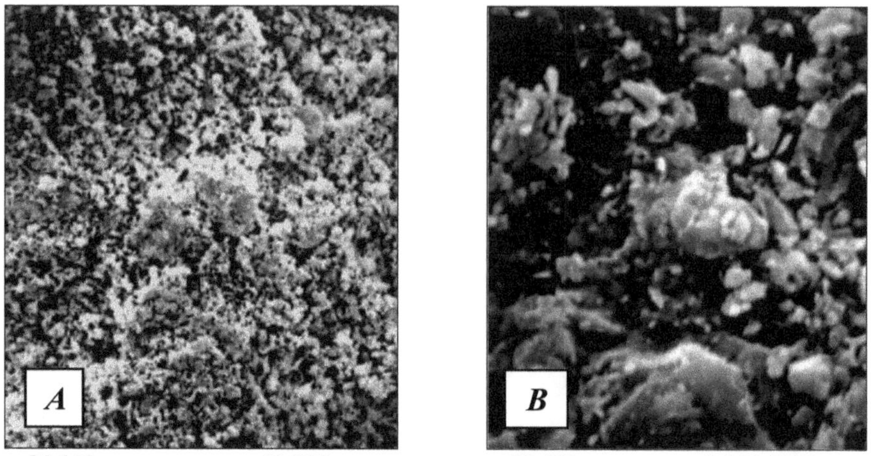

20 kV X 710 ***20 kV 5.000 K X***

P. 74 A - B (S.76)

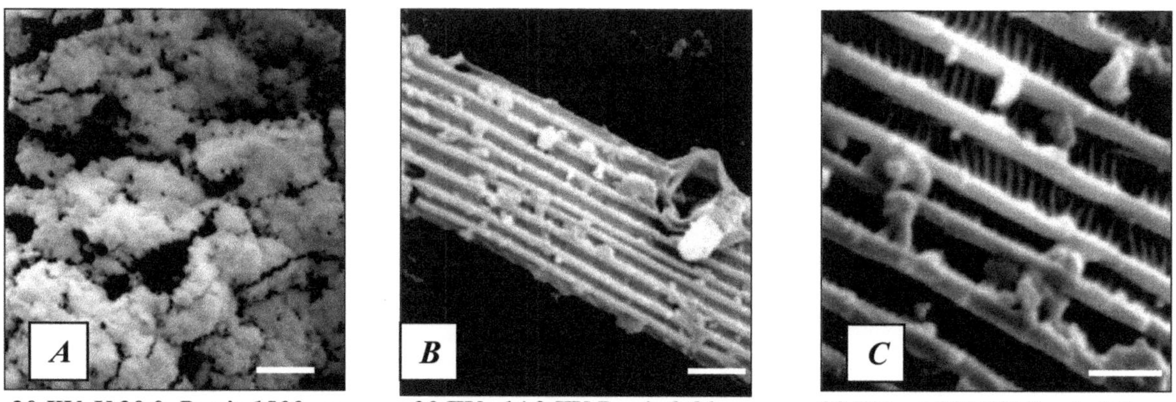

20 KV X 20.0 Bar is 1500 µm ***20 KV 14.9 KX Bar is 2.01 µm*** ***20 KV 4.000 KX Bar is 7.5 µm***

P. 75 A- C (S.67)

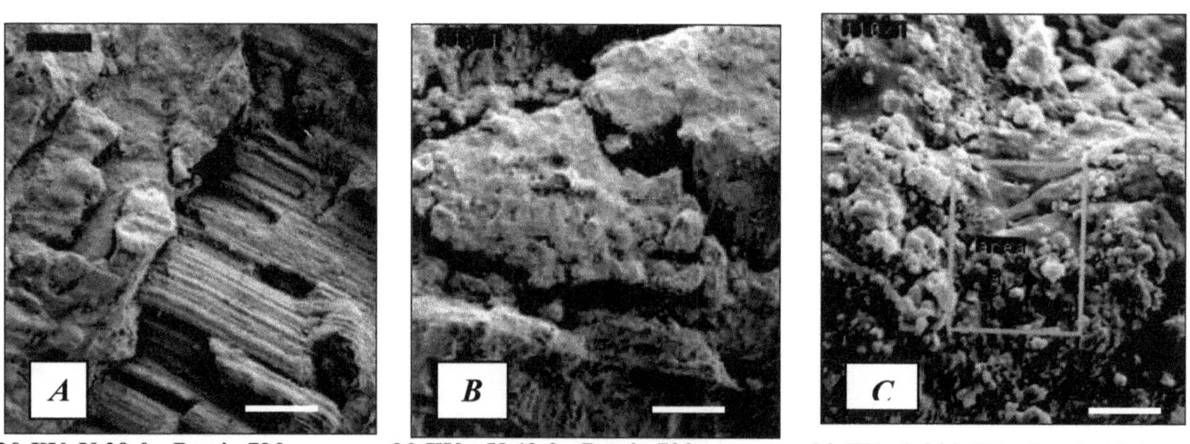

20 KV X 38.0 Bar is 790 µm ***20 KV X 43.0 Bar is 700 µm*** ***20 KV 3.000 KX Bar is 10.0 µm***

P. 76 A-C (110/1)

15 KV X 43.4 Bar is 230 µm

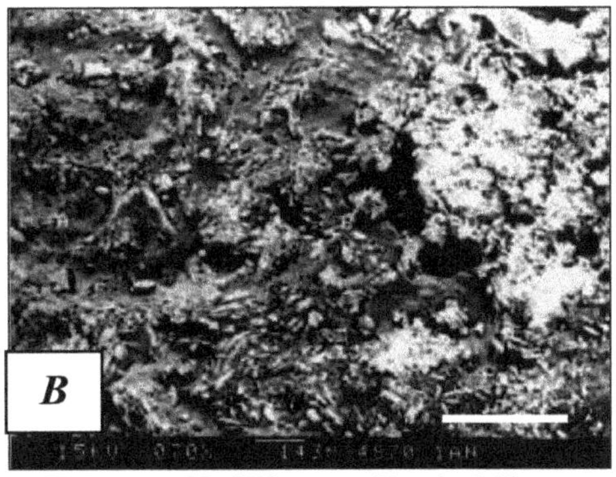

15 KV X 070 Bar is 143 µm

P. 77 A -B (S.1 A)

15 KV X 09.6 Bar is 1.04 µm

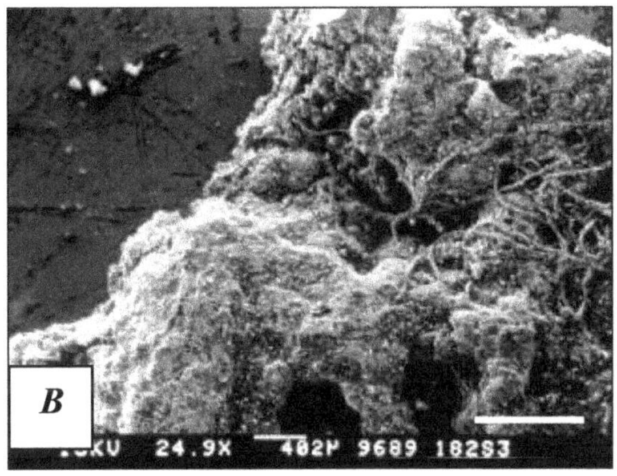

15 KV X 24.9 Bar is 402 µm

15 KV X 093 Bar is 100 µm

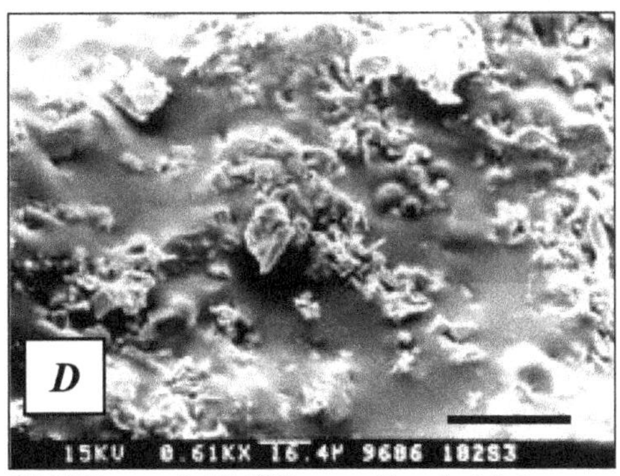

15 KV 0.51 KX Bar is 16.4 µm

P. 78 A- D (S.18)

Bibliography

Abd El Salam S. 1992. Mural Design in the Ancient Egyptian Architecture, 3100-1085 BC, Alexandria University, Faculty of Fine Arts, M.Sc. Thesis.

Abd El Salam S. 1999a. Analytical techniques used in the examination of materials of wall painting from Alexandrian sites. In *3rd Annual Symposium on Mediterranean Archaeology* (SOMA), Abstracts, Faculty of Art, University of Birmingham.

Abd El Salam S. 1999b. Analytical techniques used in the examination of materials of wall paintings from the Ancient sites of Mustafa Pasha and Anfushi in Alexandria. In C. Parisi, S. Galiardi, G. M. Parisi and G. Torcinaro (eds), *6th international conference on Non-Destructive Testing and Microanalysis for the Diagnostic and Conservation of the Cultural and Environmental Heritage,* Vol. 1: 449-46, Rome: I.C.R. and AIPnD.

Abd El Salam S. 2000. Scientific examination, analysis and cleaning of wall painting fragments from the 4th century AD Roman villa at Piddington, Northamptonshire U.K. In *26th Annual Conference CAC-ACCR. Conference Abstracts:* 33, Ottawa, Canada: Canadian Association for Conservation of Cultural Property.

Adriani A. 1936. *La Nécropole De Mustafa Pasha, Egypt*, Alexandria: Whiteread Morris Limited.

Allag C. 1982. Roman wall painting: Technique of restoration and mounting (Translated by Roger Ling). In Liversidge J. (ed), *Roman Provincial Wall Painting of the Western Empire,* B A R International Series 140*:* 85-89, Oxford: B A R.

Arnold A. and Zehnder K. 1987. Monitoring wall painting affected by soluble salts. In S. Cather (ed.), *The Conservation of Wall Paintings:* 103-135, London: Proceedings of a Symposium at the Courtauld Institute of Art and Getty Conservation Institute.

Arnold A. 1984. Determination of minerals salts from monuments: *Studies in Conservation*, Vol. 29, No. 3: 129-138.

Arnold A. 1998a. Determination processes: Salt weathering, Handout, unit 4: 1-21, Rome: ICCROM.

Arnold A. 1998b. Methodology of the study on decay weathering and conservation of monuments, Handout, unit 4: 1-9, Rome: ICCROM.

Baines J. 1989. Techniques of decoration in the hall of Barques in the temple of Sethos 1 at Abydos, *Journal of Egyptian Archaeology*, Vol. 75: 13-30.

Baines J. 1985. Color terminology and color classification Ancient Egyptian color terminology and polychromy, *American Anthropologist*, No. 87: 282-297.

Baines J. and Malek J. 1980. *Atlas of Ancient Egypt*, Oxford: Phaidon.

Barbet A. 1982. The diffusion of the third Pompian style in Gaul (Translated by Ann Gruaz). In Liversidge J. (ed), *Roman Provincial Wall Painting of the Western Empire,* B A R International Series 140*:* 75-84, Oxford: B A R.

Bayliss P., Erd D., Mrose M., Sabina A. and Smith D. 1986. *JCPDS, Mineral Powder Diffraction File: Data Book*, U.S.A: International Centre for Diffraction Data.

Béarat H., Fuchs M., Maggetti M. and Paunier D. (eds) 1997. *Roman Wall Painting: Materials, Techniques, Analysis and Conservation*, Proceedings of the International Workshop, Fribourg: Institute of Mineralogy and Petrography.

Beck W. and Koller M. 1980. Problems of heating within historic building of Austria. In *Conservation within Historic Buildings*, Preprint of the Contributions to the Vienna Congress: 22-29, London: International Institute for Conservation of Historic and Artists Works.

Bläuer-Böhm C. 1996. Assessment of quantitative salt analysis by the water extraction methods on lime mortars. In *8th International Congress on Deterioration and Conservation of Stone*, Vol. 3: 1505-1519, Berlin.

Bläuer-Böhm C. and Jägers E. 1997. Analysis and recognition of dolomitic lime mortars. In Béarat. H. , Fuchs M. , Maggetti M. , and Paunier D. (eds), *Roman Wall Painting: Materials, Techniques, Analysis and Conservation*: 223-235, Proceedings of the International Workshops, Fribourg: Institute of Mineralogy and Petrography.

Black M. I965. Coccoliths, *Endeavour*, Vol. 24, No. 93: 131-137.

Borrelli E. and Laurenzi-Tabasso M. 1996. Which binder for the Egyptian plasters of the Pharaohs monuments? A critical review of archival/bibliographic documents and some new experimental evidence, In *8th International Congress on Deterioration and Conservation of Stone*, Vol. 3: 1447-1455, Berlin.

Böttger K. G and Knöfel D. 1993. Development and testing of injection mortars of high sulfate resistance. In M.-J. Thiel (ed.), *Conservation of Stone and other Materials: Prevention and Treatment*, Proceeding of the International RILEM/Unesco Congress, Vol. 2: 1629-636. Paris, London: E & En. Spon.

Botticelli G., Danti C., and Giovannoi S. 1984. Twenty years of application on mural painting: methodology of application. In *7th Triennial Meeting*: 84.15.6-84.15.11, Copenhagen: ICOM Committee for Conservation and J. Paul Getty Trust.

Bowman A. K. 1986. *Egypt After the Pharahos 332 BC-AD 642: From Alexander to the Arab Conquest*, London: British Museum Publications.

Brown R. B. 1957. *Ptolemaic Painting and Mosaics and the Alexandrian Style*, U.S.A, Cambridge Archaeological Institute of America.

B. S. [British Standards]. 1927 No 890; 1975 No 812 part 1; 1980 No 4551; 1985 No 812 part 119 & 103.

Bugini R. and Folli L. 1997. Materials and making techniques of Roman republication wall paintings (Capitolium, Brescia, Italy). In Béarat. H., Fuchs M, Maggetti M., and Paunier D. (eds), *Roman Wall Painting: Materials, Techniques, Analysis and Conservation*: 121-127, Proceedings of the International Workshops, Fribourg: Institute of Mineralogy and Petrography.

Burnstock A. 1998. Selected methods for the analysis of pigments and plasters: availability to the independent conservator. In P. Martindale (ed.), *The Analysis of Pigments and Plasters: Its Relevance to Current Wall Painting and Stone Conservation Practice:* 5-8, London: The United Kingdom Institute for Conservation of Historic and Artistic Works.

Calamiotou M., Siganidou M. and Filippakis S. E. 1983. X-Ray analysis of pigments from Pella, Greece, *Studies in Conservation*, Vol. 28, No. 3: 117-121.

Cameron S., Urquhart D., Wakefield R. and Young M. 1997. *Biological Growth on Sandstone Building Control and Treatment*, Historic Scotland Technical Conservation, Edinburgh: Research and Education Division.

Caneva G., Pia Nugari M. and Salvadori O. 1991. *Biology in the Conservation of Works of Art*, Rome: ICCROM.

Charola A. E. and Lazzarini L. 1986. Deterioration of brick masonry caused by acid rain. In R. Baboian (ed.), *Materials Deterioration Caused by Acid Rain, Symposium Series* 318: 250-257, Washington: American Chemical Society.

Colombini M. P., Fuoco R. G. and Muscatello B. 1998. Characterization of proteinaceous binders in wall painting samples by microwave assisted acid hydrolysis and GC/MS determination of Amino acids, *Studies in Conservation*: Vol. 43, No. 1: 33-41.

Conference. 1997. Piero Della Francesca's Arezzo frescoes: Techniques and conservation, Friday, 21 November, London, Courtauld Institute of Art [unpublished paper].

Constable W. G. 1954. *The Painter's Workshop*, London, New York and Toronto: Oxford University Press.

Corbeil M-C., Oleson J. P. and Foote R. 1996. Characterization of pigments used on Roman and Abbasid frescoes in Jordan: Mural Painting/Rock Art. In *11th Triennial Meeting*, Vol. 1: 423-429, Edinburgh: ICOM Committee for Conservation.

Cronyn J. M. 1990. *The Elements of Archaeological Conservation*, London: Routledge.

Davies N. and Gardiner A. 1936. *Ancient Egyptian Paintings,* London: University of Chicago.

Davis R. 1997. International academic project, conference on conservation of wall painting in Europe and the Mediterranean, *Conservation News,* No. 62: 69-71.

Davey N. 1961. *A History of Building Materials*, London: Phoenix House.

Davey N. and Ling R. 1981. *Wall Painting in Roman Britain*, London: The Society for the Promotion of Roman Studies.

De Vos M., Donati F., Fentress E., Filippi R., Panerai C., Paoletti M. L. and Pye E. 1982. A painted oecus from Settefinestre (Tuscany): Excavation, conservation and analysis. In Liversidge J. (ed), *Roman Provincial Wall Painting of the Western Empire,* B A R International Series 140: 1-32, Oxford: B A R.

De Witte E. 1975. Soluble nylon as consolidation agent fro stone, *Studies in Conservation*, Vol. 20, No. 1: 30-34.

Dodson A. 1991. *Egyptian Rock-cut Tombs*, Shire Egyptology, UK: Shire Publications Ltd.

Doerner M. 1969. *The Materials of the Artists and Their Use in Painting, with Notes on the Techniques of the Old Master*, (Revised edition), London and Aylesbury: Rupert Hart Davis Ltd.

El-Shahat A. I996. Quaternary dune carbonates from the Mediterranean coast of Egypt: petrography and diagnosis, *Facies*, Vol. 33: 265-275.

Elzinga-Tar Haar G. 1971. On the use of the electron microprobe in analysis of cross-section of paint samples, *Studies in Conservation*, Vol. 16, No. 2: 41-55.

Emery W. B. 1958. *Great Tombs of the First Dynasty*, Vol. III, Service Des Antiquités, London: Egypt Exploration Society.

The Epigraphic Survey and The Department of Antiquities of Egypt. 1980. *The Tomb of Kheruef, Theban Tomb 192*, Vol. 102, Chicago, Illinois: The Oriental Institute of the University of Chicago.

Fahd M. I. 1994. Biodeterioration of the mural paintings of the tomb of Tutankhamun: and its conservation, *Zeitschrift Fur Kunsttechnologie und Konservierung:* 143-146.

Ferragni D., Forti M., Malliet J., Mora P., Teutonico J. M. and Torraca G. 1984. Injection grouting of mural paintings and mosaics. In Brommelle N. S., Pye E. M. , Smith P. and Thomson G. (eds), *Adhesives and Consolidants*: 110-116, London: The International Institute for Conservation of Historic and Artistic Works.

Ferretti M. 1993. *Scientific Investigation of Works of Art*, Rome: ICCROM.

Filippakis S. E., Perdikatsis B. and Assimenos K. 1979. X-Ray analysis of pigments from Vergina, Greece (second tomb), *Studies in Conservation*, Vol. 24, No. 2: 54-58.

Friendship-Taylor R. and D. 1982. Piddington, *Current Archaeology*, No. 82: 348-349.

Friendship-Taylor R. and D. 1989. Piddington, *Current Archaeology*, No. 117: 316-321.

Friendship-Taylor R. and L. 1996. Piddington, *Current Archaeology*, No. 146: 57-64.

Gabra S., Drioton S., Perdtizet P. and Waddle W. G. 1941. Rapport Sur Les Fouilles d' Hermopoulis Ouet, Tuna el-Gebel, Caire: LI.

Gettens R. J. and Stout G. L. 1966. *Painting Materials A Short Encyclopedia*, New York: Dover Publications Inc.

Gettens R. J. and FitzHugh E. W. 1966. Azurite and blue Verditer, *Studies in*

Conservation, Vol. 11, No. 2: 54-60.
Gettens R. J., Kuhn H. and Chase W. T. 1967. Lead white, *Studies in Conservation*, Vol. 12, No. 4: 125-139.
Gettens R. J., Feller R. J. and Chase W. T. 1972. Vermilion and cinnabar, *Studies in Conservation*, Vol. 17, No. 2: 45-69.
Gettens R. J., FitzHugh E. W. and Robert L. F. 1974. Calcium carbonate whites, *Studies in Conservation*, Vol. 19, No. 3: 157-184.
Gettens R. J., FitzHugh E. W. 1974. Malachite and green verditer, *Studies in Conservation*, Vol. 19, No. 1: 2-23.
Getty Museum. 1992, *In the Tomb of Nefertari: Conservation of Wall Paintings*, J. Paul Getty Museum & the Getty Conservation Institute, Santa Monica, Calforina.
Giovanni and Massari I. 1993. *Damp Buildings Old and New*, ICCROM: Rome.
Giovannoni S., Matteini M. and Moles A. 1990. Studies and developments concerning the problem of altered lead pigments in wall painting, *Studies in Conservation*, Vol. 35, No. 1: 21-25.
Goldstein J. I., Newbury D. E., Echlin P., Joy D C., Romig A. D., Lyman C E., Fiori C. and Lifshin E. 1994. *Scanning Electron Microscopy and X Ray Microanalysis*, 2nd Edition, New York and London: Plenum Press.
Green L. R. 1995. Recent analysis of pigments from ancient Egyptian artefacts. In C. E. Brown, F. Macalister and M. M. Wright (eds), *Conservation in Ancient Egyptian Collections*: 85-91. London: Archetype Publication.
Gunstone F. D., Smith D. M., and Sharp J. T. 1970. *An Introductory Course in Practical Organic Chemistry*, London: Methuen & Co. Ltd.
Haggar R. G. 1962. *A Dictionary of Art Terms*, London: Oldourue Press.
Hammer I. 1995. Preservation of wall paintings and other surfaces of architecture affected by salts, interdisciplinary remarks to interpretation of the physical system and the damage processes. In R. Pancella (ed.), *Prevention and Restoration of Cultural Heritage; Stone, Materials, Air Pollution, Murals: Scientific Research Work and Case Studies*: 575-583, LCP Congress, Montreux: Unesco.
Hornblower S. and Spawforth A. (eds). 1996. *The Oxford Classical Dictionary*, 3rd Edition, Oxford: Oxford University Press.
Hassen M. A. 1963. *History of Alexandria and Its Civilization since Old Centuries*, Egypt: Alexandria's Council (In Arabic).
Hawass Z. 1993, The Egyptian monuments problems and solutions: Giza Pyramids and Sakkara. In M.-J. Thiel (ed.), *Conservation of Stone and other Materials: Prevention and Treatment*, Proceeding of the International RILEM/Unesco Congress, Vol. 1: 19-26, Paris, London: E & En. Spon.
Helwig K. 1997. A note on burnt yellow earth pigments: Documentary sources and scientific analysis, *Studies in Conservation*, Vol. 42, No. 3: 181-188.
Hey M. 1957-8. The analysis of paint media by paper chromatography, *Studies in Conservation*, Vol. 3: 183-193.
Historic Scotland. 1995. *Preparation and Use of Lime Mortars, An Introduction to the Principles of Using Lime Mortars*, Edinburgh: Historic Scotland.
Hodges H. 1989. *Artifacts: An Introduction to Early Materials and Technology*, London: Duckworth & Co Ltd.
Holmstrőm I. 1982. Mortars, cements and grouts for conservation and repair: Some urgent needs for research. In ICCROM: *Mortars, Cements and Grouts Used in Conservation of Historic Buildings*: 19-24, Papers from a Symposium Held at Rome, Rome: ICCROM.
Hutchison C. 1974. *Laboratory Handbook of Petrographic Techniques*, London: John Wiley and Sons, Inc.

Jackson H. F. 1904. *Mural Painting*, London: Sands & Company.

Jaeschke H. F. 1995. The cleaning of New Kingdom temple wall paintings at Luxor and Abydos. In C. E. Brown, F. Macalister and M. M. Wright (eds), *Conservation in Ancient Egyptian Collections*: 147-157, London: Archetype Publication.

Járó M. 1997. Comparison of the painting materials used for wall painting in four sites of the Roman Province Pannonia. In Béarat. H. , Fuchs M. , Maggetti M. , and Paunier D. (eds), *Roman Wall Painting: Materials, Techniques, Analysis and Conservation*: 75-84, Proceedings of the International Workshops, Fribourg: Institute of Mineralogy and Petrography.

JCPDS. 1974. *Joint Committee on Powder Diffraction Data*, Swartmore, Pennsylvania.

Jedrzejewska H. 1960. Old mortars in Poland: A new method of investigation, *Studies in Conservation*, Vol. 5, No. 3: 134-136.

Jedrzejewska H. 1982. Ancient mortars as criterion in analysis of old architecture. In ICCROM: *Mortars, Cements and Grouts Used in the Conservation of Historic Buildings*, 311-329, Papers from a Symposium held at Rome, Rome: ICCROM.

Johnson M. and Packard E. 1971. Methods used for the identification of binding media in Italian paintings of the fifteenth and sixteenth centuries, *Studies in Conservation*, Vol. 16, No 4: 145-164.

Jones M. P. 1987. *Applied Mineralogy: A Quantitative Approach*, London: Graham & Trotman Ltd.

Jones P. L. 1962. Some observation on methods for identifying proteins in paint media, *Studies in Conservation*, Vol. 7: 10-16.

Kakoulli I. 1997. Roman wall paintings in Cyprus: A scientific investigation of their technology. In Béarat. H., Fuchs M., Maggetti M., and Paunier D. (eds), *Roman Wall Painting: Materials, Techniques, Analysis and Conservation*: 131-141, Proceedings of the International Workshops, Fribourg: Institute of Mineralogy and Petrography.

Kakoulli I. 1998. Hellenistic painting techniques, personal communication, 14 May 1998, Rome: ICCROM.

Kempe D. R. C. and Templeman J. A. 1983. Techniques. In K. D. R. C. and H. Anthony (eds), *The Petrology of Archaeological Artefacts*: 26-52, Oxford: Clarendon Press.

Kleiner L. M. 1995, *Ancient Binding Media, Varnishes and Adhesives*, Rome: ICCROM.

Kockaert L., Gausset P. and Dubi-Rucquoy M. 1988. Detection of ovalbumin in paint media by immunofluorescence, *Studies in Conservation*, Vol. 34, No. 4: 183-188.

Koob S. P. 1986. The use of Paraloid B-72 as an adhesive: Its application for archaeological ceramics and other materials, *Studies in Conservation*, Vol. 31, No. 1: 7-14.

Lal B. B. 1996. *Conservation of Wall Painting in India*, New Delhi: India Association for the Study of Conservation of Cultural Property.

Laue S. 1995. Climate controlled behaviour of soluble salts in the crypt of St. Maria im Kapitol, Cologne. In R. Pancella (ed.), *Prevention and Restoration of Cultural Heritage: Stone, Materials, Air Pollution, Murals: Scientific Research Work and Case Studies*, 447-453, LCP Congress, Montreux: Unesco.

Lauer J. P. 1976. *Saqqara, The Royal Cemetery of Memphis*, London: Thames & Hudson Ltd.

Laurie A. P. 1910. *The Materials of the Painter's Craft in Europe and Egypt from Early Times to the End of the XVVII TH Century, with Some Account of Their Preparation and Use*, London and Edinburgh: T. N. Foulis.

Laurie A. P. 1914. *The Pigments and Mediums of Old Masters With a special Chapter on the Microphotographic Study of Brushwork*, London: Macmillan & Co. Ltd.

Lazzarini L. 1982. The discovery of Egyptian blue in a Roman fresco of the Medieval period (ninth century AD), *Studies in Conservation*, Vol.

27, No. 2: 84-86.

Lazzarini L. 1998. Pigments and dyes, Handout: 1-15, Rome: ICCROM.

Lewin S. Z. 1982. X-Ray diffraction and scanning electron microscope analysis of conventional mortars. In ICCROM: *Mortars, Cements and Grouts Used in Conservation of Historic Buildings*: 101-131, Papers from a Symposium Held at Rome, Rome: ICCROM.

Ling R. 1991. *Roman Painting*, UK: Shire Publications Ltd.

Lucas A. 1915. *Disintegration and Preservation of Building Stones in Egypt*, Cairo, Ministry of Finance of Egypt, Survey Department: Government Press.

Lucas A. 1932. *Antiques: Their Restoration & Preservation*, 2nd Edition, London: Edward Arnold & Co.

Lucas A. 1989. *Ancient Egyptian Materials and Industries*, 4th Edition, London: History & Mysteries of Man Ltd.

MacKenzie W. S. and Guilford C. 1980. *Atlas of Rockforming Minerals in Thin Section*, London: Longman Group Ltd.

Malaguzzi-Valerj V. 1970. Ancient fresco technique in the light of scientific examination. In W. Y. Jonathan (ed.), *Application of Science of Works of Art*: 164-188. Boston: Massachusetts the Research Laboratory Museum of Fine Arts.

Mancinelli F. 1994. Documentation system applied to the restoration of the Sistine Chapel. In S. Laviald and C. E. Bernardelli (eds), *Restoration relationship between language and image*: 109-124. Palazzo dei Papi, Piazza Salorando Viterbo, Italy: World Scientific Co. Ple. Ltd.

Mairinger F. and Schreiner M. 1982. New methods of chemical analysis: A tool for the conservator. In N. S. Brommelle and G. Thomsom (eds), *Science and Technology in the Service of Conservation*: 5-18, London: The International Institute for Conservation of Historic and Artists' Works.

Martin E. 1977. Some improvements of analysis of paint media, *Studies in Conservation*, Vol. 22, No. 2: 63-67.

Martinez-Ramirez S., F. Puertas F., Blanco-Varela M. T., Thompson G. E, Haneef S. J. and Wood G. C. 1996. Durability of a new lime mortar with Biocide properties. In *8th International Congress on Deterioration and Conservation of Stone Proceeding*, Vol. 3: 1565-1571, Berlin.

Massari I. 1998. Laboratory exercises: Measurements of humidity in walls, Handout, unit 4, Rome: ICCROM.

Masschelein-Kleiner L. 1974. An improved method for thin-layer chromatography of media in tempera paintings, *Studies in Conservation*, Vol. 19: No. 4: 207-211.

Matteini M. and Moles A. 1984. Twenty years of application of "Barium" on mural painting: Fundamentals and discussion. In *7th Triennial Meeting, Copenhagen*: 84.15.15-84.15.19. ICOM Committee for Conservation and J. Paul Getty Trust.

Matteini M. 1987. In review: an assessment of Florentine methods of wall painting conservation based on the use of mineral treatments. In S. Cather (ed.), *The Conservation of Wall Painting*, Proceeding of a Symposium, Courtauld Institute of Art and Getty Conservation Institutes: 137-148, London: The Getty Conservation Institute.

Mayer R. 1973. *The Artist's Handbook of Materials and Techniques*, 3rd Edition, London: Faber and Faber Ltd.

Mellon F. A. 1991. Liquid chromatography/mass spectrometry, *VG Monographs in Mass Spectrometry*: No: 2.

Miller E. G., Lee N. J. and Ellam D. R. 1987. Remounting and conservation of archaeological wall paintings at the British museum. In Black J. (ed), *Recent Advances in the Conservation and Analysis of Artifacts*: 289-295, London: Archetype.

Moore R. C., Lalicker C. G. and Fischer A. G. 1952. *Invertebrate Fossils*, New York, Toronto and

London: McGraw Hill Book Company, Inc.
Mora P. and Mora L. 1984. *Conservation of Wall Paintings*, Rome: ICCROM.
Mora P. 1974. *Causes of Deterioration of Mural Paintings*, Translated by H. J. Plenderleith, Rome: International Center for the Study of the Preservation and the Restoration of Cultural Propriety.
Mora P. and Philippot P. 1968. Conservation of wall painting. In International Centre for the Study of the Culture Property, *The Conservation of Cultural Property*: 169-189, Paris: Unesco.
Morgan G. 1992. Romano-British Mortar and Plaster, University of Leicester, School of Archaeological Studies, Ph.D. Thesis.
Morgan G. 1998. Plaster Analysis and The Conservator or restorer. In P. Martindale (ed.), *The Analysis of Pigments and Plasters: Its Relevance to Current Wall Painting and Stone Conservation Practice:* 16, London: The United Kingdom Institute for Conservation of Historic and Artistic Works
Mural Painting Conservation. 1998. *Preparatory Manual for Course Participants, International Course on the Conservation of Mural Paintings & Related Architectural Surface,* Rome: ICCROM.
Murnane W. J. 1983. *The Penguin Guide to Ancient Egypt,* London: Allen Lane.
Newman R. and Halpin S. 1992. *Developments in the Analysis of Material Abstracts* Washington (D. C.), New York: The American Institute for Conservation Buffalo.
Noshy I. 1937. *The Arts in Ptolemaic Egypt: A Study of Greek and Egyptian Influences in Ptolemaic Architecture and Sculpture*, London: Oxford University Press.
Paine S. 1993. The effects of bat excreta on wall paintings, *The Conservator*, No. 17: 3-10.
Pascoe M. 1980. Toxic hazards from solvents in conservation, *The Conservator*, No. 4: 25-27.
Penkala B. and Bralewska E. 1981. Comparative analysis of Antique gypsum mortars from the walls of the theatre in Alexandria, the Pyramid of Cheops and early Medieval walls of the Relics at Wislica and Gniezno in Poland. In *6th Triennial meeting*: 81-10-10-1-81-10-1-7, London: ICOM Committee for Conservation.
Perander T. 1982. Mortar study in Finland: Maintenance of historic buildings. In ICCROM: *Mortars, Cements and Grouts Used in Conservation of Historic Buildings*: 141-144, Papers from a Symposium Held at Rome, Rome: ICCROM.
Petersen K., Krumben W. E., Häfner L. E., and Mieth A. 1993. Aspects of biocide application on wall paintings: report on Eurocare project EU 489, biodecay. In M.-J. Thiel (ed.), *Conservation of Stone and Other Materials: Prevention and Treatment*, Proceeding of International RILEM/Unesco Congress, Vol. 2: 597-604, Paris, London: E & FN Spon.
Peterson S. 1982. Lime water consolidation. In ICCROM: *Mortars, Cements and Grouts Used in Conservation of Historic Buildings*: 53-61, Papers from a Symposium Held at Rome, Rome: ICCROM.
Petushkova J. P. and Lyalikova N. N. 1986. Microbiological degradation of lead containing pigments in mural paintings, *Studies in Conservation*, Vol. 31, No 2: 65-69.
Pinnska J. and Rabie A. H. 1999. Study of the building materials used in construction of Maadi town temple, Fayoum, Egypt. In C. Parisi, S. Galiardi, G. M. Parisi and G. Torcinaro (eds), *6th international conference on Non-Destructive Testing and Microanalysis for the Diagnostic and Conservation of the Cultural and Environmental Heritage* Vol. 1: 621-631, Rome: I.C.R and AIPnD.
Pique F. Dei L. and Ferroni E. 1992. Physicochemical aspect of the deliquescence of calcium nitrate and its implications for wall painting conservation, *Studies in Conservation*, Vol. 37, No. 4: 217-225.
Plesters J. 1955. The preparation and study of paint cross section, *Museums Journal*,

Vol. 54, No. 4: 97-101.
Plesters J. 1956. Cross-sections and chemical analysis of paint samples, *Studies in Conservation,* Vol. II: 110-157.
Plather L. E. 1997. Konserveringen av takmaleriet fra Ål Stavkirke i 1965: Et tibakeblikk på Soluble Nylon som konserveringsmiddel. In F. Hanssen-Baner and K. Kollandsrud (eds), *Konserveringsmidler & Konserveringsmetoder* (*Consolidants and Conservation Methods*): 115-123, Nordisk Konservatorforbund XIV Kongress, Oslo: NKF-N.
Plather U. 1997. Testing av Soluble Nylon i Ål-taket. In F. Hanssen-Baner and K. Kollandsrud (eds), *Konserveringsmidler & Konserveringsmetoder* (*Consolidants and Conservation Methods*): 125-131, Nordisk Konservatorforbund XIV Kongress, Oslo: NKF-N.
Porter B. and Moss R. L. B. 1960. *Topographical Bibliography of Ancient Egyptian Hieroglyphic Texts, Reliefs and Paintings, The Theban Necropolis*, Part 1, Privet Tombs, 2nd Edition, Oxford: Clarendon Press.
Porter B. and Moss R. L. B. 1964. *Topographical Bibliography of Ancient Egyptian Hieroglyphic Texts, Reliefs and Paintings, The Theban Necropolis*, Part 2, Royal Tombs and Smaller Cemeteries, 2nd Edition, Oxford: Clarendon Press.
Preusser F. 1987. Scientific and technical examination of the tomb of Queen Nefertari at Thebes. In S. Cather (ed.), *The Conservation of Wall Paintings*, Proceedings of a Symposium at the Courtauld Institute of Art and the Getty Conservation: 1-12, London: The Getty Conservation Institute.
Profi S., Weier L. and Filippakis S. E. 1974. X-Ray analysis of Greek Bronze age pigments from Mycenae, *Studies in Conservation*, Vol. 19, No. 2: 105-112.
Profi S., Perdikatsis B. and Filippakis S. E. 1977. X-Ray analysis of Greece Bronze age pigments from Thera (Santorini), *Studies in Conservation*, Vol. 22, No. 3: 107-115.
Rakowski S. 1973. *The Tomb of Queen Nefertari*: *Problems of Conservating Wall Paintings*: Diagnosing the State of Preservation and Conservation Proposals the Working Group of the State Ateliers for the Preservation of Historical Property, Warsaw: Conservator's Information Center.
Rees-Jones S. 1990. Early experiments in pigment analysis, *Studies in Conservation*, Vol. 35, No. 2: 93-101.
Reller A. and Wild P.–M., Wiedemann H.G., Hauptmann H. and Bonani G. 1992. Comparatives studies of Ancient mortars from Giza, Egypt and Nevali Cori, Turkey. In P. B. Vandiver, J. R. Druzik, G. S. Wheeler, and I. C. Freestone (eds), *Materials Issues in Art and Archaeology* III, Vol. 267: 1007-1011, San Franciso, California: Materials Research Society Symposium Proceedings.
Rickerby S. 1991. Heat alterations to pigments painted in the fresco technique, *The Conservator*, No. 15: 39-44.
Robins G. 1997. *The Art of Ancient Egypt*, London: The British Museum Press.
Rollo B. A. L and Hartley K. 1994. Iron age and Roman Piddington: The Mortaria 1979-1993. Including a brief interim report on the site from 1989-1993, *Upper Nene Archaeological Society, Fascicle 2*.
Rose M. E. 1990. Modern practice of gas chromatography/mass spectrometry, *VG Monographs in Mass Spectrometry*, Vol. 1, No. 1.
Rossi-Manaresi R. and Tucci A. 1991. Pore structure and the disruptive cementing effect of salt crystallization in various types of stone, *Studies in Conservation*, Vol. 36, No. 1: 53-58.
Rostovtzeff M. 1919. Ancient decorative wall painting, *The Journal of Hellenic Studies*, Vol. 39: 144-163.

Rozenberg S. 1997. Pigments and fresco fragments from Herod's palace at Jericho. In Béarat. H., Fuchs M., Maggetti M., and Paunier D. (eds), *Roman Wall Painting: Materials, Techniques, Analysis and Conservation*: 63-72, Proceedings of the International Workshops, Fribourg: Institute of Mineralogy and Petrography.

Ruffle J. 1977. *Heritage of Pharaohs, An Introduction to Egyptian Archaeology*, Oxford: Phaidon.

Said R. 1962. *The Geology of Egypt*, New York: Amsterdam.

Schafer J. and Hilsdorf H. K. 1993. Ancient and new lime mortars: The correlation between their composition, structure and properties. In M.-J. Thiel (ed.), *Conservation of Stone and Other Materials: Preservation and Treatment*, Proceedings of the International RILEM/Unesco Congress, Vol. 2: 605-612, Pairs, London: E & FN Spon.

Schiegl S., Weiner K. L. and El Goresy A. 1989. Discovery of copper chloride cancer in ancient Egyptian polychromic wall paintings and Faience: A developing archaeological disaster, Vol. 24: 1-15, Max-Planck Institut Für Kernphysik Heidelberg.

Schiegl S., Weiner K. L. and El Goresy A. 1992. The diversity of newly discovered of deterioration painting in Ancient Egyptian pigments: Consequences to entirely new restoration strategies and to the Egyptological colour symbolism. In P. b. Vandiver, J. R. Druzik, G. S. Wheeler, and I. C. Freestone (eds), *Materials Issues in Art and Archaeology* III, Vol. 267: 831-858, San Francisco, California: Materials Research Society Symposium Proceeding.

Shekede L. 1998. The effect of Laser radiation on polychromed surface. In P. Martindale (ed.), *The Analysis of Pigments and Plasters: Its Relevance to Current Wall Painting and Stone Conservation Practice*, 23-25, London: United Kingdom Institute for Conservation of Historic and Artistic Works.

Schmid W. 1993. (Translator) Considerations on the techniques of wall painting conservation in situ. In *Frescoes Restoration Experience: Methods and Criteria of Conservation*: 91-96, Italy, Brescia: Irecoop Emilia Romagna.

Schmid W. 1998a. The study of mural painting techniques, Unit 3, Handout: 1-5, Rome: ICCROM.

Schmid W. 1998b. Aesthetic presentation, Handout: 1-5, Rome: ICCROM.

Schmid W. 1998c. Treatment methods and materials: Detachment, Hand out, Rome: ICCROM.

Science for Conservators. 1992. *Cleaning*, Vol. 2, The Conservation Unit of the Museum and Galleries Commission, London and New York: Routledge.

Segal I. and Porat N. 1997. Composition of pigments from the Hellenistic wall in Acre. In Béarat. H., Fuchs M., Maggetti M., and Paunier D. (eds), *Roman Wall Painting: Materials, Techniques, Analysis and Conservation*: 93-104, Proceedings of the International Workshops, Fribourg: Institute of Mineralogy and Petrography.

Shurkri N. M., Philip G. and Said R. 1956. Geology of the Mediterranean coast between Rosetta and Bardia, Part II, Pleistocene sediments: Geomorphology and microfacies Egypt *Bulletin-Institute Egyptian*: Vol. 37: 395-433.

Simpson [..] and Brown [..] [Initial not state]. 1994. *Conservation of Plasterwork*, Historic Scotland Technical Advice Note 2, Edinburgh: Crambeth Allen.

Smeaton J. E. and Burns G. 1988. Physicochemistry of the tomb of Nefertari, Egypt. In E. V. Syare *et al.* (eds), *Materials Issues in Art and Archaeology*, Vol. 123: 299-304, Reno, Nevada: Pittsburgh, Pa, Materials Research Society Symposium Proceeding.

Smith W. 1946. *A History of Egyptian Sculpture and Painting in the Old Kingdom*, London: Oxford University Press.

Sneyers R. V. and De Henau P. J. 1968. The conservation of stone. In International Centre for the Study of the Preservation and Restoration of Cultural Property, *The Conservation of Culture Property*, 209-235, Paris: Unesco.

Speroni S., Tersigni C., Torraca G., Cerea S., Fort M., Guidobaldi F., Rossi-Doria P., De Rege A.,

Picchi D., Pietrafitta F. J. and Benedetti G. 1982. Lime basd mortars for the repair of ancient masonary and possible substitutes. In ICCROM: *Mortars, Cements and Grouts Used in Conservation of Historic Buildings*: 63-99, Papers from a Symposium Held at Rome, Rome: ICCROM.

Stewart J. and Moore J. 1982. Chemical techniques of historic mortar analysis. In ICCROM: *Mortars, Cements and Grouts Used in Conservation of Historic Buildings*: 297-310, Papers from a Symposium Held at Rome, Rome: ICCROM.

Tabasso M. L. 1993. Use of traditional materials for the conservation of stone and wall painting in Italy: Real advantage or "Romantic" Utopia?, 4th Seminar on the Conservation of Asian: 52-61, Nara Cultural Heritage: Traditional Materials and Techniques in Conservation.

Tabasso M. 1998. Pigments alteration, Handout: 1-5, Rome: ICCROM.

Teutonico J. M. 1988. *A Laboratory Manual for Architectural Conservators*, Rome: ICCROM.

Theophilus. 1961. *The Various Arts*. Translated by Dodwell, C. R., London, New York: Thomas Nelson and Sons LTD.

Tite M. S. 1972. *Methods of Physical Examination in Archaeology*, London and New York: Seminar Press.

Torraca G. 1970. Deterioration processes of mural paintings. In W. Y. Jonathan (ed.), *Application of Science in Examination of Works of Art*: 169-174, The Research Laboratory, Boston: Massachusetts Museum of Fine Arts.

Torraca G. 1988. *Porous Building Materials, Materials Science for Architectural Conservation*, 3rd Edition, Rome: ICCROM.

Tubb K. W. 1987. Conservation of lime plaster statues of ′Ain Ghazal. In Black J. (ed), *Recent Advances in the Conservation and Analysis of Artifacts*: 387-391, London: Archetype.

Valentini M., Sala G. and Trovati A. 1994. Investigation on internal microclimate of a confined historical building: Evaluation of the results in relation to the conservation of frescoes. In V. Fassina, H. Ott and F. Zezzo (eds), *The Conservation of Monuments in The Mediterranean Basin*, 3rd International Symposium: 263-268, Italy: Venice.

Van Asperen De Boer J. R. J. 1984. Some practical aspects of the preparation and study of cross sections. In *7th Triennial Meeting*: 84.1.5-84.1.6, Copenhagen: ICOM Committee for Conservation and J. Paul Getty Trust.

Van Hees R. P. J. and Koek J. A. G. 1995. Treatment of rising damp: A laboratory evaluation methods: *Methods of Evaluation Products for the Conservation of Porous Building Materials in Monuments*, Rome, ICCROM: International Colloquium.

Vasari G. 1907. *Vasari on Technique*, London: J. M. Dent & Company.

Verhoeven U. 1987. Eine technologische Rarität: Das Brenne von Ziegeln in der Grabdekoration des Mittleren und Neuen Reiches, *Mitteilungen des Deutschen Arcäologischen Instituts, Abteilung Kairo*, No. 43: 259-266.

Vitruvius. 1960. *The Ten Books on Architecture*, Translated by Morris Hicky Morgan, New York: Dover, Publications, Inc.

Wallert A. and Elston M. 1997. Fragments of Roman wall painting in the J. Paul Getty museum: A preliminary technical investigation. In Béarat. H., Fuchs M., Maggetti M., and Paunier D. (eds), *Roman Wall Painting: Materials, Techniques, Analysis and Conservation*: 93-104, Proceedings of the International Workshops, Fribourg: Institute of Mineralogy and Petrography.

Watkinson D. 1972. *First Aid for Finds*, 2nd Edition, The British Archaeology Trust: United Kingdom Institute for Conservation of Artistic Works.

Weatherhead F. and Buckley A. 1989. Artists: pigments from Amarna. In B J. Kemp (ed.), *Amarna Reports V*: 40-202, London: the Egyptian Exploration Society.

Weatherhead F. 1995. Two Studies on Amarna Pigments. In B J Kemp (ed.), *Amarna Reports VI*: 98-384, London: the Egyptian Exploration Society.

Welton J. E. 1984. *SEM Petrology Atlas, Methods in Exploration Series*, Tulsa, Oklahoma: American Association of Petroleum Geologists.

Wilimzig M. and Bock E. 1995. Biological material testing, methods of evaluating of porous building materials in monuments: 87-92, International Colloquium, Rome: ICCROM.

Zehnder K. 1996. Gypsum efflorescence in the zone of rising damp, monitoring of slow decay processes caused by crystallizing salts on wall paintings. In *8th International Congress on Deterioration and Conservation of Stone Proceeding*, Vol. 3: 1669-1678, Berlin.

Egyptian and Græco-Roman Wall Plasters and Mortars: A Comparative Scientific Study

Safaa A. Abd El Salam
University of Leicester
School of Archaeological Studies

Abstract

The development of methods of examination and analysis for painted plaster allows us to identify and determine not only the original materials used, but also to define the causes of subsequent alteration, which has affected both the painted layers and the rendering.

The aim of this research is to examine and analyse wall plasters and mortars from different sites and periods, in order to identify composition, structure and mineral content, and to define the main forms of deterioration and decay affecting the wall paintings. The strategy of examination of these materials is divided into three stages with the following analytical methods.

1-Preliminary analysis: In the initial examination a microscope was used at 10X magnification and to look at polished cross-sections, to identify the structure of the mortar and the painted layers.

2-Chemical and micro-chemical analysis: Microanalysis (spot tests). Standard methods were used to identify the quantitative and qualitative nature of the composition of plasters and mortars, including the measurement of the calcium carbonate ($CaCO_3$) content, of the layers and in some instances the deposit which covered the painted surface, and the analysis of any water-soluble salts. The analysis of pigments was carried out using micro-chemical tests.

3-Physical methods: X-ray diffraction and X-ray powder diffraction (X-RD & X-RPD) confirmed the mineralogical compounds in the plasters and pigments. Induction coupled plasma spectroscopy (ICP) detected the other metal ions present in the materials. Polarized light microscopy (PLM) revealed internal structures. Scanning electron microscopy and dispersive X-ray microanalysis (SEM & EDS) were used for surface structure and to define deterioration and decay factors.

The analytical methods involved in the examination and analysis of the plasters and mortars of wall paintings helped to identify their compositions and structures. Analysis also elucidated the manufacturing techniques of the materials and the main problems of the deterioration and decay, which is useful in the conservation and restoration processes.

Ägyptischer und griechisch-römischer Wandverputz und Mörtel: Eine vergleichende naturwissenschaftliche Studie

Safaa A. Abd El Salam
University of Leicester
School of Archaeological Studies

Zusammenfassung

Die Entwicklung von Untersuchungs- und Analysenmethoden für bemalten Verputz ermöglicht es uns, nicht nur die Ausgangsmaterialien zu erkennen und zu bestimmen, sondern auch die Gründe einer im Laufe der Zeit eintretenden Veränderung der bemalten Schichten und des Putzes festzulegen.

Das Ziel dieser Studie ist es, Wandverputz und Mörtel von verschiedenen Fundplätzen und Zeitperioden zu untersuchen und zu analysieren, um ihre Zusammensetzung, Stuktur und ihren Mineralgehalt festzustellen und um die Hauptformen der Zersetzung und des Zerfalls zu bestimmen, die die Wandmalereien befallen. Der Untersuchungsplan für diese Materialien ist in drei Stufen aufgeteilt, die die folgenden Analysenmethoden einschließen:

1-Voranalyse: Für die Erstuntersuchung wurde ein Mikroskop mit einer 10-fachen Vergrößerung benutzt, um polierte Schliffe zu untersuchen und die Struktur des Mörtels und der bemalten Schichten zu bestimmen.

2-Chemische und mikro-chemische Analyse: Mikroanalyse (Punktanalyse). Mittels Standardmethoden wurde die quantitative und qualitative Zusammensetzung der Putze und Mörtel bestimmt, einschließlich des Kalziumkarbonat-Gehalts ($CaCO_3$) der einzelnen Schichten und in einigen Fällen auch der Ablagerungen auf den bemalten Oberflächen, sowie einer Analyse aller wasserlöslichen Salze. Die Analyse der Farbstoffe wurde mittels mikro-chemischer Tests durchgeführt.

3-Physikalische Methoden: Röntgenbeugungsanalysen (XRD), einschließlich Röntgenpulverdiffraktrometrie, bestätigten die mineralogische Zusammensetzung der Verputze und Farbstoffe. Induktiv gekoppelte Plasma-Atomemissions-Spektroskopie (ICP) konnte die anderen metallischen Ionen in den Materialen nachweisen. Polarisationsmikroskopie ermittelte die inneren Stukturen. Rasterelektronenmikroskopie (REM) mit energiedispersiver Röntgenstrahlmikroanalyse (EDX) wurde für die Untersuchung der Oberflächenstruktur benutzt, und um Zersetzungs- und Zerfallsfaktoren zu bestimmen.

Die für die Untersuchung und Analyse der Verputze und Mörtel von Wandmalereien eingesetzten Analysenmethoden halfen, ihre Zusammensetzung und Struktur zu bestimmen. Die Analysen erklärten zudem die Herstellungstechniken der Materialien und die Hauptprobleme der Zersetzung und des Zerfalls, was für Konservierungs- und Restaurierungsprozesse nützlich ist.

www.ingramcontent.com/pod-product-compliance
Lightning Source LLC
LaVergne TN
LVHW071203210926
848482LV00015B/715
* 9 7 8 1 8 4 1 7 1 3 8 8 5 *